AUTO MANIA

AUTO MANIA

Cars, Consumers, and the Environment

Tom McCarthy

Yale University Press
New Haven & London

Published with assistance from the foundation established in memory
of William McKean Brown and from the income of the
Frederick John Kingsbury Memorial Fund.

Set in Electra and Trajan types by Tseng Information Systems, Inc.
Printed in the United States of America.

Library of Congress Cataloging-in-Publication Data
McCarthy, Tom, 1959–
Auto mania : cars, consumers, and the environment / Tom McCarthy.
p.   cm.
Includes bibliographical references and index.
ISBN 978-0-300-11038-8 (clothbound : alk. paper)
1. Automobile industry and trade — United States — History — 20th century.
2. Automobiles — Environmental aspects — United States — History — 20th century.
3. Consumer behavior — United States — History — 20th century. I. Title.
HD9710.U52M427 2007
338.4'762922209730904 — dc22   2007014691

A catalogue record for this book is available from the British Library.

The paper in this book meets the guidelines for permanence and durability
of the Committee on Production Guidelines for Book Longevity of the
Council on Library Resources.

10   9   8   7   6   5   4   3   2   1

For Mom and Dad
For Rory and Mac
And for Lisa

CONTENTS

Acknowledgments

Many rivers to cross . . . .

Researching and writing history is a solitary endeavor that requires a lot of help along the way. Words cannot express my gratitude to those who helped me write this book. I could not have done it without them.

First, I thank Susan Strasser, who encouraged me to take on the automobile. My misgivings about the magnitude of this project were well founded, but she was right that it needed to be done.

This book began as a Yale dissertation. Committee members Jean-Christophe Agnew and the late Robin W. Winks were both incredibly helpful. Special thanks are due my dissertation adviser, John Mack Faragher. His help with the dissertation and professional advice over the years have been invaluable, and I am deeply grateful. I also thank Thomas E. Graedel for allowing me to audit his graduate industrial ecology course at the Yale School of Forestry and Environmental Studies.

I have the best colleagues in the world in the History Department at the U.S. Naval Academy. I thank you all—especially those who have helped me with various chapters over the past five years. Thanks are also due to the reference librarians at Nimitz Library, each of whom contributed to the book, and to Flo Todd, academia's fastest and most patient interlibrary loan administrator. Summer grants from the Naval Academy Research Council and Faculty Development Fund grants to acquire photographs and artwork smoothed my way down the homestretch.

My debts to the reference librarians and curators in the collections where I researched this book are many. All have truly been a researcher's best friend. I began at the Automobile Reference Collection of the Free Library of Philadelphia, where I thank the late Stuart McDougall and Bob Rubinstein. Thank you, Bob Casey, Judith E. Endelman, Terry Hoover, Andrew Schornick, Linda Skolarus, and the other great people at the Benson Ford Research Center, The Henry Ford, for hosting and helping me over extended visits to Dearborn in four summers. I especially thank the museum for a Clark Travel-to-Collections Grant.

I thank the Huntington Library in San Marino, California, for an Andrew W. Mellon Foundation Fellowship that enabled me to study the automobile side of the California Dream for two months. I also thank Ellen G. Gartrell and Jacqueline V. Reid at the John W. Hartman Center for Sales, Advertising, and Marketing History at Duke University for a 1999 J. Walter Thompson Research Fellowship to use the J. Walter Thompson Archives and the Kensinger Jones Papers. Thanks to Francis X. Blouin, Jr., and William K. Wallach at the Bentley Historical Library, University of Michigan–Ann Arbor, for a Mark C. Stevens Researcher Travel Fellowship. A big thanks to the folks at the Smithsonian Institution's National Museum of American History who supported my project with a Smithsonian Predoctoral Fellowship. Special thanks to Jeffrey Stine, whose on-going support for this project I deeply appreciate, and to Pete Daniel, Charles McGovern, Suzanne McLaughlin, Shelley Nickles, Wendy Shay, and Roger B. White. A Woodrow Wilson Postdoctoral Fellowship in the Humanities made possible a productive and delightful year of writing and teaching at Elon University. Special thanks to Nancy Midgette, Jim Bissett, and my great colleagues in the Department of History and Geography there. Support from the National Endowment for the Humanities and the American Historical Association as part of the summer institute "Rethinking America in Global Perspective" made possible a month of research in the photograph collections at the Library of Congress.

Many research librarians, archivists, and curators helped in shorter visits to other collections. I extend special thanks to Dace Taube, curator at the Regional History Collection, University of Southern California; Matthew W. Roth and Morgan P. Yates at the archives of the Automobile Club of Southern California, Los Angeles; Bill Holleran, curator at the Kettering/GMI Alumni Foundation Collection of Industrial History at Kettering University, Flint, Michigan; Mark A. Patrick, curator at the National Automotive History Collection, Detroit Public Library; Mike Smith, reference archivist at the Walter P. Reuther Library, Wayne State University, Detroit; Kevin P. Manion at Consumers Union, Yonkers, New York; and to the librarians and curators at the Michigan Historical Center (both library and archives), Lansing; the Special Collections Department of the University of Virginia Library, Charlottesville; the U.S. Department of

Transportation Library, Washington, DC; the Chemical Heritage Foundation, Philadelphia; the National Archives, College Park, Maryland; and the Library of Congress.

Thanks to *Environmental History, Michigan Historical Review,* and *Progress in Industrial Ecology* for permission to reprint material that first appeared in their pages.

Thanks to Cindy Eckard for giving the manuscript a rigorous read-over and to Colin Myers for doing the same and for help in organizing the photographs. The technical assistance of Shane M. Bowler sharpened many of the cartoons scanned from old magazines.

Special thanks to the whole team at Yale University Press, whose invaluable professional expertise helped me to turn an idea and a manuscript into this book. Thank you, Jean Thomson Black, Laura Davulis, Laura Jones Dooley, Jessie Hunnicutt, and Matthew Laird.

My greatest debt is to my family. I deeply appreciate the love, interest, and support of my parents, Tom and Ruth McCarthy. I thank my kids, Rory and Mac, who always lift my heart. My greatest regret is that writing this book took so much time away from your early years. Like most historians, I hope that in trying to understand how this world came to be as it is I can make a small contribution to making it a better world for you and your children. My biggest thank you is for Lisa Dumont. Lisa has seen this project through with me from beginning to end. She has gracefully tolerated reduced circumstances, extended absences, and my long hours of unnatural intimacy with the computer while pursuing her own career and raising our kids. Her help with the manuscript and photographs has been invaluable. She has shouldered more than her share to keep our lives going while I worked to finish this book. Thanks, kiddo.

Finally, I thank my car. Like most activities in American life, I could not have researched this book without it. We have traveled nearly two hundred thousand miles together. I get more than forty miles per gallon on the highway, and my car has always tested well within emissions standards. American automakers can make a damn good, affordable vehicle when they set their minds to it, and my Ford Escort is proof. I'm grateful for all the good service it has given me. I'll miss it when we finally part.

This book is finished. But there are still many rivers to cross.

Introduction: The Great Experiment

The American love affair with the automobile sheds important light on what may prove to be the most important question of the twenty-first century: Can six to nine billion human beings find contentment through the unending acquisition of material possessions without irreparably harming the planet they depend on for life itself? Historian J. R. McNeill called this test of the planet's human carrying capacity "a gigantic uncontrolled experiment." Given the huge stakes involved and the uncertainty of the outcome, it seemed to me that more historians ought to be exploring the intersection between consumer behavior and environmental impact.

We can learn much about the Great Experiment by examining the century-long relationship between Americans and their automobiles. First, the environmental impacts of the automobile are more varied and substantial than is commonly recognized. Second, these impacts have occurred because Americans have consciously and unconsciously pursued important personal and family agendas that have involved practical and psychological satisfactions offered by automobile ownership and operation. In other words, Americans have used automobiles for more than just transportation. Third, the potent mix of practical and symbolic satisfactions that automobiles provide, as well as the significant economic stakes they represent for both consumers and producers, have made it hard, not so much to recognize their environmental impacts (many were recognized early on), but to address them.

The environmental impacts—on human health and on the air, water, and natural resources necessary for life—are presented here in the larger framework of the *product life cycle*. This perspective reminds us that consumer goods can touch the natural world at four points in their existence: *raw material extraction, manufacturing, consumer use,* and *disposal*.

The twentieth-century American experience with the automobile provides great environmental stories at each of these four points. Clear-cut hardwood forests and open-pit iron mines, industrial pollution, and abandoned automobiles are all examples of the environmental impact of the automobile that go beyond the commonly recognized problems of smog and global warming. Each phase of the product life cycle has a unique environmental history that has included important and often unheralded environmental success stories, such as the development of automobile recycling and the adoption of the catalytic converter, that permit cautious optimism about the automobile's future. The automobile also required a supporting infrastructure to provide fuel for the vehicles and roads on which to drive them. But I do not discuss the environmental impacts of petroleum drilling, refining, and distribution or road building. Nor do I explore the environmental impact of sprawl, although the automobile certainly encouraged twentieth-century Americans to spread out and convert far more land into human-defined landscapes than would otherwise have been the case.

All of the automobile's environmental impacts occurred in a larger context shaped by consumers as much as (and sometimes more than) producers. To make this claim is not to downplay or excuse the negative impact of producers. The automakers have had extraordinary power to shape the nature and degree of the automobile's environmental impact through choices about product and manufacturing design as well as through marketing. Yet consumer choices have also influenced the nature and degree of environmental impact. Buyers and sellers have a codependent relationship, and they bear joint responsibility for the side effects of their behavior.

American automobile consumers challenge the historian as much as they challenge automobile marketers. It often seems that we know everything and nothing about them. In this book I describe what consumers have done with automobiles (behavior). I also explore possible reasons for their actions (motivations). In a more speculative vein, I occasionally suggest connections between the larger American social context and consumers' motives and behavior. Historians are often reluctant to venture explanations where the evidence is not unambiguously persuasive. I am no different. But I share these thoughts so that the reader may ponder them, too.

The automobile was America's glamour product for much of the twentieth century, so evidence about the behavior of automobile consumers is plentiful. Although few people confided their innermost thoughts to diaries when they went out and bought a car, a veritable army of interested people observed and explained their behavior. The evidence that observers left behind in mass-circulation periodicals, business publications, trade journals, market research, memoirs, occasional open company records, and academic studies allows historians today to look over their shoulders and learn a great deal about automobile consumer behavior.

Explaining why consumers behave as they do is another matter. While writing this book I thought often about the insights of a pioneering group of psychologists that includes Timothy D. Wilson, Daniel Gilbert, Amos Tversky, and Daniel Kahneman. The upshot of their findings, which scholars in other fields and policymakers are only just starting to appreciate, is sketched here. A rational decision, whether defined by method or by optimal outcome, is rare. Thus, historical explanations that presume that a human action sprung chiefly from rational calculation are suspect. Powerful motives strongly influence much of our behavior. But they often operate at a level beyond our consciousness. These motives are very difficult for us to discern directly and are, therefore, largely outside our conscious control. Emotions—conscious and unconscious—often stimulated by external factors such as the behavior of others, also play an important role in human decisions. So do the promotional efforts of companies trying to sell products, but only in a general way, an influence that paradoxically is strongest when people are not paying close attention. Although powerful influences are at work, most of us do not really know why we do what we do most of the time. When asked why we bought a product, we make up plausible stories that seem true to us. Because people generally have a hard time discerning their own motives from introspection alone, motives must be inferred from behavior. But individuals also have a tough time accurately observing their own behavior.

The good news is that outside observers appear to be as good as, and often better, at spotting the motives behind our behavior than we are ourselves. And that is mostly what the observers and commentators on American automobile consumers have done from the beginning. It is what I do as a historian in this book. There are challenges in interpreting this evidence (often explained by the frustrated sales and marketing professionals who were paid to make sense of it in the first place), but there is no shortage of material. The vantage point, as any marketing professional will affirm, is not perfect, but with the automobile it is often very good.

Given the basic relationship between consumers, producers, and environmental impact, the options to reduce harm to the natural world seem straightforward. Consumers could demand environmentally friendly cars, automakers could produce the least harmful cars possible, or concerned citizens could use the law to regulate the detrimental behavior of consumers and producers. The twentieth-century American experience with the automobile shows just how hard each approach has been to pull off in practice, even when scientists, engineers, hunters and anglers, homeowners, conservationists, environmentalists, and journalists have called attention to the problems.

The story begins with what may have been the greatest class provocation in American history. The public antics of wealthy young sportsmen in the first automobiles triggered a strong emotional reaction from other Americans—part enthusiasm, part anger, part envy—that soon made the ownership of large, fast, powerful automobiles synonymous with successful belonging in middle-class America. This emotional context and other factors frustrated the early efforts of engineers and scientists to address the two problems that have been central to the automobile's environmental impact ever since, noxious emissions and gasoline dependence, which they immediately recognized would become far greater issues once the automobile was widely adopted.

The Model T arrived at this charged moment. Its success is best explained, less by its low price (which came later), and more by how well the car's practical and symbolic attributes spoke to the conflicting emotions surrounding the automobile. Mass production and consumer financing made automobiles more affordable. Searching for effective ways to cope with anxieties caused by the impact of sweeping economic and social changes on their sense of status, dignity, and respect, Americans quickly turned automobile ownership into a self-affirming "community" of the successful and made their country the world's first mass automobile society, which dramatically magnified the impact of consumer behavior on the natural world. Under pressure from the perceived threat that runaway consumer demand posed to finite oil supplies, the surgeon general and the federal government refused to block efforts by General Motors and Standard Oil to put tetraethyl lead in gasoline to improve engine efficiency, despite forceful warnings from medical experts about widespread atmospheric lead poisoning.

Mass production meant massive environmental impact in other areas of the product life cycle. Nowhere was the impact more evident than in the great flows of raw materials and the manufacturing processes at Ford's River Rouge complex. In the 1920s Ford actually chose to market its cars by telling consumers how they were made. Yet when Ford publicized the Rouge, using the greatest concentrated source of industrial pollution from automobile manufacturing that ever existed to make the greatest effort ever made by a major corporation to show consumers how its product was made, consumers saw an "industrial epic" rather than an environmental nightmare. Ford's extraordinary efforts at waste reduction, which did reduce the environmental impact of its manufacturing processes, helped in this regard.

Mass automobile ownership brought the American automobile market to the "saturation point" crisis of the mid-1920s, as consumers clamored for greater variety in cars so that they could distinguish themselves from one another and producers sought better ways to sell replacement vehicles sooner. Much to Henry Ford's distaste, consumers encouraged wary automakers to accept planned obsolescence and develop the styling-centered annual model change. Mass production and de facto acceptance of planned obsolescence also created mass disposal problems that led to the birth of the automobile junkyard and the beginnings of automobile recycling. At Henry Ford's insistence Ford experimented with a high-volume disassembly line at the Rouge, which brought the company to the threshold of designing its cars so that they could be taken apart. But further pursuit of these consumer and industrial developments was constrained by the fifteen years of depression and war that followed.

Elvis Presley's purchase of a pink Cadillac in 1955 was a moment that in retrospect seems the apogee of the American auto industry's golden decade, 1945–55, when automakers and consumers together found much to love in chrome and tailfins. The direction of consumer tastes and producer offerings epitomized by the Cadillac exacerbated the automobile's impact across the entire product life cycle and marked the 1950s as the "dark ages" of the automobile's environmental impact. Yet automakers found that they could not pollute with impunity. Industrial pollution from plants like the Rouge affected consumers, especially recreational sportsmen and homeowners, and smog caused by automobile exhaust affected Los Angeles homeowners and businesspeople alike. Their complaints set in train a thirty-year process of strengthening environmental regulations to protect their larger consumer, leisure, and business agendas.

The American consumer's love affair with the postwar automobiles of the Big Three did not last. After 1955 some Americans stood the Cadillac-dominated status hierarchy of automobiles on its head by buying the Volkswagen Beetle. At the same time, critics who ridiculed Detroit's cars as overpowered, oversized, overstyled, overpriced, and oversold found a public audience for the first time in nearly half a century. Sales fell 46 percent in three years, precipitating the Buyers' Strike of 1958, which did in the Edsel and forced Detroit to respond with smaller, plainer cars. Environmental concerns were not yet national concerns in this important transitional moment, although postwar trends toward greater size, weight, embellishment, and speed all made the automobile's environment impact worse. Still, the sudden shift in consumer tastes toward smaller cars raised the possibility that consumers and critics could nudge producer offerings in environmentally friendly directions using explicit environmental concerns.

One postwar development that went uncriticized was the effort to convince American families to buy a second car. More than any other factor, the growing number of cars on the road finally made the automobile's environmental effects a subject of national debate in the 1960s. More-car-per-car and more cars all added up to a greater disposal challenge for America's automobile wreckers. Unfortunately, the scrap industry's greatest customer, the American steel industry, stopped buying auto scrap at this very moment, which resulted in overflowing junkyards and the abandoned automobile problem of the 1960s. Booming auto production also outstripped efforts in the 1950s and 1960s to bring air and water pollution under control at the Rouge, which fueled further protests by sportsmen about water pollution and triggered a grass-roots fight against air pollution in Dearborn's South End. Californians, meanwhile, waited for a technological fix for smog in the form of factory-installed emission control devices from Detroit. Detroit's lethargy led to the controversial 1969 smog antitrust suit that alleged that the automakers conspired to withhold the technology. Growing frustration with these automobile-related problems, articulated and orchestrated by the national media, led to the creation of the Environmental Protection Agency (EPA) and the passage of the Clean Air Act amendments of 1970, one of the toughest and most controversial environmental laws ever passed in the United States.

Over the next five years, the EPA forced American automakers and consumers to confront the environmental implications of their relationship, and in turn, automakers and consumers forced regulators and environmentalists to recognize the limits to regulation. GM's Ed Cole embraced the catalytic converter and convinced the oil industry to provide the unleaded gasoline necessary for the catalytic converters to work effectively. Installation of catalytic converters on 1975 model cars (with subsequent improvements) is the single biggest step ever taken to reduce the automobile's environmental impact. At the same time, the 1970 Clean Air Act amendments forced the EPA to require of the states implementation plans that demonstrably reduced pollution to mandated levels by 1975. The law and the courts stipulated that these plans must include land use and transportation controls to meet the air quality standards in urban areas if necessary. When drivers and commercial landowners in Los Angeles and other cities rebelled at the prospect, the EPA backed down, and everyone agreed that they could live with smog until Detroit could produce effective emission control devices. The transportation control plan controversy in 1973 marked the furthest reach of federal efforts to regulate human behavior on behalf of the environment.

The energy crisis arrived amid these efforts. The twin oil shocks in 1973–1974 and 1979 that sent gasoline prices to record highs strongly reinforced a trend to small cars that started in 1967. By 1980 there was universal agreement among automakers, government regulators, and industry analysts that a revolution was at hand. The future belonged to the small car. The revolution that followed was not the predicted one. As the energy crisis faded in memory, it became clear that small cars had been stigmatized by their association with a downer decade. It was also clear that the power of critics and regulators to force the automakers and consumers to find more environmentally friendly common ground had crested. If consumers had limited producers' impunity to harm the environment, they also limited the ability of regulators to impinge on their consumer agendas.

The Baby Boomers who bought cars in the 1980s and 1990s had been raised on Ralph Nader, smog, and the energy crisis, yet millions embraced the sport utility vehicle (SUV), a vehicle that posed safety, emissions, and fuel economy problems. To many this behavior was as inexplicable as it was disappointing. Yet in the larger historical context of America's experience with the automobile, the Boomers' vehicle choices were not particularly surprising. Like their predecessors in 1910–1924 and 1945–1955, they experienced a powerful need to create a "community" around ownership of a particular kind of vehicle. Yet there was an important difference this time. Consumers knew a great deal about the automobile's environmental impact, and commentators regularly reminded them that their developing taste for SUVs aggravated this impact. The message for those concerned about the relationship between consumer behavior and environmental impact was painful. Although knowledge of environmental harm may be a necessary condition for consumers to change their behavior, American choices in automobiles in the two decades after 1980 proved that it was far from a sufficient condition when more important personal agendas and human limitations were involved.

When we recognize that the automobile has provided Americans with important psychological satisfactions that go beyond the convenient personal transportation provided by four wheels and an engine, we can see that the environmental impact of the automobile is a symptom of deeper things in the experience of modern life. Americans' hunger for respect and other satisfactions, and the difficulty that an economic system premised on unceasing change has had in giving them the reassurance and affirmation they need, has made it difficult for them to care for the natural world to the degree that they might have, even after they began to recognize the harm that they were doing. Although the typical American automobile became more environmentally friendly in the last third of the twentieth century and may well become more so in the future, the message for those concerned about the relationship between consumer behavior and environmental impact remains sobering.

This is not an angry book. We don't need another angry book about automobiles, a perspective neither new nor helpful. There were many wonderful

things about being an American in the twentieth century. Having cars and driving were certainly among them. Automakers that work hard to sell their cars to
consumers are not evil. Self-interested and oblivious consumers are not evil. But
the two together do pose problems for the environment, and on this score both
could do better. Rather than celebrating or criticizing automobiles, we need
to understand the broader and deeper connections between consumer behavior and environmental impact, subjects too often treated separately. This book's
special contribution is linking consumer agendas with environmental impacts
in the context of the century-long American relationship with the automobile.
With this larger aim in mind, I have written the book for individuals who are
curious about these issues, not for academic or professional experts. The stories
told here are based on both primary and secondary sources. Some have been
told before by others. But many are presented here, to my knowledge, for the
first time. Nearly all involved primary research that made it possible to form my
own conclusions. Covering this ground in a book of readable length required
some sacrifice of detail and nuance. Interested readers can explore specifics in
greater depth by delving into the large literatures on automobiles, automakers,
consumer behavior, and environmental impact. The causes, connections, and
ramifications of the Great Experiment are important. *Auto Mania* is my take
on it.

—•————•—

The Arrogance of Wealth

The public knew William K. Vanderbilt, Jr., Foxhall P. Keene, Albert C. Bostwick, David Wolfe Bishop, Edward R. Thomas, and Harry S. Harkness from the society and sports pages, their social lives in New York and Newport, and their exploits as sportsmen. These young heirs to the great Gilded Age fortunes owned and raced yachts, bred and ran thoroughbred, steeplechase, and harness racing horses, and played competitive polo, golf, and tennis against one another. When Vanderbilt and his friends seized the automobile and took to the roads to have fun, they became Thorstein Veblen's "conspicuous leisure" and "conspicuous consumption" on wheels. Less fortunate Americans gaped in awe, envy, and exasperation. Americans later lionized the automobile's inventors and early manufacturers, but they forgot the men who actually showed them what the automobile was all about. No group did more to popularize the gasoline-powered automobile, stimulate the imaginations of their fellow citizens, and give substance and trajectory to the relationship between Americans and automobiles.[1]

Vanderbilt and his fellow sportsmen first demonstrated publicly the greatest pleasure of the automobile—speed. The newspapers and automobile magazines of the time make clear that no appeal approached it in importance in the first years of the automobile's existence. Speed had nothing to do with getting somewhere sooner. Driving fast was simply fun. The human experience of speeds greater than those provided by the horse dated only to the advent of the railroads in the 1840s, but early automobile drivers gushed with enthusiasm over the new thrill of having the speed and direction of an open-air vehicle entirely in their own hands. "When one sees several miles of clean road ahead," wrote an early enthusiast, "he enjoys shooting it at the highest speed possible." New York

Speed. *Courtesy of the Smithsonian Institution, NMAH/Transportation.*

City automobile owners quickly found the flat, well-maintained country roads of Long Island ideal for this purpose.[2]

Commentators soon wrote of "the speed craze" and "speed mania." "Those who indulge in abnormal speed simply for its own sake . . . are mostly degenerates, devoid of all self restraint and having absolutely no control over themselves," wrote the editors of the automobile enthusiasts' magazine *Horseless Age* in 1903. "The speed habit resembles the alcohol and morphine habits . . . in the last stages the victim indulges it with complete abandon." These editors, who took a more utilitarian view of the automobile, worried that the overexuberance of sportsmen like Vanderbilt would provoke an excessively negative reaction that would cripple the new technology in its infancy.[3]

William K. Vanderbilt, Jr., enjoyed driving fast and showing off, but he also played for a larger audience and sought more tangible markers for his exploits. In June 1900 he bought an imported, twenty-eight horsepower Phoenix-Daimler for ten thousand dollars with the expressed intent of having the largest and fastest car in America. In July he used his newly dubbed White Ghost to drive the dirt roads from Newport to Boston in two hours, eighteen minutes, half the time of the previous record, and almost as fast as the railroad. For reporters, Vanderbilt coyly painted the dash as a pleasure trip but noted that the speed register he carried on the car had shown sixty-five miles an hour on the run into Boston on

Speeding Sportsman William K. Vanderbilt, Jr., c. 1906. *Library of Congress, Prints and Photographs Division, LC-USZ62-99747.*

Blue Hill Avenue. He allowed that, if he hadn't been stopped for speeding, he would have beaten the railroad's time.[4]

Not to be outdone, in August 1900 Albert C. Bostwick brought back a prize-winning twenty-four horsepower Panhard-Levasseur from France that he had purchased from Count Rene de Knyff for thirteen thousand dollars. A showdown was in the offing. When Bostwick pleaded the need for new tires, though, Vanderbilt in the White Ghost won the automobile races held at Aquidneck Park outside Newport a month later, on 6 September. He repeated his triumph in 1901 before seven thousand spectators when he vanquished Foxhall Keene in a new Mercedes that Vanderbilt christened the Red Devil. Though he had passed on Aquidneck, Bostwick soon dominated rivals in races held in the New York City area. In these and subsequent vehicles, Vanderbilt and Bostwick won races and set world speed records at various distances over the next few years.[5]

Newspapers around the United States widely reported these exploits. Vanderbilt, Bostwick, and their automobiles became sensations. The first national automobile show, held in New York in November 1900, featured Bostwick's Panhard as an attraction. Vanderbilt later in life recalled his early trips on the open road in the White Ghost. "We couldn't rest when we wanted to. We even were forced to go hungry because every time we stopped at a restaurant we were forced to fight our way through a dense crowd to get back to it." This publicity colored Americans' early consciousness of the automobile. "My dreams at night," one early automobile purchaser recalled, recounting how he had come down with "automobile fever," "were visions of Red Devils, White Ghosts and Black Phantoms."[6]

Using their expensive imported European cars to set speed records between cities and to race one another on the track and in hill climbs, Vanderbilt and his friends generated publicity that influenced consumers. Automobile buyers, *Horseless Age* noted, sought vehicles with "such high motor power that the car will 'romp right up' any hill on the high gear." For many prospective buyers, hill-climbing power constituted the acid-test of any particular model. "Most of the larger cities," wrote the magazine's editors in 1905, explaining the auto-purchasing public's decided preference for high-powered internal combustion engine vehicles, "have some well known test hill on which customers require cars to be demonstrated to them, and which must be climbed at a considerable speed, and sometimes on the high gear, if the prospective purchaser is to be satisfied." The hill climbs, more than any other context, showed people the superiority of gasoline-powered automobiles over electric and steam-powered vehicles.[7]

By 1904, when vehicles such as Vanderbilt's ninety-horsepower Mercedes proved too powerful for the annual hill climb at Eagle Rock, New Jersey, the hill climbs had made their point. "The public has been educated by the commonness of large and powerful cars," *Horseless Age* wrote, "to expect a certain lavishness of power and size in everything called by the name automobile." When Vanderbilt and his fellow sportsmen lent a glamour and prestige to large gasoline-powered automobiles that made irresistible copy for the press, they shaped American tastes in automobiles that persisted. People think of Henry Ford as the father of the American automobile, but Ford was actually the midwife to mass ownership. William K. Vanderbilt, Jr., and his fellow speeding sportsmen fathered the American love of large, fast, powerful cars.[8]

More than thirty years later, Alden Hatch remembered the excitement of the first automobiles, as he watched Foxhall Keene lead the inaugural Vanderbilt Cup Race:

> Beyond the curve of the narrow, macadamized road there sounds a tremendous, staccato roar that swells awesomely with the passing of each tense second. Everyone climbs onto the rail fence and strains his eyes in the direction of the uproar. I need Father's steadying touch as I balance precariously on my perch, half insane with excitement.

Then it comes, tearing the morning mists apart. Orange flames shoot from its flanks; dense black smoke swirls in its wake. It leaps toward us with unbelievable speed. Seated in the midst of that streak of fire and smoke, I can see a slim, calm figure, like a god riding the cosmic storm. Father shouts in my ear, "That's Foxie."[9]

Vanderbilt and his fellow sportsmen were not alone among wealthy Americans in popularizing the automobile. A much larger number helped introduce the automobile to the country through the "touring craze." The automobile freed people from having to travel with other passengers by a fixed route on the railroad's schedule. Best of all, it enabled people to explore the vast spaces between the railroads. Early automobilists found with excitement that in many ways rural America had changed little in the nineteenth century. The countryside also gave the early automobilists a chance to get closer to nature. History and nature offered escapes from the daunting new realities of turn-of-the-century urban America and represented an earlier innocence that the automobile now placed within reach.[10]

Promoters quickly saw the potential in touring. "The automobile," *Horseless Age* predicted in 1901, "will, in a measure, bring back the times of yore when the wayside inn flourished and the stage coach was the vehicle of interurban communication. To-day stage coach travel would have no fascination for anyone, although it has its good points. The bad points are practically all absent when traveling in private automobiles, and for this reason we expect to see automobile touring become one of the most popular pastimes in America within the next few years." This prediction was fully realized. The first purchasers of automobiles invariably took off for the countryside, not only to speed where they could, but also to explore a new world. Speeding and touring—often combined—constituted the twin pleasures that sold wealthy Americans on the first automobiles. Both pursuits provided those who could afford automobiles an irresistible source of pleasure as well as adventure.[11]

Along with the thrill of speed, touring confirmed early American automobile owners' preference for large, powerful cars. They found large cars necessary to carry passengers, of course, but the longer wheelbase also made for a smoother and safer ride on the wretched backcountry roads. The greater weight required greater engine power, and the conditions of the roads demanded even more. But the American preference for large, fast, powerful touring cars, especially of the popular open tonneau style with low sides, is a reminder that the point of touring was not merely to see but also to be seen. Whether they drove cars or fixed them, other people saw them with their autos. No doubt, they preferred to be seen riding, if not speeding, in them, but being seen fixing them provided

Touring. *Library of Congress, Prints and Photographs Division, LC-USZ62-7119.*

satisfactions as well. Only a public failure to fix the vehicle and the need to ask for help brought embarrassment.

Automotive historian Gijs Mom has called the automobile an "individual adventure machine," an apt description and reminder that the automobile was far more than merely an improved form of transportation for its earliest users. To appreciate this description requires underscoring the link between adventure and a new conception of manhood at the turn of the twentieth century. Noting that adventurous young men pioneered both bicycles and automobiles, Mom showed that bicycling had its subcultures of racing and touring, as well as its mechanical challenges to test the rider's skill. The automobile offered intensified experiences in each of these areas for men to distinguish themselves further from their more timid fellows. There could be no adventure without risk and no distinction without accepting the risk and successfully demonstrating oneself the master of the technology in public.[12]

Bicycle and electric vehicle maker Albert Pope famously derided the gasoline-powered automobile, arguing that "you can't get people to sit over an explosion." But he misunderstood the people who bought automobiles. Internal combustion engines appealed to Vanderbilt and the speeding sportsmen for precisely that reason. They used large, powerful gasoline-fueled automobiles to set themselves apart and to communicate the message that they were privileged men—that it took money, skill, and "balls" to drive an automobile. Americans got the message.[13]

"No American sport . . . has ever enlisted so much power and money," the *Washington Post* wrote of the automobile in 1902. "No other sport, bar none, has ever aroused such direct personal antagonism. No other sport has come so surely to stay, and no sport in its very beginning has ever promised so surely to become a matter of vital concern for the whole country." The introduction of a major new consumer product can cause turbulence. When it is a highly appealing, expensive, and publicly used product with strong class and gender associations, it can also be socially destabilizing. Suddenly, one group in society enjoys a new set of pleasurable experiences while everyone else does not. For the have-nots the result can be discomfort, since the high cost of the product prevents them from acting on their desire to have it. Class distinctions suddenly become painfully obvious. The speeding millionaire sportsmen so effectively demonstrated and publicized the speed and power of the automobile that its introduction had an "in-your-face" quality. Their behavior aroused strong emotions in other Americans, provoking a bitter reaction while also stoking the desire of millions to own an automobile, too.[14]

A century later, Americans, if they had any inkling of a negative reaction against the automobile in its first years, knew only the jibe "Get a horse!" that called to mind shortsighted, antimodern rubes who failed to recognize the automobile's obvious potential utility. These skeptics, so the usual story goes, soon came around. Yet in fact, many people reacted to the automobile with intense anger, and even acts of violence—often tinged with class hostility—that were conveniently forgotten in the post–Model T democratization of American automobility. "The aggressive and drastic opposition . . . ," observed the *Washington Post* at the time, "is unique in the history of the development of any new sport in this country." This reaction had little or nothing to do with a preindustrial mindset or the knee-jerk rejection of a new technology. Far from it. Nearly all Americans initially reacted favorably to the automobile, saw its potential, and viewed its future with optimism. But many resented how the new technology was introduced by the wealthy. More to the point, they disliked what being automobileless said about their status. Americans have repressed this part of the story.[15]

The automobile may have been the plaything of the wealthy, but it was not something that they played with at home. To indulge in the twin pleasures of speeding and touring, they took their vehicles out on the public streets and roads with everyone else. Here the problems began. By far the biggest one derived from the automobile's greatest pleasure: speeding. As early as November 1900, the editors of *Horseless Age* warned their readers that "the need of reasonable consideration on the part of automobilists in the matter of speed is a thing that cannot too often be insisted on. Both on account of runaways of horses and because of the apparent danger to pedestrians, the public sentiment aroused by the apparent disregard of others' rights on the road is very apt to find expression in legislation quite unnecessarily oppressive." The magazine's warnings made little headway; the mania among wealthy Americans for automobile speed waxed unabated.[16]

More than anything, the attitude that accompanied the speeding outraged the public. "There is plenty of reckless driving which endangers no one but the foolhardy chauffeurs [that is, drivers] themselves," the editors at rival magazine *Automobile* noted, "but with it is an appalling amount of cold-blooded, deliberate disregard . . . of the rights of other users of the road. . . . Every suburban road has seen the motor hogs who refuse to give half the road, and force horse drivers, and even cyclists, into the ditch." "The abuse of powerful signaling horns," wrote *Horseless Age*, "has perhaps done as much as anything to stir up opposition to the automobile in country and fashionable residence districts. Who has not at one time or another seen an automobilist speeding along the road at an illegal rate, continually tooting his horn as a signal for other road users to clear the way? The practice savors of a spirit of arrogance or domination, and is resented by all who are annoyed by it." Legislators responded quickly by imposing speed limits in the northeastern states. This legislation, as many of the automobile's boosters rightly sensed, was a kind of "class legislation" intended to put the wealthy speeders in their place, but that task proved exceedingly difficult. If William K. Vanderbilt, Jr., can be called the father of Americans' love of automobile speed and power, he should also be remembered as the father of the American speed limit, traffic fine, license plate, and revocable driver's license—successive steps taken in an escalating battle to curb speeding.[17]

Speeding angered the public, not only because the speeders flouted the law with near impunity, but also because speeding and reckless driving were the leading cause of automobile accidents. "The word 'fatality,'" the *Washington Post* noted in 1902, "is linked in the popular American mind with the name 'automobile.'" Had wealthy Americans simply been killing themselves, as they soon began to do with some regularity, the public might well have been content to let it go at that, but wealthy drivers began to kill and injure middle- and working-class pedestrians. The public usually presumed that speeding caused these accidents, and frequently it did, thereby fusing in the popular mind two

Pleasure meets social responsibility. *Courtesy of the Smithsonian Institution, NMAH/Transportation.*

potent reasons for opposing the automobile—speeding and accidents—with inflammatory consequences. When a child was run over in an eastern city in June 1900, a mob quickly formed in search of the automobile owner—only to learn that the vehicle involved was not an automobile. Incidents that did involve automobiles, however, resulted in mob violence directed at drivers. In this climate, drivers who had been in accidents became circumspect about the causes, rarely admitting to speeding. Worse, with public sentiment running against the automobile and mob violence a distinct possibility, many drivers calculated that it was not worth stopping to inquire about the welfare of their victims, which, of course, only further enraged the public. "The scorching chauffeur is hard to deal with," *Horseless Age* lamented, and if his speeding was not problem enough, "when an accident is caused through his recklessness he, as a rule, increases his pace to escape the law, instead of stopping to offer assistance to those he has injured."[18]

William K. Vanderbilt, Jr., might easily have become a lightning rod for this hostility. "Mr. Vanderbilt's famous 'White Ghost,'" the *Washington Post* noted, "has done as much as any automobile in the world to emphasize the dangerous side of the horseless carriage." But Vanderbilt led a fairly charmed life behind the wheel. Although he had several accidents, the worst of these occurred in Europe, and none involved the loss of human life. (The animals in his path, it should be noted, fared less well.) His fellow sportsman Edward R. Thomas, who had the misfortune of hitting children and elderly women in several well-publicized accidents, presented another story.[19]

The toll from automobile accidents grew with the number of automobiles on the road. As the "speed craze" grew in the 1900s, so did the public outrage. From the outset, it made wonderful copy for the daily newspapers, which both publicized the accidents and criticized the automobile. In New York, the *New York Times* took the lead in this regard. It regularly reported speeding arrests and accidents involving wealthy automobilists on the front page and excoriated the perpetrators and automobiles generally on the editorial page. The automobile press greatly resented this unwelcome publicity, and *Horseless Age*, although resolutely opposed to speeding and reckless driving, took the daily press to task for magnifying the public outrage. But sensationalizing traffic accidents and criticizing the automobile sold as many papers and perhaps more than simply celebrating the exploits of sportsmen such as Vanderbilt and hyping the automobile. By 1908, even *Horseless Age* resigned itself to "the annual slaughter."[20]

Given the flagrant speeding, the arrogant disdain for pedestrians and other road users, the open flouting of the law, the failure of legal remedies to curb the abuses effectively, and the inflamed climate created by the daily newspapers, some people, not surprisingly, took justice into their own hands. Sometimes this reaction took mild if pointed forms. "Our attention has been called by men prominent in the [automobile] industry in this city [New York], who have occasion to drive on Long Island roads," *Horseless Age* commented in July 1901, "that tacks, glass and other tire puncturing instruments have of late been distributed over these roads in great quantities, presumably by parties who consider their interests affected by speed excesses." Significantly, rather than expressing anger, the magazine's editors conceded that these practices were merely "retaliation for aggression actually committed [by automobile drivers]." "Nothing could show better than this how the entire fraternity of automobile users and the whole industry are compelled to bear the sins of inconsiderate drivers."[21]

Stonings, which occurred when automobile owners drove through rural districts or urban neighborhoods where they were not welcome, constituted the most pervasive and shocking form of opposition. A stoning involved more than a single stone being thrown at an automobile. Rather, vehicle, driver, and passengers came under sustained assault from a group typically consisting of boys, male teenagers, and young men. They threw not only stones but mud, sticks, beer bottles, lead pipes, and decaying vegetables. These incidents occurred around the nation throughout the automobile's first decade. As early as 1897, youths threw stones at automobiles in New York City, and Rhode Island farm boys stoned the vehicles of wealthy Newport automobilists when they ventured into the countryside. By December 1903, reader letters in *Horseless Age* suggested that "sufferers from this annoyance are many" and urged that stone throwing be made a felony offense.[22]

Sometimes the popular anger that Americans directed at automobile drivers took stronger forms. Farmers in New York State and Wisconsin threatened motorists with shotguns. In Wisconsin, according to a report in *Automobile*, "one farmer is reported to have said that he derives great pleasure in shooting from some secluded spot at the tires of passing autos." James Gilmore horsewhipped Mr. and Mrs. W. L. Gould when they passed Gilmore's horse and buggy just outside Watertown, New York, causing Gould to lose control of the automobile and crash into a ditch. A boy with a twenty-two-caliber rifle shot and wounded William A. Graff while he drove from Pittsburgh to Butler, Pennsylvania.[23]

Although these violent acts were unquestionably motivated in part by class hostility, they also had approval from elites. The anti-automobile crusades in the popular press constituted one important sanction. The *Washington Post* went so far as endorsing highway robbery when it reprinted a story from the *Philadelphia Record* that noted that "rural constables, magistrates, and farmers, as well as suburban people who work in town and cannot afford automobiles to carry them thither, heartily approve of the holding up of automobile parties by bands of hungry and thirsty hoboes who pose as constables and threaten the travelers with the calaboose [jail] and a fine unless they shall 'whack up' [pay]."[24]

The anti-automobile laws passed by the state legislatures and the broad public outrage that these efforts reflected also served to countenance extralegal action. Most editors and legislators did not explicitly endorse violence or vigilante actions, but some people in positions of authority did, as when a New York City magistrate in 1902 urged pedestrians to carry firearms to shoot any motorist who threatened to run them down. The following year a New York City daily quoted the noted minister and municipal reformer Dr. Charles Henry Parkhurst as saying: "If I could be Czar of New York for thirty days I would kill the first automobilist caught exceeding legal speed in the streets of the city." Such extreme statements represented a minority viewpoint, but even the middle-class *New York Times* editorialized that "there is one thing that travels faster than the swiftest automobile yet invented and that is a bullet. It would not be well for the drivers of automobiles to induce the belief that firearms offer the only really efficient protection against them."[25]

The *Times* anguished editorially over the stoning problem, alternately condemning and rationalizing it. "The automobile has come to be regarded as a dangerous vehicle under all conditions, and boys throw stones at it partly from a sense of public duty and partly because it is 'fun' to do so . . . ," ran one editorial. "The millionaire who finds his racing machine dented and scarred may think he has a grievance, but considering what people of his class or their imitators have done and are daily doing to make the automobile terrible, he would do well to devoutly give thanks that the dents and scars are not on his own head, where, to tell the plain truth, they would usually do most good." Acts of hostility, such as when Edward R. Thomas and his wife were stoned in their open car by a group of working-class boys on a Manhattan street in 1902, drew support that cut across class lines. In response to this incident one clergyman asked, "What more natural, then, that the hoodlum of the street should blindly strike at the hoodlum of the automobile?"[26]

On 27 February 1906, Woodrow Wilson, then president of Princeton University, made the most famous anti-automobile statement in remarks before the North Carolina Society of New York City. "I think that of all the menaces of today the worst is the reckless driving in automobiles," news reports quoted Wilson as saying. "In this the rights of the people are set at naught. When a child is run over the automobilist doesn't stop, but runs away. Does the father of that child consider him heartless? I don't blame him if he gets a gun. I am a Southerner and know how to shoot. Would you blame me if I did so under such circumstances?" He continued: "Nothing has spread Socialistic feeling in this country more than the use of automobiles. To the countryman they are a picture of [the] arrogance of wealth with all its independence and carelessness." Wilson's remarks seem shocking, especially coming from a future U.S. president, but in 1906 they were typical of anti-automobile statements made by many elites and compelling evidence, not of a cavalier attitude on the part of Americans toward vigilante violence, but of the emotional intensity of the reaction to the automobile, anger that merits scrutiny, not casual dismissal or awkward silence.[27]

These incidents suggest a deeper dimension to the anger that went beyond speeding, accidents, and the obvious dangers to life and limb, a sense that an immense social transgression was taking place. Automobiles that interrupted children's games in the street caused the stonings, said some. No doubt, drivers did

Who Owns It, Anyway?

"Who Owns It, Anyway?" *Life*, 1902.

occasionally interrupt children's games. Yet the theme of territoriality pervaded, not just accounts of stonings and other assaults, but accounts of opposition to automobiles generally. People often treated drivers who passed through their neighborhood as unwelcome trespassers, regardless of the drivers' legal rights to the public road. Drivers were seen as interlopers, as wealthy, powerful, hyper-masculine transgressors who did not always behave with the expected deference of an awkward guest.[28]

Rural Americans, who in most states were forced by highway laws to bear a disproportionate share of the cost of the upkeep of the roads that ran through or by their property, experienced an acute sense of transgression. The roads were open to all, it was true, but previously all had meant mostly oneself and one's neighbors, not wealthy male strangers from elsewhere. Rural landowners had a legitimate proprietary interest in their roads. They found it especially galling to have wealthy drivers speeding up and down their roads, frightening their horses and raising dust, or to have affluent tourists come and gawk at them, flaunting their wealth and privilege. As a result, average Americans gave many a wealthy early automobilist a rough lesson in manners, rights, and respect.[29]

Although acts of popular resistance to the automobile added further "adventure" to driving, in fact they took the pleasure out of the experience because they robbed the driver of a sense of mastery, which hurt the early automobilists where it counted most. Animosity, often of an overt class nature, accompanied these anti-automobile acts. These actions cannot be attributed to shortsighted cranks who missed the obvious advantages of the automobile. On the contrary, people saw the aggression of early automobile drivers as a threat to their social standing, masculinity, and dignity, and they responded with understandable anger.[30]

Accompanying the enthusiasm and social turbulence of the introduction of the automobile were other issues. The emerging conservation movement forced Americans for the first time to grapple publicly with issues of diminishing natural resources and increasing waste. At the same time, the modern concept of pollution began to surface in popular consciousness. Were the unpleasant byproducts of America's rapidly expanding industrial enterprises merely nuisances or were they public health threats? How should these problems be regulated, if at all? To understand the automobile's reception in the larger context of such concerns, it is instructive to examine two issues—the use of gasoline as a fuel and the related problem of automobile exhaust—the two environmental issues that late-twentieth-century Americans most readily associated with the automobile.[31]

Engineers like Peter M. Heldt and Albert L. Clough, editors at *Horseless Age* who saw their responsibility as smoothing the way for the mass adoption of the automobile by solving problems that lay in its path, found conservation ideas and their implications for the future of the automobile and the automobile industry unavoidable and obvious. In his article "Conservation Idea in Its Application to Motoring" (1910), Clough, an MIT-trained engineer and a longtime technical writer at *Horseless Age*, argued for the importance of extending the current ideas about natural resource conservation to the resources most important to the automobile industry, since "as the supply of any natural product begins to become restricted on account of its approaching or prospective exhaustion the current price of the product goes up and the quality is very likely to drop." Commentators in automobile columns and magazines regularly expressed concerns about the supply of rubber and hardwood. But the adequacy of gasoline supplies concerned scientists, engineers, and industry observers the most. Thanks to Vanderbilt and the speeding sportsmen, American automobile buyers by 1904 had decisively chosen gasoline-powered automobiles over electric and steam-powered alternatives. Concerns about gasoline dependence reflected this fact and the growing conservation consciousness among America's technical elite, who expressed their worries in language virtually indistinguishable from that used to express the same concerns over the next century. "So far as is known," Clough wrote, "there are no petroleum supplies in process of formation, at least at such a rate as to interest this or many succeeding generations."[32]

Unlike some contemporaries, Clough was not an early-twentieth-century doomsayer, and he explicitly cautioned against "taking the position of the extreme alarmist." "It often happens," he wrote, "that an excessively severe demand opens new and unexpected sources of supply, and it may be that this will happen in the petroleum industry." Nonetheless, he argued that the issue merited the attention of both automobile manufacturers—who, he said, should design carburetors that more readily vaporized the denser grades of gasoline, thereby wasting less fuel—and consumers—who should ensure that they did not use a air-gasoline mixture that was overly rich in gasoline.[33]

Clough had no illusions that it would be easy to convince consumers of the merits of conservation. He feared that people would simply ignore the warnings about the gasoline supply with the blithe certitude that it would outlast their time. But he remained optimistic that his conservation message would resonate with many. "There is a large number of conscientious people who have a horror of waste," he wrote, "who are pained to see any of the natural resources which are a part of the birthright of the race carelessly thrown away, and who are willing to take some pains that posterity be not robbed by their carelessness."[34]

Here at the close of the automobile's first decade, Clough demonstrated a nuanced appreciation of one of the automobile's chief environmental liabilities: its dependence on petroleum for fuel. Moreover, he related this concern from the perspective of generational stewardship, the idea that the present generation should not borrow or plunder the patrimony of future generations. Then, to top it off, he concluded, "If these grounds are not sufficient it may be remarked that our streets smell badly enough without the addition to the atmosphere of vast quantities of unburned gasoline." Latter-day environmentalists concerned about the automobile really could not ask for more early prescience than this. But how widely were such concerns shared? As Clough realized, the issue was whether the concerns he expressed would have an impact, not just on technical people, but on people with more influence on the direction of automobile developments: manufacturers and consumers.[35]

2

FORESIGHT AND EMOTION

Before conservation reached the national stage and even before the automobile age really got started, industry observers, following the lead of farsighted scientists and engineers, issued warnings about relying on a nonrenewable fuel. *Horseless Age* commented on these worries in an article entitled "Oil Famine Bugaboo" in January 1897. Two years later *Motor Age* editorialized: "Even conservative engineers, think it the part of wisdom to look around for a substitute for gasoline." These concerns intensified as the new century began. "One of the great problems of the near future in connection with the popularization of the automobile will be that of an adequate and suitable fuel supply," predicted engineer Thomas J. Fay in *Horseless Age* in 1905. "The available supply of gasoline, as is well known, is quite limited, and it behooves the farseeing men of the motor car industry to look for likely substitutes."[1]

Not everyone expressed these concerns in pessimistic terms. E. P. Ingersoll, the publisher and first editor of *Horseless Age*, dismissed them with a technological optimist's flourish: "The great needs of civilization are quite certain to be supplied as they arise, and as to future generations we can safely trust them to settle their own difficulties and satisfy their own wants." The less sanguine, including engineers like Albert Clough, focused on possible solutions. "Many persons feel that the present restriction of the internal combustion engine to gasoline as a fuel is likely to seriously limit its development and increase its cost of operation to a prohibitive figure," he wrote in 1904, "but fortunately it is not too much to expect that ere long these engines will be successfully operated upon kerosene and even upon less refined products of petroleum, if not upon other gaseous or liquid fuels of low cost and great abundance."[2]

Gasoline is a petroleum derivative. It is not a single substance but a mixture of the lighter hydrocarbon compounds separated from the heavier substances in petroleum by the process of fractional distillation—by heating the petroleum and allowing the lighter compounds to rise to the top of the heated mixture before drawing them off. These lighter hydrocarbons constituted between 10 percent and 20 percent of a barrel of crude oil, depending on the oil's source. Gasoline began its history, not as a valued resource, but as an industrial waste left over when refiners distilled kerosene from petroleum. The principal commercial product of the American oil industry during the second half of the nineteenth century was kerosene for use in lamps. Kerosene manufacturers simply dumped the worthless gasoline back into the natural environment wherever expedient.[3]

But people noticed that gasoline both evaporated easily and had an explosive force when ignited. Inventors and oil companies seized on these qualities and experimented with gasoline in internal combustion engines. As a result, they found a profitable and arguably less polluting use for one of the oil industry's industrial wastes. Kerosene refiners could stop dumping their gasoline and begin to sell it to the new automobile owners. When electricity provided a new basis for household illumination, kerosene became the oil industry's troublesome by-product.[4]

Given its original status as an unwanted by-product, refiners sold gasoline at a low price—about seven cents a gallon in 1899. But demand from automobile users grew rapidly, and within a year the price doubled to fifteen cents a gallon. In 1902, the price jumped more than 50 percent from twelve to about eighteen to twenty cents. The price then fluctuated between twenty and twenty-five cents a gallon for the balance of the decade. Even at this price, automobile owners found the price of gasoline reasonable, particularly when viewed in relation to the overall expense of purchasing, operating, and maintaining an early automobile. Given the wealth required to own an automobile and the pleasures to be had from driving one, early automobile owners generally showed little concern about cost. In this context, industry observers viewed the sharp rise in the price of gasoline, not as problem in itself, but as a sign of growing scarcity. With recent discoveries of oil in Texas and California, the industry could accept scarcity as a temporary problem to be solved as the oil industry tapped these new sources and increased production to meet demand. But simultaneous indications that the Appalachian oil fields of Pennsylvania and Ohio were approaching the point of exhaustion and the realization that the oil from the new fields did not yield anywhere near the percentage of gasoline as the older fields raised the prospect that these fields might prove insufficient to meet the demand. As a result, people continued to worry whether it made sense for the new industry to use gasoline as its chief fuel.[5]

John D. Rockefeller's Standard Oil monopoly was another concern. "To the commercial leaders of the [automobile] industry, who assume the responsibility for the millions of capital invested in it," *Horseless Age* editor Peter M. Heldt noted in 1905, "the fact that the great industry which they are engaged in building up is entirely dependent upon a fuel the supply of which is in the hands of a monopoly, and is also limited by natural conditions, must be somewhat disquieting." While he acknowledged that a practical alternative was not quite at hand, he concluded that "in view of the possible limitation of supply it would be highly important if only a tolerably satisfactory substitute for gasoline could be found or developed."[6]

By 1905, scientists and engineers had devoted considerable attention to the search for an acceptable substitute. They quickly focused on kerosene and ethyl alcohol. With substantial petroleum production relative to the rest of the world, Americans initially leaned toward kerosene, but kerosene, though more plentiful than gasoline, still presented the same problem of finite supply. Europeans, in contrast, lacked substantial petroleum reserves and immediately sought an alternative fuel that could be produced at home from renewable sources. Lacking the large domestic petroleum deposits of the Americans and Russians or the extensive overseas empires of the British and French, the Germans worried that petroleum was a nonrenewable resource to which they lacked guaranteed access. Pushed by Kaiser Wilhelm himself, the German government and distillery industry actively pursued the alcohol alternative. Within a few years American engineers also concluded that alcohol held the most promise as a substitute to gasoline. Even better, it could be mixed with gasoline. "This fuel," Thomas J. Fay enthused, "far from being controlled by a monopoly, is the product of the tillers of the soil."[7]

Many Americans found this link with agriculture compelling. Federal agencies, solicitous of the needs of agricultural America, played an important role in exploring the use of alcohol as an alternative to gasoline. The Department of Agriculture and Geological Survey researched alcohol as a fuel, and the State Department ordered American consuls abroad to collect information on its commercial and industrial uses. Support for alcohol extended beyond scientists, engineers, and federal agencies. Even the automobile's greatest boosters gave support. The Automobile Club of America, which included the speeding sportsmen in its membership, reported strong interest in alcohol from its members and sponsored a competition for alcohol-powered vehicles. "Alcohol is not only a decidedly satisfactory substitute for gasoline as a motor fuel," *Motor* argued in 1904, expressing the strong consensus among experts, "but is much cleaner, less odorous, and freer from danger of explosion. Slightly more alcohol is required for the development of a given power than is required of gasoline, but this difference is more than offset by the lower cost—13 cents a gallon in Germany at the present time." By 1906, people concerned about gasoline had settled on alcohol as the preferred alternative. They knew that both fuels had pros and cons, but they concluded that alcohol had no rival as the best alternative to gasoline.[8]

It did not matter. Automakers and consumers ignored warnings about the gasoline supply and the advice of the experts. They refused to develop, let alone purchase, internal combustion vehicles that ran on alcohol. Pricing, competing social concerns, marketplace decisions by potential producers, technology, timing, and power all contributed to this outcome. In 1904, the price of alcohol suitable for use in internal combustion engines in the United States was, not thirteen cents a gallon, but about fifty cents a gallon, more than double the price of a gallon of gasoline. And this price did not include an excise tax placed on alcohol by the federal government, which raised the price of 94 percent commercial alcohol to $2.50 a gallon.[9]

In 1862 Congress and the Lincoln administration had imposed the alcohol excise as a "sin tax" to raise revenue to pay for the Civil War. Lawmakers retained it after the war to encourage temperance. By the early 1900s the United States found itself the only major industrial nation that did not exempt alcohol used for industrial purposes from such taxes. Efforts to repeal the tax or relieve industrial users of its burden had been made for nearly two decades and predated the automobile. But lobbyists from the petroleum and wood alcohol industries successfully rebuffed these assaults on the tax. Spokesmen for these two industries naturally argued that repeal would cause an increase in the social misery caused by drunkenness.[10]

Led by scientist Elihu Thomson and General Electric, lobbyists for several industries eventually turned congressional sentiment in favor of repeal. With evidence available that alcohol could be produced for less than twenty cents a gallon, the proponents of "free alcohol" believed that if the tax was removed, the price of alcohol would fall, especially as producers responded to the market opportunity with increased output. With the price of gasoline rising, they confidently expected that automobile manufacturers and consumers would soon switch to alcohol. The pro-repeal efforts received a crucial twin boost from Standard Oil's fall from public grace at the hands of muckraker Ida Tarbell, whose published exposé of Standard's anticompetitive practices appeared serially in *McClure's Magazine* from 1902 to 1904, and from President Theodore Roosevelt's strong support for repeal, which he explicitly linked to punishing the oil monopoly. Despite early opposition from Senator Nelson W. Aldrich of Rhode Island, an oil industry investor whose daughter had married John D. Rockefeller,

Jr., the Free Alcohol Act won congressional approval. Roosevelt signed it into law on 9 June 1906, permitting the sale of denatured ethyl alcohol excise-free after 1 January 1907.[11]

Despite the strength of the consensus behind alcohol as the preferred alternative, the removal of the excise and reliance on market forces failed to bring about the switch from gasoline to alcohol. Timing played a decisive role in this outcome. By 1907, the automobile industry and American automobile owners were too heavily committed to gasoline. Only a quick and substantial reversal in the prices of the two fuels at this point might have encouraged the switch. But prices did not fall enough because potential alcohol producers lacked either the will or the resources to make the investments to drive down the cost of the fuel. The "whiskey trust" Distillers Securities Corporation created the United States Industrial Alcohol Company to dominate the new market, but with monopoly profits in prospect, it saw no reason to drive down the price of alcohol rapidly. Farmers' cooperatives lacked the capital to enter the new industry on competitive terms. What's more, no obvious distribution system existed for the fuel, unless one contemplated tanking up at the saloon in more ways than one. So long as the petroleum industry continued to find and develop new sources of oil to keep the price of gasoline from rising too sharply, the oilmen effectively discouraged others from making the large investments necessary to produce low-cost alcohol.[12]

Alcohol faced technical hurdles as well. It could be used in existing internal combustion engines if manufacturers modified the carburetor to vaporize alcohol better during cold starts, but automobile owners still needed 25–33 percent more alcohol by volume to produce the same power or to cover the same distance as with the cheaper gasoline. This problem could be solved with higher compression-ratio engines, since alcohol could withstand compression of two hundred pounds per square inch, whereas engine "knock," a pounding, clanking noise, accompanied by overheating, a loss of power, and damage to the engine, occurred in gasoline engines when compression reached about eighty pounds per square inch. Columbia University professor Charles E. Lucke concluded that with high-compression engines, alcohol fuel could sell for double the price of gasoline and still be superior to gasoline-powered vehicles on a cost-per-mile basis. But the engineers and craftsmen involved with American metallurgy, machine tool, and automobile manufacturing at the time lacked the ability to build a high-compression, alcohol-powered car that was not so heavy as to make the vehicle lethargic yet strong enough to take the pounding delivered by the primitive suspension systems of early automobiles on America's ungraded and unpaved roads. Given these issues, manufacturers stayed with gasoline engines, opting for the challenge of meeting demand for these vehicles rather than developing competitive alcohol engines. And with sportsmen such as Vanderbilt demonstrating the joys and the symbolic cachet of gasoline-powered automobiles, consumers saw no compelling reason to push for a change either.[13]

The failure of alcohol disappointed many. "It will be some time before denatured alcohol will be more economical than other fuels," Peter Heldt reluctantly concluded in 1908. Still, he saw reason for hope in the long run, especially given the large quantity of agricultural waste that begged to be used. In the late twentieth century, some people looked back wistfully on electric and steam-powered vehicles as opportunities that were missed—options that were not in the running once it became clear that internal combustion engines delivered more of what consumers wanted—speed, power, and adventure with a declining level of inconvenience. Alcohol provided the best viable opportunity to switch to a renewable alternative fuel, a prospect that seemed promising to many in 1906 but one that fell short and then dimmed with the mass adoption of the gasoline-powered automobile.[14]

Scientists and engineers spotted another environmental problem with the automobile remarkably early. Automobiles produced an exhaust with a noxious odor that was often visible in the form of smoke. "Smoky exhaust," as observers called it, was an unwelcome by-product of using gasoline. However, unlike the choice of gasoline, automobile exhaust more directly involved the nondriving public. Americans at the beginning of the twentieth century for the first time grappled with the issue of industrial smoke problems in a serious way. Consequently, they viewed automobile exhaust in the larger context of what was then termed the "smoke problem" or the "smoke nuisance."[15]

Although people worried about the impact of smoke on public health, they remained uncertain as to which smoke nuisances presented such a problem. They defined nuisances as sights, sounds, or smells that offended the senses, not necessarily as threats to health. The automobile and its emissions arrived amid this fundamental uncertainty. Nonetheless, under Anglo-American common law, people did not need to demonstrate that a nuisance posed a threat to public health to act against it. In theory, they just decided that the nuisance offended community norms. In reality, acting against nuisances increasingly required lawmakers and judges who did not see tolerance for the nuisance as a necessary price of progress. But the automobile added several wrinkles to the nuisance problem. Apart from taxicab and livery services, automobile use and exhaust arose largely in the pursuit, not of business, but of recreation and pleasure. Complicating the issue, the automobile smoke nuisance involved, not private activity on a fixed

WHEN THE GASOLINE AUTO PASSES BY.

Smoky exhaust. Edward W. Kemble, *Life*, 1902.

piece of real estate, but the use of numerous mobile pieces of private property on public roads.[16]

Urban public health concerns highlighted both benefits and problems with the automobile. In a prevailing climate of near paranoia over germs (a recent discovery), many people noted the public health benefits of replacing horses with automobiles. But optimism about the health benefits of automobiles went further. Some physicians gushed about the benefits of motoring for all sorts of ailments, including baldness and dandruff. One English physician, considering airborne bacteria, even argued that automobile exhaust was an excellent disinfectant. These outlandish views remind us that the tendency to view exciting new technologies as panaceas often worked to the automobile's advantage. But the potential public health benefits of the automobile constituted only a minor strand in the larger discourse about the benefits and problems of the automobile that occurred before 1910. The benefit of eliminating horse manure certainly did not obscure or overshadow the problem of exhaust or the biggest public health problem posed by the automobile—the number of people being killed by them, already the principal cause of the animosity directed at early automobile owners.[17]

Observation led to investigation by mechanics, engineers, and scientists, who soon concluded that two factors caused "foul exhaust." The first was relatively straightforward. "Smoky exhaust," they concluded, was simply a matter of too much crankcase oil in the cylinders where it was partially consumed with the gasoline during combustion. The second factor proved more complex. Direct experience quickly acquainted automobile owners and mechanics with one of the gaseous by-products of internal combustion, carbon monoxide. At sufficient concentrations, this odorless, colorless gas caused asphyxiation and death. But the early alarm passed when it became clear that carbon monoxide was not life-threatening if dissipated in the open air. The presence of carbon monoxide in automobile exhaust underscored the fact of imperfect combustion. Engineers and scientists did not fully understand the reasons for this failure, but they believed that imperfect combustion accounted for black smoky exhaust (as opposed to blue smoke from overlubrication) as well as the bad smell. They knew that the complete combustion of hydrocarbon compounds, consisting of carbon and hydrogen atoms, in the presence of oxygen should yield two naturally occurring by-products, carbon dioxide gas and water. When they recognized that the chemistry of internal combustion fell short of this outcome, they began the quest, which continued across the twentieth century, to achieve perfect combustion in gasoline-powered engines in order to eliminate the undesirable by-products in exhaust.[18]

In England, the noted Scottish mechanical engineer Dugald Clerk conducted an important series of early experiments into the problem. The automobile press in the United States reported his results in early 1908. Perfect combustion, mechanical engineers believed, required the right mixture of gasoline and oxygen in the cylinder for the given power that the engine was being asked to generate. To complicate matters, this mixture varied at each power level required of the engine. Clerk's experiments proved that the mixture for each level could be optimized but that it was impossible to produce perfect combustion at any level with current engine technology. His work also made plain that most automobiles were not running at anywhere near the optimal mixtures of gasoline and oxygen and were, therefore, emitting carbon monoxide and other noxious gases to a greater extent than they might. Most important, Clerk found that the gasoline-oxygen mixture that minimized noxious emissions coincided with the mix that optimized fuel economy. From the standpoint of late-twentieth-century concern about the automobile's environmental impact, it is hard to imagine a more important finding. As early as 1908, people concerned with either conserving gasoline supplies or the problem of noxious exhaust could aim at this goal. But an appreciation for this point never made it beyond the technical literature.[19]

The public viewed smoky exhaust as a bad odor rather than a significant public health threat, much less an environmental problem. But several groups, animated in part by foresight, refused to let go of the exhaust issue. Heldt and his editorial colleagues at *Horseless Age* remained alert to the possibility that it might be something more. They paid special attention to the possible public health ramifications of automobile exhaust, reporting the concerns of Dugald Clerk about the threat from carbon monoxide as the number of automobiles in urban areas increased. Albert Clough saw a direction to these developments. "As the use of motor vehicles increases," he wrote, "it may be expected that the emission of carbon monoxide may be recognized by law as dangerously unsanitary, especially in crowded city districts, where automobiles congregate in large numbers." Although the editors of *Horseless Age* called attention to evidence of the possibility of harm to human beings and the natural environment, they knew that a convincing case could not yet be made on this evidence. Still, they called for regulation, not because scientists or public health authorities had determined conclusively that automobile exhaust harmed people or plant life, but simply because it offended the human sense of smell.[20]

Many people beyond the editorial offices of *Horseless Age* shared a willingness to resort to regulation to address smoky exhaust. A number of American cities passed or flirted with passing ordinances directed at the problem. New Yorkers took these efforts furthest. After several years of sporadic protest that smoky automobiles were ruining the experience of nature in Central Park, reformers circulated petitions in 1908 that successfully pressured the New York City Board of Parks into passing an ordinance that permitted the authorities to expel offending vehicles from the city's parks after 1 August 1908. Despite challenge, the courts upheld the ordinance, and people generally deemed it a success.[21]

Municipal reformers led by Edward S. Cornell next mounted pressure to extend regulation to smoking vehicles on the city's regular streets. "New York is approaching . . . the highly desirable condition of being a horseless city," the *New York Times* editorialized in support of the proposed law, "but as the number of self-propelled vehicles increases the evil of smoke emission from these machines is becoming steadily more serious, and, unless stopped in time, this new form of the old 'smoke nuisance' will make us regret the passing of the horse, dangerous and troublesome beast as he is from the standpoints of cleanliness and hygiene."

After failing to persuade the city's aldermen, reformers prevailed on the Board of Health to amend the city's sanitation code on 27 April 1910.[22]

Unlike the park ordinance, however, the citywide ordinance ultimately failed. Despite intermittent efforts that met with some success over the next few years, the ordinance simply proved too difficult to enforce. The law called for distinguishing light smoky exhaust from dark exhaust. But an automobile emitted exhaust of varying density depending on the amount of work the engine did — when it was going up a hill, for example. People began to acknowledge that the problem might not be entirely preventable by automobile owners' careful maintenance.[23]

Even Peter Heldt at *Horseless Age*, who initially blamed the problem on careless owners, conceded that smoky exhaust involved more than encouraging owners to be careful about engine lubrication. "No doubt some of the blame for the smoking nuisance rests with the operators of cars," he wrote in 1910, "but we are inclined to think that those most responsible are the makers. We are led to this conclusion by the fact that certain makes of vehicles are exhausting smoke whenever they are seen, while other makes are not so affected. . . . The threat of a crusade against smoking automobiles in this city [New York] should serve as a warning to both builders and users." The failure of the New York City ordinance marked the beginning of an important shift in perception of the emission problem from being primarily the responsibility of the owner to being that of the manufacturers.[24]

By 1910, urban Americans certainly recognized that automobile emissions presented a problem. They knew that some doctors claimed that emissions threatened public health and might get worse as more people purchased and drove automobiles. Already, strong supporters of the automobile, such as Heldt and *Horseless Age*, and many sectors of the public had indicated a readiness to regulate auto emissions on fundamentally nonscientific grounds simply because smoke and odor offended the senses. But ultimately, Americans decided not to pursue further regulation of automobile exhaust, a decision made in the evolving context of their larger reaction to the automobile.

Americans viewed the automobile with enthusiasm, but many reacted angrily to how it was introduced. The automobile aroused other powerful emotions as well. Envy is an elusive emotion for the historian. People don't like to get caught publicly envying others' superior wealth, power, or automobiles, much less have their envy recorded for posterity, because the person who feels it assumes an inferior relationship with respect to the person he or she envies. Accounts of

"I WISH WE HAD ONE OF THOSE!"

"I Wish We Had One of Those!" Paul Reilly, *Life*, 1916.

Americans' early reactions to the automobile provide less evidence of envy than either enthusiasm or anger, but observers still noted plenty of it, and it helps to explain why people generally left gasoline and smoky exhaust to conservation-minded engineers and municipal health reformers.[25]

"Almost the principal motive in buying an automobile," one 1904 observer noted, "is to excite the envy of friends." One Southern California nature lover admitted in 1906 that he "felt a few sharp arrows from the quiver of envy shoot through his soul whenever an automobile whizzed by," but quickly added that "he has outlived his foolishness and has learned to thank his lucky stars that he can't afford to keep one." In March 1903 the *New York Times* commented editorially on a speech made to an automobile group in Berlin by Prince Henry of Prussia. "Prince Henry . . . tells the automobilists that their unpopularity is due in part to envy of their expensive joys and in part to their own lack of consideration for the safety and convenience of less fortunate users of the public highways. Almost everybody yearns to own an automobile, and comparatively few can do more than hope—not too confidently—that in the future they too may have a chance to rush along the highway in a private locomotive. It may be, therefore, that much of our indignation when we see others do it is a mingled emotion."[26]

Probably the most interesting evidence of envy arose in connection with what was called "the chauffeur problem." "Among the multitude of problems which are facing the automobile trade and sport there is perhaps none which is more perplexing than the proper estimate of the chauffeur as a class," A. B. Tucker wrote in 1903. "He is at present far from being a fixed quantity. As an element in automobile life he is looked upon with varying degrees of importance. His exact position is less firmly fixed in the public mind now than it ever has been." Tucker worried about more than the chauffeur's class status. "Nine-tenths of the fiction which has been written on automobiling throws a romantic glamour about the person of the chauffeur," he noted. "He is either a lord in disguise or a gentleman athlete under a cloud or something of that sort, and the always beautiful young lady whom he drives invariably falls in love with him." It was one thing for the wealthy to use an automobile to impress members of the opposite sex but quite another for a man who could not afford an automobile to be reaping sexual advantages that he did not deserve simply because he drove one.[27]

The chauffeur problem came to a head over "joy riding," the practice of chauffeurs borrowing their employer's cars without permission and taking friends for rides, which *Horseless Age* defined as "generally a combination of 'scorching' and 'boozing.'" The magazine was not exaggerating. Accounts in the nation's newspapers about joyrides that resulted in accidents had a depressing similarity. Joyrides were mostly nocturnal pleasures where the chauffeur "borrowed" the car after the owner was through with it for the evening. Often ordered to drive the car to a public garage where it was housed, the chauffeur instead picked up a few friends and went speeding or drinking or both—eventually encountering unseen curves, embankments, brick walls, and telephone poles. "Late last night Faust invited five friends to take a . . . joy ride . . . ," read a typical account. "The car was going 60 miles an hour, when one of the girls protested at the speed. Faust laughed, and added 10 miles to the speed. The car became unmanageable, jumped the road, plowed through a ditch, and struck a palmetto tree. . . . Faust's neck and skull were broken." Newspaper accounts suggested that joyriding was widespread, and for obvious reasons. The chauffeurs simply could not resist the temptation of driving the cars on their own, showing off for friends, and appropriating some of the masculine prestige conferred on the early masters of the new technology.[28]

In short order, chauffeurs ceased to be the subject of romantic glamour and became the scapegoats for many automobile problems, particularly speeding, accidents, and—a newly discovered problem—drunk driving. States quickly passed laws that made it illegal to take an owner's car without his permission. Of course, many of those taken for illicit rides were women, so the lawmakers had more on their minds than just speeding. One joyriding San Bernardino woman told Los Angeles police that just before her high-speed accident on Wilshire Boulevard she had been riding on the chauffeur's lap! By 1908, the editors of

Horseless Age had no question in their minds about the chauffeur's social status: he was a servant and would remain one. He had been put in his place. Most middle-class Americans could not yet afford an automobile. They readily supported steps that prevented an undeserving working class reaping the benefits that they envied. Thus, the boundaries of the privilege would be well policed until they could afford automobiles. In the meantime, a chauffeur was to be seen driving only when clearly subservient to his master, a clear example of envy in action.[29]

Americans recognized the automobile's practical advantages as transportation, but the emotionally charged issues raised by how wealthy Americans (and their chauffeurs) introduced the automobile powerfully affected American consciousness and life. On one hand, Americans felt the immense appeal of the automobile—the exhilaration from the machine's speed and power, the pure thrill of going fast, the sense of discovery and adventure in exploring the countryside, the challenge in keeping the vehicle functional, the satisfaction of being seen by others and oneself as sufficiently wealthy and masculine enough to be the owner and master of this marvelous new invention.

On the other hand, many also found the advent of the automobile upsetting. Before World War I, most people could not afford an automobile, certainly not one with the power to turn heads the way their own heads had been turned. The inescapably public nature of automobile use meant that those who did not possess one instantly suffered public loss of face and status. Intentionally as well as unintentionally, the wealthy flaunted their possession of the automobile in a manner that made it one of the greatest class provocations in American history. The speeding sportsmen also used the automobile to assert a hypermasculinity that accentuated the gender anxieties of middle- and upper-middle-class men already gripped with worry about their masculinity. These actions gave great offense and provoked an intense, widespread, sometimes class-tinged reaction that also drew support from wealthy elites who were embarrassed and appalled by this breach in democratic social etiquette by others of their class.[30]

This emotion-charged atmosphere overwhelmed other problems with the automobile, such as gasoline dependence and smoky exhaust, that already worried observers with foresight. Moreover, adding these concerns to the already potent list of grievances that the public had with automobile owners was not what the public really wanted. To regulate the automobile in a manner that protected people from its most obvious problems and to punish the transgressors was one thing, but the revenge that most people wanted, the only acceptable balm for their envy and for their damaged sense of self-respect, was automobile ownership itself. The pleasure that the wealthy so obviously enjoyed in large, fast, powerful, gasoline-fueled automobiles made these vehicles potent symbols of wealth, masculinity, and skill. Thus, from the beginning, Americans used automobile ownership, especially of large, powerful automobiles, in a quest for respect. This quest ultimately brought social and environmental ramifications that in America significantly changed the relationship between people and the natural world.

3

———————•———————

A Monstrously Big Thing

The emotions that the speeding sportsmen aroused in Americans sparked the automobile revolution of the 1910s and 1920s and made the United States the first automobile society in the world. Judging from their behavior, Americans found few questions more important during the first quarter of the twentieth century than whether they owned an automobile. Were you among the haves or the have-nots? Americans rushed to answer this question affirmatively. To do so required what popular author and historian Philip Van Doren Stern called "an instrument of historic destiny," the respectable and affordable Ford Model T. Between 1908 and 1927, Americans bought fifteen million Model Ts, while the total number of automobiles on the road jumped from two hundred thousand to twenty million. The makers of "instruments of historic destiny" are commonly lauded as visionaries. So it was with Henry Ford. From the outset, he envisioned making an inexpensive, well-made car that most Americans could afford. And he delivered it.[1]

Like the engineer-editors at *Horseless Age*, Ford saw the automobile as an improved form of transportation destined to become an everyday necessity once the price became affordable. He was not alone in recognizing the tremendous business opportunity for an automaker who could profitably sell an automobile for under five hundred dollars, the magic price at which it was thought millions of middle-class Americans could stretch their budgets to afford one. Many people grasped this prospect early. But the early low-priced offerings failed to make the grade because consumers found them underpowered, unreliable, and unremarkable. None offered the cachet to rival the appeal of the large, powerful, expensive vehicles of the wealthy. So the question remained: Who would be first to offer an affordable *and respectable* automobile?[2]

The goal could not be reached, it seemed clear, without high-volume production to substantially lower per-unit costs. But building this capacity was a fantastically expensive proposition. How did an automaker build it without selling a large number of cars to pay for it? What investors would finance a company without assurance that it had a car that would sell in the numbers necessary to recoup the investment? This catch-22 stood between the automakers and the obvious. Without a revolutionary breakthrough, the only way forward was with a very well designed product that appealed to enough people to earn the profits needed to reinvest in production and to drive down costs gradually but continuously. This car did not have to sell for five hundred dollars immediately, but it had to sell for less than the typical automobile while providing substantially more car per dollar than any other make on the market.

In 1905 the market consisted of wealthy Americans. But here an automaker faced the fiercest competition. More than two hundred automakers in the United States fought mainly for the elite's automobile dollars. Many wealthy people no doubt would appreciate a car that offered more value per dollar and buy it. And yet, there were good cynical reasons to believe that many others might be less interested in a car that stood apart in terms of value as opposed simply to high price. Indeed, the average price of a car sold in America before 1908 was rising. Although many wealthy Americans had yet to buy an automobile, the number of potential wealthy customers was limited. The prospects for building sales quickly to the size necessary to lower production costs sharply by selling autos to the wealthy therefore seemed small.[3]

For Henry Ford the reasons for disdaining this market ran deeper than the economic rationale. Many early automakers derived vicarious status from selling expensive, well-appointed automobiles to the wealthy. Henry Ford did not share this aspiration. In fact, hobnobbing with the wealthy and catering to their automobile needs interested him not at all. Like many people who had grown up on farms in the second half of the nineteenth century, Ford harbored a suspicion, even hostility, toward wealthy Americans. "We do not want to cater to the rich man . . . ," Ford told visitors to his small Detroit factory in 1906. "It is my whole ambition to build a car that anyone can afford to own, and to build more of them than any other factory in the world." Ford had his eye on another market, people like himself—the farmers and small townspeople of rural America. Nearly 60 percent of Americans still lived in rural communities. No group of Americans had a stronger practical need for motorized personal transportation. If this market seemed in retrospect like an obvious one for Ford to choose, he must be given credit for focusing on a group whose hostility to the automobile—or at least to the behavior of the first owners—still ran strong. But Ford knew this market. He knew its needs and its psychology. His challenge was to build a car that suited

Henry Ford and his Model T, 1921. *From the Collections of The Henry Ford, P.O.3015.A.*

both. He knew, too, that a car that sold well in rural America would also sell well with many other middle-class Americans. Henry Ford later derived much personal satisfaction from the popularity of the Model T with rural Americans. He may even have felt some glee at putting farmers on the same plane as the wealthy.[4]

No one had a greater hand in the design of the Model T than Ford himself. He worked side by side with his designers day after day for nearly two years. The key to the car that they made was combining lightness with power and durability. To achieve these goals the company pioneered the use of vanadium steel, which provided greater strength with less weight. The result was a tough, simply made and easily repaired, twelve-hundred-pound car with a very respectable twenty-horsepower, four-cylinder engine that was ready to take a beating and come right back for more.[5]

As well designed as it was, the Model T's success depended on some important early publicity. Much would later be made of Henry Ford's antipathy to advertising. In fact, for much of the Model T's life Ford did not advertise the car. But Ford did not have this luxury at first. Advertising helped establish the Model T as the car that stood apart from the rest. On 3 October 1908 Ford ran a full-page advertisement for the new Model T in the *Saturday Evening Post*, the largest-circulation magazine of the day. The ad prominently touted the car's $850 price but in a context that clearly stressed the car's value in comparison with higher-priced vehicles. "The one real automobile value among all the 'season sensation' announcements," the ad claimed, "is this big, roomy, powerful five-passenger touring car at the hitherto unheard of price of $850.00. A car that possesses at least equal value with any '1909' car announced, and at the same time sells for several hundred dollars less than the lowest of the rest." The "rest" that the ad referred to were the larger, high-priced cars, but the ad really worked by separating the Model T from the twenty-four models on the market that already sold for nine hundred dollars or less. The ad suggested that here, finally, was an affordable car that was more than an affordable car. It caused a minor sensation, and prospective customers deluged Ford's Detroit offices with mail inquiries.[6]

Inquiries do not always result in sales, let alone make a sales sensation. The company and its local dealers worked hard to create favorable publicity based on the Model T's actual performance. To use the idiom of a later age, they set out to prove that the car "walked the talk." Racing was the most important activity in this regard. A major national boost came when the Model T won the June 1909 transcontinental race between New York and Seattle that the young mining heir M. Robert Guggenheim sponsored to promote that year's Alaska-Yukon-Pacific Exposition in Seattle. But the Ford Motor Company, Ford dealers, and Ford owners raced the new Model T all over the country and in all types of races — on speedways, in endurance runs, and in hill climbs. The company shared the news of the car's successes with its dealers, who in turn spread the word to potential customers, often through the automobile pages of local newspapers. These accumulating successes in varied settings proved that the Model T was special. In fact, by giving drivers like Barney Oldfield and Ralph DePalma in their high-horsepower specialized racing machines real competition, the Model T sparked a revived public interest in racing, especially by "stock cars," automobiles designed for the public rather than for racing. Organizers even barred the Model T from the inaugural Indianapolis 500 in 1911 when drivers balked over a requirement to carry several hundred pounds of lead as a weight penalty. The car's racing success depended as much on its reliability as its speed and power, and it soon became clear that the Model T was durable. Within thirty months of introduction, the press touted the first hundred-thousand-mile Model Ts. When it was reported that, in a reversal of the usual roles, the Model T had pulled a horse free after it became mired in mud, it was clear that an affordable car that commanded respect had arrived. By March 1911, before even 0.3 percent of its lifetime sales had been made, commentators called the Model T "this famous car."[7]

With a great product, well publicized, and offered at moderate cost, Ford began well with the Model T. But even as "the greatest money value ever yet put into an automobile," as the *Christian Science Monitor* described it, the Model T did not automatically sell itself. Nor was the vaunted sales organization that Norval A. Hawkins created sufficient to sell the car. It always takes buyers as well as a seller to make a market. What did Americans discover in the Model T that pleased them, and how did they find these satisfactions? Since the benefits and pleasures of owning a car were so publicly obvious, the general idea of owning a car pretty well sold itself—once people could see around their anger with the early speed maniacs, a reaction often smoothed by a little neighborly envy. The first-time buyer's primary desire, early owner K. P. Drysdale noted, "is simply to own an automobile—something to ride in." Specific makes of automobiles, by contrast, had to be actively sold by dealers because consumers who had made up their minds to buy a car generally made some effort to choose the best one they could afford. The sheer number of cars on the market made this choice hard. Until consumers collectively established a car as a make to buy, both sellers and buyers had to put considerable thought and effort into the selling process before consummating a transaction. Consumers faced a real risk where a low-priced car was involved. As important as it was to be able to afford something to ride in, a poor choice might advertise one's poverty and stupidity.[8]

For all that has been written about the success of the Model T, surprisingly little has been said about the deeper subtleties of the car's appeal. Most historians have pointed to the Model T's practical virtues as a machine, combined with its ever lower price, to account for its popularity. True enough. But other facets of the Model T made it appealing, too, and these are worth reviewing to understand who first bought the Model T and why it became the choice for them. The Model T appealed chiefly to two markets. The rural market of farmers and small townspeople, as intended, provided most buyers in the car's first decade. The urban and suburban white-collar middle class provided a second market, although somewhat later. Farmers found the practical appeal of the Model T, indeed of any car at all, overwhelming. Time was money on the farm as elsewhere, and owning a car meant the difference between an all-day wagon ride to town and back and an hour's automobile trip instead. In the early twentieth century rural Americans lacked access to high-speed transportation. This shortcoming not only posed a handicap to economic life, it also placed severe constraints on social life. For the farm wife especially, life on a farm was practically a prison. The automobile promised to address this hardship. For no other group of Americans did the automobile have anywhere near as great a practical appeal.[9]

When a rural American grasped this benefit, the question became getting the most car for one's money, which meant in many regions the car that coped most reliably with wretched roads. Ford designed the Model T for precisely these conditions. The car's high road clearance allowed it to travel deeply rutted country ways. The car's high power-to-weight ratio made it nimble on hills and maneuverable in mud. "The only reason for the existence of the Ford automobile," Henry Ford later said in reference to the car's success, "is that we have bad roads in America." The car also got from twenty to thirty miles per gallon depending on road and driving conditions, an important consideration for rural people who regularly needed to drive longer distances. Farmers naturally found a hundred and one uses for the car around the farm.[10]

Ford designed the Model T to be not only durable but simple to fix. When the company created a widespread parts distribution system, which it did early on with just this goal in mind, it offered a car that owners could easily and inexpensively repair. The car almost could not be totally wrecked. No matter the damage, replacement parts could be purchased—from the Sears, Roebuck catalog if need be—to fix it up. This virtue certainly made the Model T cheaper to operate for farmers who were already comfortable doing repairs on farm machinery. For many Ford owners the repairs brought not annoyance but a deep attachment to the car. The Model T, historian Reynold Wik recalled from firsthand experience, "meant personal mechanical involvement." "We worked on the machine," he remembered, "we took it apart, made adjustments, added accessories, pampered it, tinkered with it almost daily, worried and brooded over it, cussed it when it failed and bragged about it when it performed superbly." And, he might have added, fell in love with it. The Model T was not an instant hit with farmers, though. Despite the car's early fame, farmers often remained careful buyers, so even in rural areas several years went by before the car really began to sell. But the Model T's virtues slowly made themselves obvious, encouraged by a larger economic context of rising farm product prices and land values that meant greater farm income.[11]

These virtues of the Model T appealed to other Americans, only less so. While middle-class Americans in urban and suburban areas were certainly attracted by the low price of the Model T, they generally lived in areas with ready access to high-speed transportation that they already used to get to work and to shop. Car ownership improved their transportation options but not as dramatically as it did for rural Americans. For the urban middle class a car remained largely a recreational possession used on Sundays, holidays, and vacations. Moreover, white-collar Americans on salary experienced lost purchasing power in 1916–1920 due to the inflation brought on by World War I. Nor did it help that urban Americans, at least initially, were less accustomed to doing their own repairs on machinery and often lacked buildings in which to garage the vehicle, a necessity when cars were not yet closed to the elements. City people thus paid more in operating expenses for a Model T. These were strong enough impediments that ownership of cars by middle-class Americans in urban areas lagged behind rural areas for awhile.[12]

One of Ford's dealers, Roscoe Sheller, remembered that in the early years he had to give a lot of demonstration rides and driving lessons to make a sale. He compared the Model T to a missionary, and like conversions, no sales were more important than the first ones in a community, which got the T out on the roads, where it could be observed and promote itself. Country doctors, who had perhaps the strongest practical need for faster and more reliable transportation in rural America, often played this important role. People who lived in rural areas, where it was impossible simply to drive anonymously past a neighbor, also promoted the Model T by giving people lifts. Ford advertised extensively in national media until 1911, but personal observation, experience, and word-of-mouth endorsements from friends and neighbors sold the car.[13]

In 1911, when Ford asked two thousand owners why they had bought their Model Ts, 85 percent said word-of-mouth recommendations of other Ford owners. People experienced some peer pressure in these discussions from family, friends, and neighbors. Recalling the purchase of his first automobile, Minor Hinman said simply, "The pressure got too great." If men generally negotiated the transaction with the dealer, women and children in a family often influenced the choice of car as much or more than the father and proved just as eager to drive it. "The people in Ford cars are not the only ones that enjoy them," Gerald Stanley Lee wrote in March 1914. "They fly through the streets addressed to all of us—happy valentines about the world and about the way things are going."[14]

Dealers like Sheller appreciated that the company continued to promote the car beyond advertising. Ford plowed profits from growing sales back into production to lower the cost of making the car, which resulted in regular and often dramatic price reductions. The magic five-hundred-dollar barrier was reached in August 1913 and then bettered the following year. Then there was the increasing publicity generated by the company's production miracles at Highland Park, Michigan, the three-hundred-thousand total annual sales challenge with fifty-dollar rebate for Model T buyers on 31 July 1914, and, above all, the an-

Miracle of mass production. A day's output at Ford's Highland Park Plant, c. 1913.
From the Collections of The Henry Ford, P.833.99165.2.

nouncement of the five-dollar day in January 1914, which kept the Ford name —
now legend — before the public eye. Ford, the man and the company, became
national sensations that kept people observing, talking, and thinking about the
car.[15]

Sometime between 1910 and 1912, national Model T sales reached a tipping
point and took on a self-accelerating momentum that turned the nation's seven
thousand Ford dealers into order takers more than salesmen. By 1913, every third
car on the road was a Model T. Of the first million Model Ts that Ford sold, 64
percent went to the farm and small town market. By 1920, nearly a third of farm
households in America owned automobiles, but the rate of ownership was mark-
edly higher in the Midwest (53 percent) and Far West (42 percent), where the
farmers were wealthier and the distances greater, than in the South (14 percent).
In Sheller's Washington territory the tipping point occurred in spring 1917. In
one memorable day, to his amazement one customer led him through a word-
of-mouth network to four more sales with no more effort on his part than tag-

ging along. At day's end, he had more sales than he had ever made in a month. "Without anyone appearing to know what was happening," he recalled, "mass antagonism toward the 'auto' was gradually melting. The change was contagious and spread like measles in a kindergarten."[16]

Soon enough Model T owners constituted sizeable communities locally and nationally. The car was unique, its idiosyncrasies were legion, and its owners were so numerous that an immense popular lore grew up around it. Shared exasperations like hand-cranking the car on a cold morning worked against snobbery or one-upmanship and created a democratic camaraderie. "Every Model T owner," Floyd Clymer remembered, ". . . was a disgruntled grouch with a thousand complaints. But they were the good-natured complaints of a proud owner, of an adoring parent." Model T jokes were part of this lore and of the same flavor as the complaints. By 1916, there were eight books of them in print and selling well. "The jokes about my car sure helped to popularize it," Henry Ford mused. "I hope they never end." Model T owners also formed Ford clubs, and county and state fairs held "Ford Days" for Ford owners. Historian Daniel J. Boorstin called these shared experiences of ownership and expertise "communities of consumption" and argued that twentieth-century Americans increasingly relied on them to replace older forms of community. They offered a subtle, perhaps largely unconscious, assurance that one fit in.[17]

If there was community in mass Model T ownership, there was less conformity than one might suspect. Although Model T owners bought them so that they could join their neighbors in the circle of respectability that automobile ownership represented, they did not buy their cars to have one exactly like their neighbors. Almost from the beginning third-party manufacturers offered accessories for the Model T. The car's design made it easy to replace, upgrade, or simply add to the factory-standard parts. Eventually, some five thousand accessories and customization options helped owners remain individuals within the larger community of T owners.[18]

Historians have given Henry Ford full credit for the practical utility of his car but less credit for its symbolic appeal—the fit between the car and the psychological needs of its buyers. No car before the Model T (and few after it) so fully realized the sales potential from both kinds of appeal. For all its practical virtues, the Model T, through its toughness and lack of luxury, possessed its own symbolism. The timing of its arrival proved as important in this regard as its design. In 1908 anger about the introduction of the automobile was still widespread; envy was still acute. Ford's car provided many Americans the means to satisfy their envy and to close the public status gap that the advent of the automobile had painfully opened between themselves and the wealthy. Just how much of the Model T's symbolic appeal was consciously designed into the automobile (then, as now, a hard thing to do successfully) and how much derived from the fact that Henry Ford was a lot like the people who bought the car can never be known. What can be said is that Ford made the T inexpensive enough, well-made enough, and, most important, just large, powerful, and fast enough that buyers could close most of the status gap between themselves and the wealthy without hypocritically aping them or leaving themselves open to ridicule for choosing a cheap, slow, poorly made car. Avoiding these twin pitfalls was important. As Henry Ford put it in a letter to *Automobile* in January 1906, his car had to be "in every way an automobile and not a toy." Americans wanted to be car owners, but they wanted to be different from the controversial owners among the wealthy. The Model T expressed this distinction perfectly.[19]

Nowhere was Henry Ford's design more attuned to his customers' psychological needs than in the car's power and speed. He made no pretense that the Model T belonged in the class of large, fast, powerful, and expensive automobiles preferred by the speeding sportsmen. It was slower and thus safer. Its owner could not be readily associated with the irresponsible excesses of the speeding sportsmen. And yet, Ford offered a car that stood apart from the smallest, slowest, or most underpowered cars on the market. The Model T's twenty-horsepower engine and twelve-hundred-pound weight gave it a good power-to-weight ratio, which meant that even though its top speed normally might be only thirty-five to forty miles per hour, the car had good acceleration and a real alacrity. There were times and places—on hills and in challenging road conditions—where the nimble Model T left some of the larger, more expensive, and powerful makes in the dust and mud, which the surprisingly strong performances of the Model T in actual competition with larger, more powerful, and much pricier automobiles underscored decisively. The symbolic resonance of the car's competitive achievements cannot be overemphasized. When the *Chicago Daily Tribune* reported in 1917 that "the conversion of a model T touring car or Ford roadster into a speedster has become the most popular of American garage sports," those men were working on more than their cars. Not surprisingly, as many more Americans bought cars like the Model T, the speeding sportsmen abandoned automobiles as a recreational pursuit in favor of more exclusive hobbies like polo, flying, and speedboating.[20]

The Model T even offered tongue-in-cheek symbolic satisfactions. "The driver of the old Model T," author E. B. White remembered, "was a man enthroned." The car's high ground clearance, like sport utility vehicles in the late twentieth century, allowed its owners literally to look down on the drivers of more expensive makes. A good, plain, speedy, rugged, no-nonsense car, the

Polishing his ticket to respect, 1913. *Courtesy of the
Smithsonian Institution, NMAH/Transportation.*

Model T embodied a satisfactory set of practical and symbolic characteristics
for consumers who wished to own automobiles yet distance themselves from the
wealthy, irresponsible, and socially transgressive automobile owners who preceded them. The Model T owner thus had it both ways: he owned a car that
was similar enough where it counted most to lay claim to the cachet that went
with automobile ownership yet different enough to assert a new and, in his eyes,
a superior basis for respect by demonstrating the practical intelligence to choose
value over ostentatious expense. "Self-respect," Abilene, Kansas, newspaper editor Charles Moreau Harger observed in 1911, "is one of the things that has come
to the farm with the motor car."[21]

    The practical appeal of the automobile, especially for rural Americans, can-
not be downplayed. But the emotional context of America's first experiences
with the automobile—the desire of average folks to experience the fun of driving, to redress a deeply felt wrong, and to reassert a claim to respect both in their
own eyes and in those of wealthy others through their own possession and use of
automobiles—also played a powerful role in the early mass adoption of the automobile in the United States. Ford's sales manager Norval Hawkins understood

this deeper reality. "A man's emotions," he wrote, "not his thoughts, control his Desires." Ownership of a good, plain, practical car like the Model T became synonymous with middle-class decency for those who prided themselves on being regular folks who disdained material pretense but who refused be left in the dust. In democratic America people bought their first cars, not just because they were useful as well as fun, but because their self-respect demanded it.[22]

During the two decades between 1908 and 1929, first-time buyers of cars like the Model T that sold for under a thousand dollars swelled the great American automobile boom. However, there were two different sales booms that occurred in this price range. The first took place between 1912 and 1917. The practical and symbolic satisfactions that Americans found in the Model T, farm prosperity, and the price reductions that came with the mass production breakthroughs at Ford's spectacular Highland Park plant, especially the introduction of the moving assembly lines in 1913, fueled this surge. Between 1908 and 1919, the average wholesale price of an American-made passenger car fell from $2,109 to $827—a reduction of more than 60 percent without adjusting for the inflation of World War I. The automakers actually met consumers halfway. As the price of the typical car fell, the income of Americans rose. Between 1890 and 1925, real per capita income grew by 85 percent, which meant substantially greater purchasing power for millions of Americans. At the same time, the average length of the workweek decreased by nearly 20 percent, from sixty to just over fifty hours per week. More Americans thus had greater discretionary income to spend on consumer goods beyond the basics and more free time to enjoy these goods. If the wealthy sparked interest in automobiles, the arrival of respect-worthy and affordable cars like the Model T, rising real income, and increasing leisure time kindled the revolution that transformed the country.[23]

A second and even greater boom in the sales of cars priced under a thousand dollars occurred after the post–World War I recession ended in 1921. Before this date, the low-end market was the Model T market. As the new decade began, the rapidly growing number of automobile "haves," though still a minority of Americans, actually consisted of two rather distinct groups, Model T owners and people who owned other makes. Dealer Sheller remembered that the word *automobile* was actually reserved for the other more expensive makes. As the 1920s progressed, however, Ford faced growing competition in its market, not only because other automakers successfully copied Ford's mass production techniques to cut costs and lower prices (often with Ford's acquiescence), but because an explosion in automobile consumer financing made cars like General Motors's slightly more expensive Chevrolet much more affordable.[24]

From 1920 to 1929 the amount of consumer debt in the United States more than doubled from $3.3 billion to $7.6 billion. As historian Lendol Calder shows in *Financing the American Dream,* "the modern system of credit for consumption has its roots in the two decades after 1915." "The driving force behind this huge expansion of debt," Calder writes, "was, literally, the driver." Between 1922 and 1929, automobile debt quintupled. With evidence of the importance of automobile ownership evident everywhere in American life in the early 1920s, the people who pioneered automobile financing knew they had a sure thing. Others were not so certain. Going into debt to buy an expensive plaything like an automobile seemed like foolish hedonism run amok. Encouraging the practice could not be good for the individuals and families involved, for the economy, or for the country. Earlier generations supposedly had shown far more discipline, deferring purchases until they could afford them from current income or from savings. In fact, as Calder strongly argues, no such golden age of consumer discipline, deferred gratification, and thrift had ever existed in America. Americans had pursued their American dreams through debt even before they stepped on Plymouth Rock.[25]

In the nineteenth century, middle-class Americans financed the purchase of a number of consumer goods that occupied pride of place in their homes—sewing machines, expensive furniture, organs, and pianos, for example—with credit extended either by the manufacturer or by the retailer. In fact, to a remarkable degree, sales and marketing practices in the automobile industry—including the use of both financing and trade-ins—closely followed those in the piano industry. Although the connection between the two industries may no longer be obvious, early-twentieth-century purchasers saw a clear relation between these two consumer goods. In their time, both the piano in the parlor and the automobile were the most prestigious consumer goods that a middle-class family owned, intimately part of the public face that a family presented to the community. Humorist Will Rogers got it right when he quipped, "Even the most experienced can't tell by looking at a car how many payments are still to be made on it." This quality put pianos in the parlors of middle-class homes in America before the turn of the century, and now it worked for automobiles. The last thing that a family wanted was to have that piano or auto repossessed and answer neighbors' questions about what happened to it.[26]

Given the automobile's expense, durability, and symbolic appeal, it is remarkable that people did not begin financing auto purchases sooner. Despite several early exceptions, including an important one in General Motors, automakers did not pioneer automobile financing. Although its customers often arranged third-party financing, Ford discouraged the practice. With demand so strong that most companies had no problem selling to dealers for cash whatever they could make, and with their own hands full with the challenge of increasing production to meet demand, the automakers had little immediate incentive to provide customers with credit to generate more sales. By the mid-1910s, Model T sales made clear the existence of a large potential middle-class market for automobiles. Yet bankers, not yet in the business of consumer lending, passed up this opportunity. The bankers' reluctance presented an opportunity for entrepreneurs—the people who created the automobile finance company. In December 1913 E. F. Weaver established the first automobile consumer finance company in San Francisco. General Motors created the General Motors Acceptance Corporation (GMAC) in March 1919 to extend credit primarily to its dealers. Consumer lending apparently began as an ancillary business and then just took off. Soon GMAC was the largest finance company. During the 1920s, automobile financing swelled to dramatic proportions, led by four large firms, GMAC, Commercial Credit Corporation (CCC), Commercial Investment Trust (CIT), and Universal Credit Corporation (Ford's belated entry to the field), tied to but not necessarily owned by automakers, and backed up by hundreds of small, local companies independent of the automakers.[27]

The standard terms for the sale of a new car with financing were one-third down and the balance in twelve monthly installments. By the early 1920s, industry observers estimated that 70–75 percent of automobiles sold in America were financed. A closer analysis conducted in 1926 suggested that the number approached 60 percent. The percentage of cars being financed was highest for cars in the lowest price class, approximately 70 percent, and higher for used cars than new cars, 65 percent versus 56 percent. About 90 percent of the credit was provided by the large firms, including GMAC. As Calder notes, "Installment credit 'trickled up' the social ladder to become part of the middle-class way of life."[28]

Competition among finance companies produced increasingly liberal terms and growing worry about the practice. More than fifteen hundred articles on the merits and morals of consumer financing appeared in national magazines and scholarly journals between 1915 and 1930, an outpouring that made installment sales one of the decade's great topics for moralizing. Retailers clamored against automobile financing, not because they were opposed to the use of credit, but because automobile owners paid their auto debts first. For retail clothiers, already losing ground to the automobile in the competition for the consumer's dollar, this development added insult to injury. "Practically every merchant," reported Allen Sinsheimer of the National Association of Retail Clothiers, "declared that since automobiles have been sold on the weekly payment plan his collections have been cut down materially." But the auto finance companies worried, too.[29]

General Motors hired the distinguished Columbia University economist Edwin R. A. Seligman to conduct a study of the practice that was published in 1927 as the *Economics of Instalment Selling*. Given carte-blanche access to GMAC records, he produced a ringing endorsement. GMAC experienced losses on consumer loans of just 0.2 percent in its first seven years. These findings were widely publicized, especially by GM itself. Although the public moralizing and hand-wringing nearly ceased after Seligman's study, the American business community remained deeply ambivalent about installment selling. One indication was that corporations did not push financing plans through advertising. After reviewing eighteen hundred ads that appeared in *Ladies Home Journal* between 1901 and 1941, economic historian Martha Olney concluded that national corporate advertising did little to encourage the practice. Businessmen recognized that only a steep business downturn would provide the ultimate test of the practice. Most dreaded the prospect. In their quest for automobiles, Americans forced consumer financing on a worried business community. They bore the lion's share of the responsibility for the consumer financing revolution of the 1920s. Liberal consumer financing, combined with the growing availability of cheap used cars, lowered the barriers to purchasing an automobile even further, so that by the mid-1920s most American families who really wanted an automobile could buy one. One survey estimated that 55 percent of American families were automobile owners by 1927. Historian Frank Stricker has ventured that perhaps even 30 percent of working-class families owned automobiles. The consequences of this massive effort to join the community of automobile "haves" rippled outward to touch nearly everyone and everything.[30]

Consumer financing also raised a deeper problem that extended well beyond the automobile. "Time-buying," wrote Ray W. Sherman in the pages of *Motor* in 1925, "has caused more intensive work than any scheme of mere money-saving ever devised." General Motors's John J. Raskob was explicit on this point: "With obligations under the installment plan, [people] find themselves in need of an income. . . . Now a man [must] put aside so much a month for an automobile, a radio, a washing machine, etc., . . . which, in the last analysis, means that he is performing a greater amount of work than would be the case if he did not have to meet payment of his obligations each month." The father of market research, Curtis Publishing's Charles Coolidge Parlin, told his corporate clients in 1932 that "the automobile furnished one of the greatest incentives to industry and sobriety labor ever had," and he suggested that the desire to purchase automobiles and the ability to do so through consumer financing encouraged a "better attitude" on the part of labor. Thus, the changes in work and consumption transforming American society, Calder concluded, were mutually reinforcing. "Once consumers step onto the treadmill of regular monthly payments," he wrote, "it becomes clear that consumer credit is about much more than instant gratification. It is also about discipline, hard work, and the channeling of one's productivity toward durable consumer goods." Automobile-inspired consumer financing intensified existing patterns of economic behavior, not just in the realm of consumption, but in the workplace as well. Americans celebrated hard work, so they viewed the uncoerced desire of people to work harder in a positive light. But there was an unrecognized downside to greater effort in the workplace. By working harder to pursue consumer goods with ever more ardor, Americans also turned up a full notch the pressure that they placed on the natural world. Had it been able to speak for itself, it might have howled in protest.[31]

As Americans bought and drove the forty million automobiles that the automakers strained to produce before 1930, the entire process caused thousands of environmental impacts in scores of industries. Miners dug iron ore from the earth on Minnesota's great 110-mile long Mesabi Range. The coal necessary to make iron and steel and to generate electricity to run the plants came from beneath the hills and mountains of West Virginia and Kentucky. Hardwood forests in northern Michigan fell to provide the lumber for automobile bodies. Rubber plantations sprang up in Sumatra, Indochina, and Malaya to supply the industry with tires. American consumers' demand for automobiles set in motion one of the greatest flows of raw materials in human history.

By the early 1920s, the U.S. auto industry was the largest industry in the largest economy in the world. The automakers led industrial America in the consumption of a wide variety of raw materials—steel, malleable iron, rubber, plate glass, nickel, lead, and upholstery leather. In 1929, the industry used 18 percent of total U.S. production of all forms of steel, while taking 60 percent of the strip steel and nearly 30 percent of the sheet and bar steel. It used 52 percent of the malleable iron, 84 percent of the rubber, 73 percent of the plate glass, 58 percent of the upholstery leather, 31 percent of the lead, and 26 percent of the nickel consumed in the United States. It also took 37 percent of the aluminum, 24 percent of the tin, 18 percent of the hardwood lumber, 16 percent of the copper, 10 percent of the cotton, and 6 percent of the zinc. Most of the basic raw materials came from the Great Lakes region of the United States and Canada. However, the automakers received materials and parts from nearly every American state as well as a number of countries around the world.[32]

Although the impacts varied, the extraction and processing of raw materials produced environmental damage. Minnesota's Mesabi Range iron ore tended to be found in soft deposits near the surface, which made it accessible by open-pit mining techniques. U.S. Steel owned mines that accounted for 72 percent of the Mesabi's output. At the company's Hull-Rust-Mahoning Mine ("the Man-made Grand Canyon of the North"), which eventually became the world's largest open-pit mine, the techniques of strip mining were pioneered and perfected. Strip mining spread great quantities of red dust on the surrounding area and left the land deeply scarred. Between 1890 and 1977, mine operators took three billion tons of iron ore from the Mesabi Range, a quantity that met nearly 60 percent of U.S. demand.[33]

When the raw materials reached the factories of the automakers and their parts suppliers, another set of environment impacts occurred. Making automobiles produced some of the most extraordinary industrial pollution ever recorded. Fly ash, iron oxide, heavy metals, sulfur dioxide, and of course, millions of tons of carbon dioxide belched from the smokestacks of the coke ovens, blast furnaces, foundries, steel mills, and plants of the American steel and auto industries. Iron, sulfuric acid, cyanide, phenols, and heavy metals poured into the sewers and rivers that served as liquid waste conduits away from the plants. The manufacturers dumped the remaining by-products—liquids, sludge, and solid wastes—that resisted ready disposal in the atmosphere or a local water body wherever an accommodating landowner could be found.

The environmental impacts continued with consumers. With every mile that they drove, they released more unburned hydrocarbons, nitrogen oxides, carbon monoxide, and carbon dioxide into the atmosphere, while using more petroleum and leaving less in the ground. Once people had automobiles they took them into the countryside to explore nature and to visit America's scenic wonders, more than fulfilling the early predictions of automobile boosters. In the 1920s automobile tourism became a major recreational activity, and governments and businesspeople began to transform the countryside to encourage it by building roads and providing tourist amenities. Historian Paul S. Sutter has showed that Aldo Leopold, Robert Marshall, Benton MacKaye, and Robert Sterling Yard, who founded the Wilderness Society in 1935 and became the leaders of the mid-twentieth-century American wilderness preservation movement, shared an acute sense of outrage about the impact that tourists, automobiles, and roads had on scenic, undeveloped nature. "Driven wild," as Sutter put it, these men eventually pushed back against these inroads by calling for the preservation of wilderness areas defined explicitly by the absence of roads and cars. By the 1920s, too, more than a million cars a year wore out, blighting the landscape and posing a complex disposal problem.[34]

Thus, at every stage of the automobile product life cycle — from resource extraction and manufacturing to consumer use and disposal — the multiple environmental impacts intensified dramatically with the explosion in the number of automobiles, as Americans bent the natural world to meet their overwhelming desire for automobiles. The accelerated sales made possible by mass production and consumer financing caused greater environmental impacts across the entire automobile product life cycle and caused these impacts to occur earlier than otherwise would have been the case.

As Americans made the automobile their chief talisman of successful belonging and purchased them in the millions, they forced a reckoning with the choice of gasoline as a motor fuel. The pressure that resulted led to some remarkable technical ingenuity but also down a path that had serious public health consequences. In 1916, a sudden spike in gasoline prices revived concerns about the adequacy of petroleum supplies. A secretary at Ford Motor Company reported in May 1916 that the firm received fifty to a hundred letters a day on the subject. By 1920, concerns about the adequacy of petroleum resources to meet the skyrocketing demand from auto owners began to verge on alarm. Between 1909 and 1920 the number of registered automobiles in the United States increased 2750 percent, but the annual production of gasoline increased just 800 percent. Studies by the U.S. Geological Survey and others predicted the depletion of known reserves within ten to twenty years, if prevailing rates of gasoline use persisted.[35]

The oil companies knew that if they failed to find new oil fields and ways to squeeze more gasoline out of a barrel of crude oil, they would surrender the automobile fuel market to ethyl alcohol. In July 1912 Standard Oil of Indiana chemist and executive William Meriam Burton filed a patent application for a process to "crack" heavy fuel oil molecules under high temperature and pressure to produce lighter molecules, or gasoline. The Burton thermal cracking process doubled the gasoline yield from a given quantity of crude oil. Burton and Standard Oil kept the discovery secret in the hope that the company could increase its gasoline output without driving the price of gasoline down, but in 1914 Walter F. Rittman, a chemical engineer working for the U.S. Bureau of Mines at Columbia University, developed an alternative thermal cracking process to break the Standard Oil patent monopoly. The press lauded Rittman as a hero at the time, but whether he did his country a service in the long run is debatable. In the decade between 1913 and 1923 petroleum refiners improved the average yield of gasoline from crude oil from one gallon of gasoline per eight gallons of oil to one gallon of gasoline per three and a quarter gallons of oil. Even so, automobile sales and gasoline consumption continued to outstrip gasoline production.[36]

The problem was not the oil industry's alone. "The automobile industry," *Scientific American* editorialized in March 1919, "can no longer afford to ignore the engine-fuel problem." Henry Ford experimented with alcohol in the years just before and during World War I, but he abandoned serious pursuit of the matter. Other automakers recognized that they could help conserve petroleum and postpone the day when they would have to run the business risk of switching to alcohol-powered cars if they designed engines that used less gasoline to produce the same or more power. The most straightforward way to do this with existing internal combustion engines was to increase compression in the engine's cylinders. Unfortunately, when compression increased, engine "knock" followed. At first, no one knew what caused knock. It might be the engine or it might be the gasoline. Tests soon showed that it was the fuel: higher compression caused gasoline to "pre-ignite" before the piston completed its downstroke. Knock posed a frustrating barrier to the more fuel-efficient engines that were desperately needed.[37]

Charles F. Kettering, who had been working on knock since 1916, moved the fuel problem to the top of the research agenda shortly after taking over as the head of the newly formed General Motors Research Corporation in 1919. Kettering decided that his research team needed to find a gasoline additive that eliminated knock, permitted the development of high compression engines, and extended the life of the gasoline supply, so that the industry could gradually position itself for the move to alcohol without undue harm to its profits or the convenience of consumers. In short, an additive provided a temporary bridge to the alcohol future.[38]

General Motors's researchers Thomas Midgley, Jr., Thomas A. Boyd, and Carroll A. Hochwalt, working under the direction of Kettering, began a trial-and-error (and eventually a more systematic) exploration of "low percentage" (small quantity) additives. After working their way through several potential solutions that had various serious drawbacks, the GM researchers discovered on 9 December 1921 that adding less than a teaspoon of tetraethyl lead (TEL) to gasoline (one part TEL to thirteen hundred parts gasoline) did the trick, making possible a doubling of engine compression ratios and halving the amount as gasoline required to travel the same distance at the same speed. Kettering later called the day the most dramatic in his career.

Smelling monopoly profits from a patent-protected formula or production process and with help from Du Pont (which then owned nearly 24 percent of GM and had de facto control over GM decisions), Kettering's company moved quickly to commercialize its discovery, marketing the gasoline in selected areas beginning in February 1923. After Standard Oil of New Jersey patented a superior method for manufacturing leaded gasoline, GM and Standard Oil created the Ethyl Gasoline Corporation in August 1924 to begin the high-volume manufacture and marketing of leaded gasoline. In a well-publicized marketing coup, the top three finishers at the Indianapolis 500 in 1923 used leaded gasoline. Consumer demand proved brisk. Within eighteen months, more than one hundred million gallons a year of leaded gasoline were being sold at more than ten thousand filling stations in twenty-seven states. GM took these steps with a speed, secrecy about production, and hype for the product that suggested a rush to profit.[39]

The environmental problem with leaded gasoline was the lead. Lead is toxic to humans and other life. Adding it to gasoline meant releasing trace amounts of lead to the atmosphere with automobile exhaust wherever automobiles were used. Suspended in the air, people inhaled the lead dust, and the metal permanently accumulated in their bodies. The lead in the exhaust also fell to earth, blanketing roads and the nearby landscape with a fine lead-laden dust. Thanks to the pioneering occupational health research of Harvard Medical School physician-scientist Alice Hamilton during the 1910s the toxicity of lead was well known in the 1920s. Midgley actually gave himself lead poisoning while working with TEL. When he wrote to Yale physiologist Yandell Henderson for an opinion on the public health impact of TEL in gasoline, Henderson replied that "widespread lead poisoning was almost certain to result."[40]

An accident at Standard Oil's Elizabeth, New Jersey, Bayway experimental manufacturing plant in October 1924 dramatically publicized the dangers of TEL. The incident killed five employees and sent another thirty-five to the hospital with severe lead poisoning. The deaths were disconcerting in the extreme. The workers readily absorbed the TEL through the skin, and it did extraordinary damage to their nervous systems. Many of the victims died hallucinating and screaming in delirium, which made the disaster front-page news around the country. Further investigation by journalists revealed that workers had died at the GM and Du Pont facilities where developmental work had been done on TEL. Hundreds, in fact, had been hospitalized. The extraordinary toxicity of TEL should not have been a surprise. Henderson and the War Department had tested it for use as a nerve gas during the war. Henderson spoke out immediately, publicly, and forcibly. "Sale to the public of gasoline containing lead . . . ," he argued, "should be stopped immediately and prohibited until the subject is fully investigated by scientific experts. Otherwise there is great danger of widespread poisoning among men at gasoline filling stations, garages, repair shops and even among the general public. Public health authorities everywhere should take immediate action." New York City and New Jersey banned the sale of leaded gasoline. Henderson continued to speak out, calling the prospect of leaded gas "the greatest single question in the field of public health that has ever faced the American public."[41]

In an effort to blunt the public hysteria, GM rushed to publicize the favorable findings in a Bureau of Mines report on TEL that it had funded. But the Bayway disaster and the continuing public controversy finally forced Ethyl to agree to a halt in sales, while U.S. Surgeon General Hugh S. Cumming, at the quiet urging of GM, Standard Oil, and Du Pont, held a conference on 20 May 1925 to hear testimony from representatives of industry, labor, science, and medicine on the matter. With ironic timing, Hamilton in this very month published her seminal *Industrial Poisons in the United States*, a book that devoted its first two hundred pages to problems with lead. Hamilton and Henderson both testified against the additive. Representatives of industry spoke in favor of it, stressing three points: that TEL offered the prospect of substantial improvement in gasoline fuel economy, that the health problem was limited to the workers who made TEL and could be corrected with greater attention to safety, and that the industrial progress of the nation depended on moving forward with the additive. Patent attorney and Standard Oil of New Jersey executive Frank A. Howard called the discovery of TEL "a gift of God." More significantly, Kettering and Midgley asserted that no alternative solutions were available (a claim that historian William Kovarik has demonstrated Kettering knew to be false). In the absence of proof that gasoline with the additive posed a direct threat to human health, the public health officials at the conference groped for middle ground between TEL's proponents and Hamilton and Henderson.[42]

Faced with conflicting assertions, Cumming and the conference participants decided to create a blue-ribbon panel of seven scientists and physicians to study the issue and report back in seven months. In a hurried study that compared service station attendants and chauffeurs who used gasoline with TEL with those who did not, the panel found that both groups had elevated blood-lead levels and that there was no appreciable difference between the two groups, which seemed to suggest that those exposed to TEL from leaded gasoline were not on a rapid road to lead poisoning. The experts cautioned, however, that further long-term studies were required. "The vast increase in the number of automobiles throughout the country," they wrote, "makes the study of all such questions a matter of real importance from the standpoint of public health, and the committee urges strongly that a suitable appropriation be requested from Congress for the continuance of these investigations under the supervision of the Surgeon General of the Public Health Service and for a study of related problems connected with the use of motor fuels." Meanwhile, Ethyl promised to do more to safeguard the health of the workers who manufactured TEL. So the burden of making a case against the additive fell on those critics like Hamilton and Henderson who believed that there would be a threat to the public at large from automobile exhaust. Although they could (and, as it turned out, did) predict the long-term health consequences, they obviously could not prove yet that there would be long-term negative consequences. Absent definite proof of harm, Cumming ruled in favor of Ethyl, a decision that in retrospect seems a craven unwillingness on the part of the federal government and the Coolidge administration to stand between General Motors, Du Pont, Standard Oil, and Ethyl and a solution that promised them profits while helping to address an urgent national need to conserve petroleum. No further studies were funded by Congress or even requested by the U.S. Public Health Service. The public became guinea pigs.[43]

Far from being deterred by the controversy, GM and Standard declared vindication, trumpeted the safety of TEL, and resumed marketing it aggressively. During the controversy, they touted TEL to scientists, engineers, and public health officials, as well as to the press and larger public, as the best available method of improving the fuel economy of automobiles in order to husband diminishing petroleum reserves. Thomas A. Boyd even bragged to the American Chemical Society that fifty-mile-per-gallon automobiles were in prospect. Afterward, the companies marketed TEL to consumers as a speed and power enhancer, and the automakers built high-compression engines that delivered greater horsepower rather than fuel economy. Faced with a choice between profits and prudential caution in the face of a strong possibility of harm to public health, senior executives at Ethyl, GM, Du Pont, and Standard Oil knowingly turned a blind eye to the health risk and chose profits. By 1940, 70 percent of the gasoline sold in America contained TEL, a percentage that rose even further after World War II. Over the sixty years from 1926 to 1986 the Ethyl Corporation produced 6.6 million tons of TEL that were burned with gasoline, and Americans inhaled lead whenever they inhaled automobile exhaust. GM made $43.3 million in patent royalties and $82.6 million in Ethyl profits before its patent expired in 1947. As management guru Peter Drucker later observed in admiration, "GM . . . made money on almost every gallon of gasoline sold, by anyone."[44]

Historian William Kovarik has argued that GM cynically ignored the possibility of using ethyl alcohol as a "high percentage" gasoline additive, when its

Profits over prudence under pressure = leaded gasoline. *Photo by John Collier.*
Library of Congress, Prints and Photographs Division, LC-USF 34-82498-C.

own research (as well as the research of others) had proven this approach to
be the superior alternative on every point save one, that alcohol did not offer
patent-protected monopoly profits. On the other hand, historian Alan P. Loeb
has argued that GM may have had legitimate reasons for concluding that technical, production, and cost problems made alcohol a less feasible additive than
TEL, at least for the short-term. More remarkably, during a visit to the German
chemical manufacturer Badische, Anilin and Soda-Fabrik (BASF) in November
1924, Kettering's hosts showed him a new nonpoisonous substance that worked
as a "low percentage" antiknock additive and cost less than TEL that he correctly guessed was iron carbonyl. Press reports about iron carbonyl as a knock
suppressant appeared as early as March 1926, only weeks after the surgeon general gave TEL the go-ahead. Scientists working on petroleum chemistry had
their own take on the TEL controversy. "I believe, and my investigations support
the idea," chemist Carleton Ellis was quoted in the same month as Kettering's
German visit, "that gasoline and kerosene are capable of sufficient modification
during the refining process to yield high compression fuels, directly, without

resort to additions such as lead ethyl." By 1927, laboratory experiments in France had proven Ellis right. But it was too late.[45]

When Americans bought their first automobiles, they made a choice with broad implications. Soaring gasoline sales due to consumer demand for automobiles posed a severe threat to finite petroleum reserves and the continued use of gasoline in automobiles. Under pressure from this perceived threat, Henry Ford continued to talk up alcohol, but there was still no follow-through from his company. GM actively pursued a solution, but it chose the first workable one that also offered the prospect of profit. Not only was another opportunity lost to switch to the cleaner-burning, renewable alcohol, but GM saddled gasoline with a new public health threat in lead, one of the automobile's worst environmental impacts. It was a bad move made under the twin pressures of corporate venality and the runaway demand of Americans for automobiles.[46]

"A time of revolution is an uneasy time to live in," editor and social commentator Frederick Lewis Allen observed in his popular 1931 retrospective on the 1920s, *Only Yesterday*. Prosperity and change offered more opportunities for economic success, but they also placed greater pressure on individuals to succeed. Success in America always came with the string of uncertainty attached, an inevitable result of living in a society characterized by a magnitude and rate of change so great that there could be no assurance of maintaining a successful or respected place in the world over time. Change in the quarter century that Americans adopted the automobile, as some of the darker developments of the 1920s suggested, strained the limits of the congenital optimism with which they resolutely fronted economic change. The simple primacy of keeping up left the overwhelming majority no time or inclination to ponder where they were heading or the associated costs.[47]

Automobile ownership meant that a person or a family stood with the community of successful Americans. To be without an automobile was increasingly a form of public nakedness in which a man, as one commentator put it, "ran the risk of being singled out among his fellows, especially on Sundays and holidays, as either hopelessly poor or perversely out of the swim." Deeper anxieties were also at work. Americans embraced the automobile, not just because it was an object of unprecedented practical and symbolic appeal made affordable by mass production and consumer financing, but because automobile ownership became perhaps the most important way that they told themselves and others that they were coping successfully with change. The 1920s marked the culmination of two generations of immense economic and social change that transformed the United States from a largely agrarian society into the most powerful and

dynamic industrial economy in the world. In a culture that lauded and often re-
warded innovation, initiative, and hard work, Americans, especially men in the
workplace, experienced extraordinary pressures to prove their worth as individuals by successfully exploiting the larger climate of opportunity and change for
their own demonstrable benefit. This cultural imperative imparted a ferocious
energy to economic activity in America, wonderfully captured by the Austrian
émigré economist Joseph Schumpeter's famous phrase "creative destruction."
It also imparted ferocious energy to material display. Americans not only made
the automobile their most cherished material possession, they seized the opportunity with relish. Indeed, they used it to proclaim their success defiantly before
the capricious and uncertain world that they unwittingly made.[48]

4

AN INDUSTRIAL EPIC

"Iron and coal form the backbone of the automotive industry, iron because it is the principal component of a motor car and coal because it is necessary both in the manufacture of iron and the production of power," the Ford Motor Company explained in *Ford Industries*, an extensively illustrated book that it published and distributed widely in the mid-1920s to explain how it made cars. "The cost of iron and coal delivered at a plant largely governs the selling price of the product. No matter how efficiently or economically a manufacturing organization may be operated, the fluctuating market prices of raw material are beyond its control. The only way to avoid price fluctuations is to control the source of raw material."[1]

By the 1920s, Ford made a million or more cars a year, a process that touched the natural world in profound ways. But Ford also needed to defend its position as America's principal producer of low-cost cars. After the dramatic productivity gains achieved at its Highland Park complex, Henry Ford and his managers realized that further substantial cost reductions in the Model T could not be found in the manufacturing process alone—at least if Ford wished to preserve the high Ford wage scale that had made the company a legend while reducing the extraordinary rates of employee turnover on the company's monotonous assembly lines. To protect and extend its market dominance, Ford had only one major remaining option if it was to stay ahead of its competitors: to reduce the cost of raw materials that fed its manufacturing operations.

Ford decided to become its own supplier of raw materials. In doing so it would eliminate the middlemen and the profits they charged. Ford's customers would pay a single profit on the finished automobile. "Backward integration," as economists call this form of vertical integration, provides a powerful competitive weapon when the barriers to entry in an industry are high and when a firm like Ford commands a large market share that permits it to take full advantage of the economies of scale that mass production permit. But it can also be a risky strategy in an industry characterized by a high rate of technological change — where the materials used, the methods of manufacture, or the product itself may suddenly change — because the extraordinary capital investments required cannot be undone easily should unforeseen developments demand it. Moreover, the strategy posed a challenge to execute because it often required a firm to manage activities outside its core competency. By the early 1920s, the automobile industry had entered a period when the rate of technological innovation was slowing, so backward integration was a reasonable strategy for Ford. By the mid-1920s, the company had redesigned its manufacturing operations around this approach. It remained important through World War II, and traces lingered into the 1980s.[2]

To implement backward integration Ford acquired sources of raw materials and built facilities to process such basic resources as iron, steel, coal, silica sand, lumber, and rubber. The challenge Ford faced with these basic industries was operating them in a manner that ensured that the raw materials could be had at a cost consistently below their market price. In addition to the cost-cutting rationale, Ford also justified backward integration by arguing that it needed to have control over some portion of its basic raw materials to be able to guarantee the continuous flow necessary for mass production. As can be imagined with a product assembled from more than ten thousand parts made from hundreds of raw materials, the Ford production men worried obsessively about continuous flow. The interruption of the flow of a single crucial material or part could bring automobile production to a halt.[3]

Steel provided the sharpest prod to Ford's backward integration. A Model T was 75 percent iron and steel by weight. U.S. Steel much earlier had shut out Ford and everyone else from the top iron ore lands on the Mesabi Range in Minnesota, so the best that the automaker could do was buy some second-quality iron ore mines in northern Michigan and settle for making iron and some of its steel in order not to be held hostage by the steel industry. Henry Ford, who retained the suspicions of nineteenth-century agricultural America toward big business even after he became the greatest industrialist in early-twentieth-century America, held similar views about rubber, coal, and timber suppliers.[4]

Ford also worried about coal, an especially important raw material for making cars in the first half of the twentieth century. Ford needed coal to make coke, one of the three ingredients (with iron ore and limestone) necessary to make iron and, ultimately, steel. It also burned coal to fire the steam engines that turned the dynamos that generated the electricity that ran much of the production machinery. Ford estimated that it used twelve thousand pounds of coal for every car that it made. So in 1923 Ford bought coal mines in West Virginia and Kentucky with deposits estimated at five hundred million tons and its own railroad, the Detroit, Toledo, and Ironton, to carry the coal north to its plants.[5]

Henry Ford's concern was well founded. Price spikes, shortages, or shutdowns in the steel or coal industry could and did wreak havoc with Ford's output and profitability. Controlling a portion of his raw materials, even if it added cost to his production, could be rationalized as insurance against these problems. But this concern with control also reflected one of Henry Ford's strongly held biases. He viewed suppliers, who made most of their profits as a consequence of surging consumer demand for his cars, as parasites, all the more so when firms like U.S. Steel were controlled by Wall Street financiers who seemed to be more interested in maximizing profits than in improving the quality and lowering the price of their product. Ford vowed to cut them out to the extent he could, or at least to deal with them from the position of strength that being his own supplier made possible.

With these motives for backward integration, Henry Ford and his senior managers decided shortly after the conclusion of World War I to bring the basic raw materials to make automobiles to a single complex located on the Rouge River in Dearborn, Michigan, a few miles southwest of Detroit. Facilities constructed on this farmland site would convert raw materials into automobile parts to ship to Ford's assembly plants around the country for final assembly into automobiles. There is no firm evidence that there was a master plan for the Rouge at the outset. The complex—essentially a steel mill, parts manufacturing facility, final assembly line, and power plant—evolved into its mature form over twenty years from 1917 to 1937. When the Rouge final assembly line began producing Fords for the Detroit area in 1927, the plant was the only one in the world where raw materials were made into finished automobiles at one site. By then the Rouge was the most famous industrial complex that had ever existed, more famous even than Highland Park. No automaker concentrated so much production at one site or pursued backward integration as far, as comprehensively, or on as massive a scale as Ford.[6]

To cultivate a mystique around how it made the Model T, Ford opened its doors and told the public about the birthplace of mass production at Highland Park, the Rouge, and the immense flows of raw materials and parts that fed the plants. "The visitor to the Ford Motor Company," the firm told *Saturday Evening Post* readers in 1924, "finds no locked doors, no secret processes, no closely-guarded methods or formulae, no restrictions of competition through patent holdings." Ford built the Rouge to accommodate plant tours, which began at the complex in 1924. Ford trained its guides to wow visitors with facts about the Rouge and its operations: twelve hundred acres, nearly ten million square feet of floor space, enough electricity generated to power a city of a million, one hundred miles of railroad track, eighty-five acres of employee parking, storage bins for two million tons of iron ore, coal, and limestone, more than a car a minute rolling off the final assembly line, and so on. "All major units and final lines," Ford's Rouge production chief Charles Sorenson recalled, "were arranged so that visitors could move along with the operations and not interfere with the workers." "With all its intricate operations," Sorenson argued, "we knew it was to our advantage to show the public how we did it. It was not considered wise to reveal the secrets of your business. That was the old practice. We reversed that idea completely. We had nothing to hide."[7]

The Rouge became the most visited industrial site that ever existed in the United States and probably in the world. In nearly six decades from 1924 to 1980, Ford took eight million visitors on tours through the complex that its publicists touted as one of the "wonders of the modern industrial world." A Rouge visit, which from May 1936 began at the Rotunda, a visitor's center built as "a fitting gateway for all visitors to the world's largest industrial permanent exposition," apparently lived up to its billing. For many years it rated as one of the top ten tourist attractions in America. In the 1920s, Ford acted on the belief that a great company with the best product and manufacturing methods could dispense with the phoniness and cost of paid advertising and thrive on fairly earned, favorable publicity alone. Henry Ford and his managers were right. Ford's plants were twentieth-century manufacturing marvels, the apotheosis of American industrial supremacy. In consequence, the company received millions of lines of free publicity in newspapers, magazines, and trade journals. Mexican muralist Diego Rivera, no special friend of capitalism, even made the Rouge the subject of his celebrated Detroit Industry murals at the Detroit Institute of Art.[8]

Not only did Ford open its plants to the public and benefit from free publicity, but in 1924 the company launched an unprecedented publicity campaign involving books, films, fair exhibits, and institutional advertising to tell Americans how it made the Model T. By the mid-1920s, the Rouge stood at the center of this publicity. The Rouge was the greatest and best-publicized example of vertical integration in twentieth-century industrial America, a facility where, as Ford's publicists bragged, raw materials were turned into finished automobiles driven from the final assembly line in just twenty-eight hours. Outside commentators, no less than Ford spokesmen, agreed that the scope and scale of the operations there were nothing short of "epic." For nearly sixty years, the "Crown

Ford's Rouge with the Rotunda in the foreground, 1939. *From the Collections of The Henry Ford, P.833.72814.*

Jewel of American Industry," as the company's publicists dubbed the complex, featured prominently in the firm's promotions. The production prowess that the Rouge epitomized was at the core of Ford's corporate identity.[9]

Ford's commitment to showing and telling the public how it made automobiles was almost certainly the greatest effort a major corporation has ever made to educate consumers about how its product was built. But the company could not do so without explaining the relations among raw material flows, manufacturing processes, and the natural world. This publicity campaign offered an unparalleled early window on the environmental impacts of making automobiles at America's largest automaker, especially the first two phases of the automobile product life cycle: the environmental impact of the industry's demand for raw materials and the pollution associated with manufacturing automobiles. By placing the Rouge at the heart of its publicity efforts, Ford made the greatest concentrated source of industrial pollution from automobile manufacturing that ever existed the basis for telling Americans how it made automobiles.

When it chose to advertise, as it did at several points in the 1920s, Ford could not resist the temptation to publicize the Rouge and the benefits of vertical integration rather than the Model T. Indeed, the company saw no reason not to. The most remarkable instance was a series of fourteen, two-page, color-illustrated institutional ads entitled "An Industrial Epic" that Ford ran in the *Saturday Evening Post* in 1924. With titles like "From Source to Service," "Mountains of Raw Material for a Nation's Transportation Need," and "Iron for Unparalleled Production," the ads all dealt with how Ford transformed raw materials and energy into automobiles, including ones that focused on the Rouge, hydroelectric power, forestry and lumber practices, coal mining, coke-making, blast furnaces, and final assembly.[10]

The fourth ad in the series, entitled "Vital Resources That Cannot Fail," provided the essence of Ford's message, as well as the company's picture of an ideal world. "No market fluctuations affect the Ford Motor Company's materials, resources, men or machines. They are free from manipulation. Controlling the five basic elements of man's productivity; iron, coal, timber, transportation and power—the sovereign essentials of industry are assured. . . . all beyond the reach of fluctuation and manipulation. All co-ordinated to one end and purpose—dependable transportation at lowest cost, for the masses of men." At a time when the *Saturday Evening Post* had a circulation of over two million, Ford's advertisements may well have told more people more about how a car was made than any other single communication before or since.

Two ads in particular shed light on the environmental implications of Ford's prodigious appetite for raw materials and the extraordinary wastes generated at the Rouge. In the seventh ad, "For the People and Posterity," Ford acknowledged concern about the impact of its timber needs—more than a million board feet a day—from the standpoint of forest conservation. Similarly, an ad captioned "Robbing Smoke of Its Waste" raised the issue of air pollution and other wastes at the Rouge. The advertisements show that Ford recognized that these processes posed environmental problems and acknowledged a need to address them, a reminder that there never was a dark age in which industrial companies wreaked maximum harm to the natural world in blindness or with impunity.

Late-twentieth-century proponents of industrial ecology argued that manufacturers should assume responsibility for the environmental ramifications of their products by adopting product designs and manufacturing processes that minimized these impacts by conscious prior intent rather than dealing with unintended consequences at the behest of angry regulators and publics. Like many concepts in the business world, industrial ecology passed for an innovation. But many of the new field's basic ideas and practices had a much longer history, a fact very evident in the case of Henry Ford, the Rouge, and Ford's raw material programs during the 1920s and 1930s. Industrial ecology provides a useful perspective to understand how Ford viewed and addressed raw materials extraction, manufacturing pollution, and the larger issue of industrial wastes generally.[11]

At the heart of industrial ecology is the same idea central to the thinking of Henry Ford and his plant engineers as they built the Rouge. Flows of materials and energy from the natural world are transformed by people into products that are used and ultimately discarded by others. But industrial ecologists viewed these flows with two aims in mind: the efficient use of raw materials and energy—that is, maximizing an output from a given amount of material or energy (or minimizing the inputs of either while holding output constant) and minimizing the negative environmental impacts from making, using, and disposing a product. The first concern involves conservation, the second, reducing pollution. These two motivations and the resulting positive environmental outcomes are central to industrial ecology. Keeping these two different aims in mind is also necessary to understand what Ford was doing with raw materials and at the Rouge.

Model Ts—and nearly all the cars since then—were made mostly from iron and steel. But Ford's "For the People and Posterity" advertisement is a reminder that before the mid-1920s the automakers used a good deal of wood in the Model T and other automobiles. In fact, as the 1920s began, the demand for wood increased, as consumer tastes shifted from open-body runabouts and touring cars to closed-body sedans and coupes. The automakers used hardwoods—maple, elm, and ash in that order—chiefly for the body frame and floorboards. In 1922 Ovid M. Butler writing for *American Forestry* reported that 250 board feet went into every Model T, which was about the average lumber content taken from a hardwood tree. Thus, at the peak of Model T production in the early 1920s, when annual output often exceeded a million cars a year, Ford needed a million or more trees—approximately twenty-five thousand acres of forest—or over 250 million board feet—a year to feed its production. The company also consumed a substantial quantity of timber to pack, crate, and ship parts and materials.[12]

As part of its efforts to cut out middlemen and ensure the flow of a vital raw material, in July 1920 Ford announced the purchase of 450,000 acres, most of it maple and birch hardwood forest around Lake Michigamme, on Michigan's Upper Peninsula. The idea was not to supply all of Ford's timber needs, just a substantial portion of them should the need arise. At the same time, Ford built an extensive wooden parts manufacturing complex at Iron Mountain, Michigan, about a hundred miles from its forests. When Iron Mountain opened in August 1924, Ford moved the making of wooden parts there from Highland Park. The new facility provided 115 million board feet of lumber a year in capacity, roughly half the company's annual needs. The goal in moving production from Detroit to the forests was to reduce long-distance freight charges on moving green lum-

Ford and forest conservation. Ford ad, 1924. *From the Collections of The Henry Ford, 64.167.19.469.*

ber south, especially since much of the lumber ended up as scrap and because water constituted a substantial portion of the lumber's weight.[13]

At Iron Mountain, Ford tried to use every part of the tree except the shade. Instead of following the usual practice of cutting boards from logs, letting the boards dry, and then cutting parts from them, Ford cut the twenty-plus wooden parts for the Model T directly from the logs and then dried them in fifty-eight drying kilns that each measured 222 by 20 feet. This step allowed Ford to cut more parts from each tree, reducing waste by 35 to 50 percent. The operation then converted the much-reduced remaining scrap wood into charcoal—the plant later became the basis for Kingsford Charcoal—and methyl (that is, wood) alcohol, as well as numerous other chemical by-products marketed to the public. The company hoped for two and a half million dollars in annual sales. Ford burned any remaining scrap or sawdust as fuel to create heat for the drying kilns and to power the plant. "The multiple waste necessary under the previously accepted practice of cutting the logs into boards, shipping the boards to various mills and burning the resulting high percentage of scrap as fuel, is largely eliminated," Ford told the public in the *Saturday Evening Post* ad. "Lumber, labor, time and by-products are saved." The company hoped to save one hundred million board feet a year as a consequence.[14]

Company publicists proclaimed all efforts to save wood as "the Ford Motor Company's contribution to national forestry conservation." Although it received no public criticism about the volume of its timber consumption, the company recognized the implications of its large demand. "Woodman spare that tree," an article in *Ford News* admonished in 1924, explaining that the "latest government reports show that, unless the command is heeded with extreme conservation and reforestation measures, timber resources of the country will be exhausted within 36 years." Ford's forest and Iron Mountain plant manager Edward G. Kingsford told the *Wall Street Journal* in 1923 that he feared the hardwood supply would not last ten years. In this context, the company repeatedly stressed the conservation methods that it used in its forests and at Iron Mountain.[15]

Ford touted two progressive forest management practices that it practiced for conservation purposes. The company bragged that in its forests no trees fewer than ten to twelve inches in diameter were cut, which is to say that it practiced selective rather than clear-cutting. The second practice was to collect and burn the remaining brush and slash to remove fuel for any future forest fires that might endanger the trees left behind. Ford garnered favorable publicity for using these methods. The *New York Times, Wall Street Journal,* and *Christian Science Monitor* ran complimentary reports. After Ovid Butler visited Iron Mountain and the Ford forests, he pronounced the operation everything that a professional forester could want. *Automotive Industries* ran a laudatory article in 1924, and followed it up with several positive editorials urging other automakers to follow Ford's example.[16]

The progressive picture that Ford painted in its publicity and that the press obligingly amplified belied a more complex reality. Even after acquiring its own forests, Ford continued to buy 80 percent of its lumber from contractors who cut elsewhere. The company did not force these lumbermen to practice selective cutting. Nor did Ford make selective cutting the general practice in its own forests, either. "When the company first started logging . . . ," recalled Ford employee Eric Stromquist, "it did it just like the others did. We cut everything clean." In fact, Ford practiced selective cutting only in a small portion of its forests from a company-run logging camp located two miles from Sidnaw, Michigan, and it did so for only several years. The Sidnaw operation served as a pilot and a show project for publicity purposes. The brush burning was not widely practiced either. Thus, when Kingsford responded to a survey of commercial forestry practices conducted by the U.S. Chamber of Commerce in 1927 by indicating that Ford did no clear-cutting and that all slash was burned, his answers were not true. Henry Ford's wishes and Ford's publicity notwithstanding, the men in the woods who cut timber for Ford mostly did it the way they always had because it was cheaper and easier.[17]

At Iron Mountain, Ford's attempt to cut parts directly from logs proved a disaster. A substantial percentage of the parts warped in the drying kilns, rendering them unusable and, ironically, increasing rather than reducing the amount of scrap. Ford lost money on the operation. Fortunately for the company, the use of wood in automobiles, which peaked at 1.1 billion board feet in 1923, declined after the mid-1920s with the introduction of all-steel, closed-body construction. The Ford Model A, introduced in 1927, averaged just fifty board feet per model. A decade later, when there were only five board feet in a Ford sedan, wood had essentially vanished from the American automobile. Thus, the importance of Iron Mountain diminished almost as soon as it became operational, a good example of the technological risk inherent in vertical integration. Ford sold its forests and the Iron Mountain plant in 1951.[18]

Ford used the same waste-reduction frame of reference in the "Saving Millions by Robbing Smoke of Its Waste" advertisement as it did in "For the People and Posterity." But the second ad, which included a full-page color illustration showing smoke being released to the atmosphere as red-hot coke was discharged from the coke ovens at the Rouge, depicted an environmental impact of automobile manufacturing at the Rouge. The auto industry distinguished between manufacturing, making an automobile's constituent parts, and assembly, putting these parts together to make a finished and functioning automobile. Ford did a substantial amount of its manufacturing first at Highland Park and then from the mid-1920s at the Rouge. Branch plants distributed throughout the United States then assembled Ford cars from the parts before shipping the cars to dealers in each region.

Manufacturing consisted largely in making metal parts. "Making cars calls for plenty of iron and steel," a Ford publicist wrote. "Producing the raw metal is hot, dusty, smoky business at best. Fiery coke ovens, plants which bake fine ore into cinder cakes, towering blast furnaces — all are essential if America is to have cars." Consequently, most of the pollution associated with making automobiles came from manufacturing the metal component parts and not the assembly of those parts. The major automakers differed in the degree to which they manufactured component parts themselves or relied on suppliers to make them. A full accounting of the environmental impact of making automobiles would necessarily entail an assessment, not just of the manufacturing done by the automakers, but also of that portion done by the automakers' suppliers.[19]

Pollution or waste? Ford ad, 1924. *From the Collections of The Henry Ford, 64.167.19.468.*

Although it relied heavily on suppliers to manufacture parts, Ford for much of the first half of the twentieth century, proved the great exception to general practice among American automakers in making more of its major component parts itself. More important, it was the only U.S. automaker to make its own steel. Ford greatly expanded its steel-making capacity at the Rouge in the 1930s and eventually became one of the top ten steel producers in the United States. Before the 1950s, most of the major parts for Ford automobiles, as well as all of the steel that it made for itself (approximately half its total need) came from one place, the Rouge. For the other automakers the environmental consequences of automobile manufacturing were widely dispersed among many plants and suppliers, but for Ford these problems and their impacts were concentrated at the Rouge, making it the best case study on the industrial pollution associated with making automobiles.

In the 1920s Ford viewed industrial pollution at the Rouge as part of a larger problem with waste, a perspective typical in American industry at the time. In practice, a manufacturing waste might involve both an excessive use of materials *and* harmful environmental impacts when the firm discarded materials. When people spoke of "waste reduction," they almost always meant using men, machines, materials, and time more efficiently rather than reducing pollution. When they spoke of "industrial waste," they usually meant the materials discarded by a company, but not always with the implication that environmental harm automatically ensued. However, the pursuit of waste reduction, even when motivated largely by the goal of efficiency or conservation, could reduce industrial waste and, consequently, environmentally harmful pollution. The question implicitly posed by the "Robbing Smoke of Its Waste" ad is the extent to which these goals—conservation and pollution reduction—animated Ford's waste reduction efforts at the Rouge. The efforts in themselves were spectacular.

Waste reduction at the Rouge was a by-product of Ford's experience with mass production. All factories that transformed raw materials into physical products created wastes in the form of extraneous gases, liquids, and solids. Engineers needed to move these away from the people and machinery involved in production in order for the factory to function. When production volume increased, the logistics involved in getting out from under the accumulating wastes could be daunting and threaten to disrupt manufacturing operations. Ford first crossed this threshold at Highland Park with mass production of the Model T. Waste disposal there posed a major headache for the company.[20]

The men who built Ford's Rouge complex therefore planned for large-scale waste management with two goals in mind. The first goal was to use existing technology simply to move wastes away from the operations. More than six hundred smokestacks, for example, drew smoke up and out of the factory and released it high enough in the air to disperse and dissipate it away from the plant. The recollections of Harry Hanson, who worked at the Rouge on laying out what was then the largest foundry in the world, provided a sense of the immediacy of this problem. "The coke from the cores and the coke from the castings," he recalled, "had separate air systems over those to pull the stuff out but it was still a hard way to get rid of it so we erected stacks about a hundred feet tall and five feet in diameter. We put them . . . close to the shakeout. In that way we could pull the dust out of the shakeout and all along the line into the air and spread it over Dearborn." Leaving aside his cavalier attitude about spreading foundry dust over Dearborn, Hanson made plain the more pressing problem that affected production and worker health inside the plant: getting smoke and dust out of the foundry.[21]

Liquid wastes presented a similar problem, and here as well Ford opted for the standard solution. The availability of water was an important consideration in the decision to build the plant on the Rouge River. Ford needed water, not only for production and to transport raw materials to the site, but also to move liquid wastes away from the plant. "The main reason for locating the plant at the Rouge is like any other large plant," William Verner recalled. "You decide on its location as to railroad, traffic, and water supply. When the Highland Park plant was built, it wasn't very long before they did have trouble with the water supply. . . . They were always bothered with sewage disposal. They had to get rid of the waste. The idea of a big plant is to have water supply available, a place to get rid of the sewage." Wastewater had to be collected and piped away from the Rouge in order for the complex to be functional. These matters were as much basic production problems as designing and laying out machinery and training the men to run it.[22]

The comments by Hanson and Verner indicate that minimizing the harmful impact of manufacturing pollution on the surrounding natural world was not a principal goal for Ford. Yet the problem of this broader, external impact was considered from the outset. Hanson was aware of it. He and other plant engineers worked with the standard technology of the time to capture as much dust as possible before it went up the foundry stacks. "We settled out as much [dust]," he noted, "as we could." Plant engineers built bag-houses to capture the dust from the Rouge blast furnaces. William B. Mayo, the senior engineer who worked most closely with Henry Ford in laying out the Rouge, warned Ford's senior management about liquid wastes in his "Report on River Rouge Location." "In the very near future," he wrote in 1916, "manufacturing plants will be prohibited from polluting the rivers with their sewage, hence—this site is such that it has sufficient space to properly take care of and treat its sewage, which in time should be led off through sewers into the Detroit River at a considerable distance below the City of Detroit." Mayo's warning, duly entered in the minutes of Ford's board of directors, was not heeded. Henry Ford and the engineers who designed the Rouge decided not to treat their industrial waste in the manner suggested by Mayo before releasing it to the Rouge River. Instead, they discharged it untreated directly to the river.[23]

This decision stemmed from the position of the plant at the lower end of the Rouge River, only two miles from its mouth at the Detroit River. Once the plant's discharges reached the Detroit River, the strong current and large volume of water in that strait effectively diluted the discharges before they reached Lake Erie, where they dispersed further and settled out. The plant's very existence was premised on the widening and straightening of the Rouge by the U.S. Army Corps of Engineers after Ford had purchased all the land bordering the river below its complex. Consequently, Ford encountered no downstream landowners aggrieved by the changed nature of the river. Decades later, Ford managers no doubt recognized the worth of Mayo's advice, but in 1917 the company concluded that, in comparison with Highland Park, the Rouge site, to paraphrase the prevailing wisdom, made dilution a viable solution for the plant's water pollution.

Although Ford's plant engineers made few direct efforts to reduce the impact of air and water pollution emanating from the Rouge before 1945, solid wastes offered an entirely different story. When the quantities of solid wastes were either large enough or valuable enough, Ford had an opportunity to recover some of the cost of production by either reducing the waste or by recycling the residual materials for internal reuse or external sale. With its focus on lowering the price of its cars, Ford missed few such opportunities. "Even a microscopic saving," as *Ford Industries* put it, "assumes impressive proportions when multiplied by a million or two."[24]

Ford's formal solid waste reduction commitment began at Highland Park in 1916. But the Rouge offered the company a substantial opportunity to design production processes with waste reduction and reuse in mind. Here, as part of the company's effort to create the ultimate modern, rational factory, the company's waste reduction activities reached their zenith. "Picking up and reclaiming the scrap left over after production is a public service," Henry Ford observed, "but planning so that there will be no scrap is a higher public service." In contrast to pollution reduction, Ford's plant engineers planned, built, and modified the Rouge with advanced thinking on waste reduction, especially the reduction of solid wastes, as a major consideration. "When certain operations produce large amounts of a certain kind of scrap which is re-used in production," L. D. Middleton, manager of the Rouge General Salvage Department, wrote in the late 1930s, "the same consideration is given to the handling of this scrap as would be given to laying out the various steps in the operations themselves. Consequently, conveyors are used and railroad facilities supplied for handling the major items which have to be forwarded to the other building[s] for re-use."[25]

Ford's Rouge waste reduction and recycling activities can be divided into two principal categories: by-products and salvage. Ford used the term by-products for waste materials that arose as a consequence of primary production that with further processing (and expense) could be sold profitably outside the company. These activities were, in fact, separate nonautomotive businesses where the firm made an ongoing commitment to market the materials. For example, in making coke from coal for the Rouge blast furnaces, Ford also produced coke oven gas, tar, ammonium sulfate, and benzol. By design, the company used the first two items as fuels in various operations around the Rouge. It sold ammonium sulfate, a fertilizer, and benzol, a fuel that could be mixed with gasoline and used in internal combustion engines, to the public. The by-product coke ovens at the Rouge produced three thousand gallons of benzol from every thousand tons of

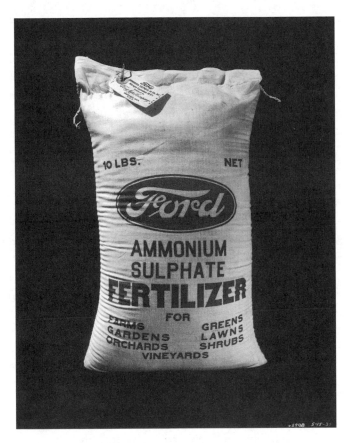

Peddling fertilizer with Fords. *From the Collections of
The Henry Ford, P.833.62988.*

coal that were coked. In 1924 Ford sold $1.15 million of these two by-products, even though Ford dealers hated being forced to sell fertilizer along with automobiles.[26]

The production of pig iron from iron ore, limestone, and coke also yielded blast furnace slag. To eliminate the cost of disposing 125 tons of this slag a day, Ford built a cement plant at the Rouge. The company's engineers designed a process that sprayed the molten slag with cold water to cool and granulate it to the size of coarse salt. They then pumped it as slurry under hydraulic pressure through thirteen-hundred-foot pipelines to a special plant to make Portland cement. The plant had the capacity to produce a thousand barrels a day, with Ford using about a quarter of the cement in its own construction activities while selling the balance on the open market. Contractors used Portland cement for many purposes, including, of course, building roads and bridges to accommodate the cars pouring from Ford's assembly lines.[27]

Salvage—the term that Ford plant engineers and publicists used then for what Americans later called recycling—generally involved waste materials reclaimed from the manufacturing process for reuse inside the company without substantial further processing beyond collection and separation by material. (Sometimes Ford sold salvaged materials outside the company—a reported four million dollars in 1924, for example—but these sales generally were ad hoc in nature.) At the Rouge, the most important salvage activities, measured by both weight and value, involved metals. After the advent of the closed, all-steel body in the mid-1920s, the amount of metal in automobiles increased. Since so many Rouge departments produced scrap metal, the general salvage department assumed responsibility for reclaiming it and maintained storage areas for ferrous (iron-based) and nonferrous metals. Reflecting its extensive use in automobiles, steel and iron by both weight and volume constituted most of the scrap metal reclaimed at the Rouge. The plant's salvage system collected and returned more than six hundred tons of steel "home" scrap to the furnaces each day. Many nonferrous metals had a value per unit of weight much greater than iron or steel, so the Ford salvage people also collected and reused copper, brass, aluminum, zinc, lead, tin (and various tin alloys), cadmium, mercury, and silver.[28]

The Ford waste reduction and recycling activities need to be set in a larger industrial context. Such activities were not especially remarkable in the 1920s, either in the United States or the leading European industrial nations. There had always been useful and profitable things to do with many wastes, and industrialization had given rise to many more, including many examples of inter-industry sales, especially as professional engineers and chemists began to play more prominent roles in designing industrial processes. In the United States these sources of interest in waste reduction and recycling received an important additional impetus from the World War I need to recycle scrap metal and a movement known as "industrial conservation."[29]

The industrial conservation movement was an outgrowth of the larger conservation movement. As with the utilitarian wing of the conservation movement, a national technical elite advanced the industrial conservation agenda—in this case engineers who worked with and within industry, especially the disciples of Frederick Winslow Taylor and those who subscribed to his principles of "scientific management." There was no popular pressure for industrial conservation. Although figures in government such as engineer and commerce secretary Herbert Hoover encouraged the movement after World War I, government officials neither used nor proposed government regulation to further the movement's goals. Engineers in industry simply pushed industrial conservation because they recognized a significant conservation and efficiency opportunity in the factory. Short of adopting pollution reduction as an explicit goal, business people in a free-market economic system that lacked direct government environmental regulation of business activities made their most important positive environmental contribution through the pursuit of efficiency.[30]

Ford did not initiate the industrial conservation movement. Much of what Ford did—using by-product coke ovens and recycling scrap metal, for example—was standard or becoming standard practice in comparable industries such as steel in the United States and Europe. Nor was Ford alone among automakers in pursuing waste reduction opportunities during the 1920s. "Nearly all of the automotive plants," William Crawford Hirsch wrote in *Automotive Industries* in June 1920, "have special departments charged with the salvage and disposal of scrap." For the most part, these efforts focused on scrap metal salvage.[31]

Nonetheless, the Ford approach to waste reduction can be distinguished in several respects. First, the scale of the Ford efforts dwarfed the activities of other firms. No other industrial firm in America during the 1920s pursued waste reduction on so many fronts involving the quantity of materials that Ford processed. Second, Ford often refused to settle for best practice when it considered that practice needlessly wasteful. Indeed, the company's engineers prided themselves considerably on their ability to do "best practice" one better. For example, Ford took waste reduction technology beyond then-current best practice in the wet process for manufacturing cement from slag, the use of hardwood scrap to produce pulp to make cardboard, and the successful large-scale use of pulverized coal in the giant Power House No. 1, the main electrical power generation facility at the Rouge.[32]

Not surprisingly, technical editors and writers often singled out waste reduction activities at the Rouge as primary examples of industrial conservation. In fact, the technical press during the 1920s featured no other industrial firm as often or as prominently for its waste reduction activities as Ford. In keeping with the leadership role that *Industrial Management* played in launching the industrial conservation movement, the journal's editor, John H. Van Deventer, used a series of thirteen articles on the Rouge that appeared monthly between September 1922 and September 1923 to focus special attention on Ford's waste reduction efforts. "The thoroughness of waste elimination at River Rouge goes as far beyond the average conception of waste elimination," he wrote in an editorial introducing the series, "as Ford's conception of a complete industrial plant, as exemplified at River Rouge, goes beyond . . . the present day industrial plant." George E. Hagemann also focused specifically on waste reduction in two articles he wrote for *Management and Administration* in 1925. "The reclamation department of the Ford Motor Company," he concluded, "is accomplishing successfully a task of immense proportions, and one profitable from the standpoint of the company, the employees, the consumers, and industry and society as a whole." A number of single articles appeared in other publications as well. Ford actively worked with editors to produce these articles as part of the larger effort to publicize how it made cars.[33]

Ironically, this publicity bolstered a mystique that also obscured important details about the company's waste reduction efforts. Company records and recollections by Ford managers suggest a more complex reality, especially when it came to Ford's accomplishments and motives in pursuing industrial conservation. As *Ford Industries* put it in 1924, "In the Ford Motor Company waste is regarded as almost criminal." The sentiment was Henry Ford's, and it was a point of pride that the company widely publicized: Henry Ford hated waste, he wanted to see how many things could be salvaged, and, if salvage was practical, whether it could be done profitably with the savings passed on to consumers. Industrial conservationists made the same point—that the efficient use of materials contributed to a firm's bottom line. But as Harrison E. Howe pointed out in 1919, "frequently . . . the real problem in waste utilization is more economic than technical. Many wastes do not occur in sufficient quantity at any one spot to make their use possible, or the cost of collection and storage defeats the project." The engineers who advocated industrial conservation certainly did not expect companies to pursue waste reduction opportunities to the point where they added to rather than reduced the cost of their operations. This economic constraint, discerning proponents of waste reduction such as Stuart Chase charged, often limited the industrial assault on waste.[34]

Ford embraced cost reduction as the chief rationale in its public waste reduction rhetoric, but in practice the company did not rigorously apply cost reduction as a test of whether to pursue or continue a waste reduction activity. Just as Howe suggested, Ford's engineers repeatedly demonstrated that most of the waste reduction and recycling activities that they undertook at the Rouge were practical in the sense that the physical amounts of waste could be reduced and uses found for many materials. But in many cases they found technical difficulties in doing so. And these difficulties sometimes frustrated efforts to save money.

Given Ford senior managers' ambivalence about (and occasional active hostility toward) rigorous cost accounting in the 1920s and 1930s, it is not possible to reach a comprehensive, definitive conclusion about the profitability of the Ford waste reduction activities. In 1925, Hagemann reported in the first of his two articles that the company saved fifteen million dollars in 1924 from waste reduction activities. Of this figure, he indicated that four million dollars came from the sale of salvaged scrap metal. Spreading the savings across the two million cars and trucks produced that year, Ford thereby reduced the selling price of a fully equipped Model T by an average $7.50, or about 2 percent. In his book *Today and Tomorrow*, published the following year, Henry Ford himself claimed that salvage generated twenty million dollars in savings each year. During the depressed 1930s, the company reported that general salvage activities at the Rouge produced a more modest four to five million dollars in profit each year.[35]

The public pose that waste reduction activities cut the cost of Ford's cars, while perhaps true in the sense of generating a modest net savings across many activities, masked a very mixed picture. "Mr. Ford abhors waste," Harold N. Denny wrote in the *New York Times* in 1933, "and will spend any amount of money to avoid it." Interviews conducted with former Ford and Rouge managers by Owen Bombard for the Ford Motor Company Archives in the 1950s are full of stories about Henry Ford's waste-cutting zeal. The recollections of Henry Ford's associates consistently indicated that he pushed waste reduction programs even when he knew they were not profitable. "I don't think the economical point always interested Mr. Ford," his personal secretary Ernest G. Liebold recalled. "Mr. Ford would insist on the by-products being salvaged," the firm's purchasing manager Albert M. Wibel added, "because he didn't want to throw them away. . . . These were usually not economical operations. We would never throw anything away regardless of whether it was economical or not. We would work out some way, and rather than set up the example of throwing something away, we would let the good ones take care of the bad ones and hope that an average would be set up."[36]

In fact, the managers involved conveyed an undertone of reluctance or embarrassment when recalling these activities. Some thought that the company's waste reduction zeal was excessive and communicated this reservation to Henry both directly and indirectly. Several quietly ignored Ford's directives. Frank Hadas went further and tried to talk Ford out of a major waste reduction project in the early 1930s on the grounds that it would not pay. This reluctance could not be characterized as open opposition, but the clear reservations expressed make plain that Henry Ford's managers and plant engineers drew a line when waste reduction activities appeared to be unprofitable, and they were not happy about pursuing them beyond this point. But Ford insisted that they give many a try and, in some cases, keep trying. "Mr. Ford's whole philosophy," John W. Thompson recalled, "was that you had to do a thing to learn it, but, you couldn't actually learn to make anything without actually making it." This approach characterized Ford's waste reduction efforts. Ford's managers suggested that the man whose obsession with lowering the cost of production to put America on wheels sometimes pursued waste reduction when he knew that it added cost to his operation.[37]

Was Henry Ford simply indulging a personal obsession, or was he struggling to accomplish something more with his waste reduction programs? The answer to this question ultimately leads back to the perspective of industrial ecology. Ford had a strong affinity for nature. He loved the outdoors and bird watching in particular. But there is no evidence that his love of nature provided the motivation for the company's waste reduction activities. Ford himself did not entirely lack what late-twentieth-century Americans considered environmental sensitivity. "It is not right," he told readers in *Today and Tomorrow*, "to put a layer of dust [from smokestack pollution] over the surrounding country and spoil its trees and plants." Nor was the company oblivious to the environmental side benefits of its waste reduction and recycling activities. And Ford's waste reduction activities produced tangible environmental benefits. "If you could see any smoke come out of the smoke stacks then there was something wrong," John W. Thompson recalled with some exaggeration. "You had to call somebody and you either captured that coal or coke, whatever it was that was coming out of there or the fires burned so you didn't get it. For a long time you couldn't see anything on those stacks. Mr. Ford wanted to save everything he could."[38]

But as Thompson's observation, Hanson's recollection about the foundry dust, and the "Saving Millions by Robbing Smoke of Its Waste" advertisement all suggested, Ford's chief motivation was not reducing pollution. "No heavy plumes of smoke roll from the tall stacks above the main power plant," the ad explained. "For smoke is waste, and waste is unpardonable by this company's standards." In the 1920s Ford viewed smoke as waste, not pollution. While burning coal or coke more efficiently and thereby reducing smoke certainly had a positive environmental impact, the company's engineers sought to cut waste, not to avoid or reduce pollution. Any beneficial environmental impact in the form of reduced pollution came as an incidental by-product of the waste reduction effort.

Like the utilitarian and industrial conservationists, Ford cared chiefly about efficiency. But instead of subordinating efficient resource use to profit maximization, which was the aim of most business firms and certainly the expectation among most industrial conservationists, he reversed the priorities and sought to maximize the physical efficiency with which the company actually used materials and energy, confident that satisfactory (although perhaps not optimal) profits would ensue. As Denny astutely observed, "Money is one of his by-products."[39]

Henry Ford's ability to pursue waste reduction past the point of profitability depended on his control of the company and its continuing profitability. The point is an important one. After he bought out the other shareholders in 1919 he no longer had to justify his decisions to anyone. Few corporate managers—given their legal responsibility to increase shareholder wealth—are ever in a position to pursue waste reduction (or any other activity) for long, unless they can make a plausible argument that it contributes to the bottom line. Ford could not always provide this justification when discussing waste reduction activities with his own managers, but then he did not have to provide it—a prerogative he chose to exercise with some regularity when it came to waste reduction.[40]

The extent and zeal with which the Ford Motor Company pursued waste reduction at the Rouge and elsewhere during the 1920s and 1930s, the degree to which it publicized these activities, the fact that it continued many activities known to be unprofitable on an open-ended basis without an explicit research and development rationale, and the lack of wholehearted support from the company's plant engineers as a result all suggest that Henry Ford's personal obsession with the matter is the single best explanation for Ford's waste reduction and recycling programs. His interest and the zeal with which he pursued these programs were likely the manifestation of a deep-seated personality need. Ford himself suggested as much when he told people that he got his related obsession with cleanliness from his "Dutch" (that is, German) mother. This idiosyncratic motivation aside, Ford showed the tremendous unilateral power that automakers had to reduce the environmental harm of their product and manufacturing processes if they so chose.[41]

Sometimes you have to work hard to see what is before your face, even when someone is obligingly connecting the dots for you. Normally, the complexity of developed industrial economies and corporate secrecy work to hide or obscure activities on the production side of the market exchange from consumers, citizens, and potential critics, but Ford in the 1920s probably did more to show the connection between its product and environmental impact than any company in American history. Four of the *Saturday Evening Post* "Industrial Epic" ads contained images of belching smokestacks that to latter-day Americans would be instant reminders of the pollution generated by industry. But before World War II, Ford could explicitly discuss raw materials, manufacturing, and finished automobiles, even with clear depictions of natural resource depletion and industrial pollution, without having to worry that consumers or the general public viewed these actions negatively. Local people knew of the harm, as did the men employed in these activities and their managers, but they lacked either the means or, more important, the inclination to complain. And the reform impulse and the investigatory journalism that fueled it were comparatively dormant during the 1920s. Above all, the automobile was America's glamour product. Down to the 1930s, Henry Ford and his company got full credit for the Model T and the mass production breakthroughs that put America on wheels. When Ford called the Rouge "the Crown Jewel of American Industry" or touted the raw material programs that fed the plant as "an industrial epic," Americans agreed, and they thronged to the Rouge and the Rotunda to see for themselves how Ford made their beloved automobiles. Ford's Rouge, raw materials, and waste reduction publicity, revealing as it was, should be seen, less as early "green-washing," and more as simply feeding and benefiting from the generally favorable impression that Americans had of all things having to do with automobiles.[42]

5

◆━◆━◆

The Death and Afterlife of Automobiles

"Any rags, any bottles, any old Fords today?" — Junk Dealer

As a tenant farmer in rural Alabama during the 1920s, Ned Cobb knew first-hand that some people worried about who had automobiles and who did not. As one of the first African-American automobile owners in his community, he heard about it from both his white and his black neighbors. "There's a heap of my race," he recalled, "didn't believe their color should have a car, believed what the white man wanted 'em to believe." But Cobb went ahead and bought one anyway. His white landlord occasionally chided him about it. Although Cobb enjoyed impressing his neighbors to a certain degree, the question of whether he was an automobile "have" or "have-not" was not of paramount importance. He simply saw that a car could be useful to him. If he could afford one, he concluded, he was entitled to have one, regardless of what others thought. "I didn't buy that car to try and get bigger than nobody else," he recalled. "I didn't buy it to show myself rich — I weren't rich; I bought it to serve my family and I knowed I was entitled to anything I wanted if I had the means to get it."[1]

Cobb found the car useful in his business activities, particularly for hauling lumber. A car also facilitated the Sunday social visiting that he considered both a duty and a pleasure. But the most important thing for Cobb was making the car available to his teenage sons so that they could pursue work opportunities, make a good impression while courting, and get on successfully in the world. "My boys, after they got big enough to drive cars," he recalled, "they come to me — they drove the car more than I did for their pleasure; that was most of the reason why I bought the car. After they got big enough to help me work, I found out, they'd talk it to me, they'd be glad if I bought a car so they could enjoy life some." The desire to help his sons, more than any other reason, drew Cobb into

77

the dynamics of the American automobile market in the 1920s. "So I . . . bought a brand new Ford car," he continued, "and it got where, after two years, them boys wanted me to change it and get a Chevrolet—a little higher grade of car."

By 1928, Ned Cobb and his family owned two cars, purchased apparently without recourse to consumer credit. Their story raises the question of whose agenda drove the American automobile marketplace in the 1920s, automakers such as General Motors or families like the Cobbs with their own plans. In the standard histories GM successfully dealt with the looming saturation point—where every family that could afford and wanted an automobile had one—and Ford's dominance through greater attention to marketing—building and motivating a strong dealer network, advertising, and, especially, adopting the styling-centered annual model change. But the Cobb boys demanded and got the latest styling in automobiles long before GM figured out how to make styling-centered annual model changes routinely profitable. The Cobbs certainly saw some automobile advertising, but it is hard to believe that their lives as African Americans in rural Alabama were awash in advertising or that their desire for a new and more stylish automobile had much to do with the automakers' promotional efforts. What is clear is that the Cobb boys paid attention to what their neighbors were doing with automobiles and how others perceived different makes and models.

Ned Cobb's life differed—in some ways profoundly—from the lives of white middle-class Americans, but when it came to automobiles, he and his sons were very like millions of other Americans in the mid-1920s. American families used their automobiles to help their older children get ahead. Making a car available to them made possible a wider range of educational, work, and social opportunities. Moreover, owning a respectable car was an important way that parents helped their children make a favorable impression in the community. Quite often, the children themselves insisted that the family purchase a good car. Where status and the future well-being of their children were involved, American parents could be quite indulgent, even those otherwise less keen to keep up with the Joneses.[2]

Americans worried about the new social practices that the automobile made possible. But the risks of unsupervised freedom were outweighed by the practical and symbolic advantages that families could provide by making an automobile available to their children. If parents feared what might happen in the backseat, they feared more that their children would not have the use of an automobile or would not be asked out by those who did. Automobile ownership meant success—or at least a healthy degree of initiative, a progressive attitude, and respectability. Parents wanted their children to share in these associations. Here Americans like Ned Cobb revealed a soft spot that made them far more susceptible to changing automobile styles than otherwise, that made them ask, after they had bought a Ford Model T for a first car, whether they wanted "a little higher grade of car," a Chevrolet perhaps, for their second.[3]

The Cobb boys' preference for Chevrolet over Ford marked an important change in the American automobile market. As the American automobile market approached the one-car-per-family saturation point in the mid-1920s, consumers found that having a car owned by half the car owners in America was unsatisfactory, a point made by the Ford joke about the motorist who met the Devil. "Help yourself to any of these cars, and take a spin around Hades," the Devil told him. "But, your majesty," the baffled motorist replied, "these are all Fords." With car ownership, their reference point changed. What now mattered more was the kind of car one had within the larger community of car owners.[4]

Consumers increasingly stigmatized the Model T as a technological and stylistic laggard. Technological innovation had slowed in comparison with the automobile's early days, but the 1920s saw its share in the electric ignition, six-cylinder and high compression ratio engines, balloon tires, four-wheel brakes, and safety glass, which all helped to render earlier models obsolete. Although some improvements eventually appeared on the Model T, most never did. By mid-1920s, the good-natured jokes about the Model T's ubiquity and its humble status gave way to jokes that had an edge of dissatisfaction. "Why put horns on Fords," asked one, "when they look like the devil without them." "The guy who owns a second-hand flivver may not have a quarrelsome disposition," went another, "but he's always trying to start something."[5]

Ironically, as the "Tin Lizzie" jokes reached their crescendo, Professor Clare E. Griffin and the Bureau of Business Research at the University of Michigan reported, in the first serious study of automobile longevity, that the Model T had an average life one-third longer than all other makes taken together, eight versus six years for all other vehicles. But as the changing nature of the jokes suggested, the Model T often died as a cherished possession in the mind of its owner long before it actually died on the road, done in not just by sheer numbers but by *perceived* obsolescence, both technological and stylistic.[6]

As the widely discussed saturation point approached, commentators began to point out that styling changes in particular offered consumers and producers common ground that met both groups' needs. Styling offered consumers a way to differentiate themselves from other automobile owners and to communicate symbolically a greater variety of things about themselves. At a minimum, having a car with the latest styling offered the consumer a chance to demonstrate that he or she could regularly afford to buy a new car. For automakers, faced with the saturation point and slowing technological change, styling meant an opportunity to encourage repeat buying sooner than would otherwise have been the case. These mutual opportunities converged in the mid-1920s to force styling and stylistic obsolescence to the fore.[7]

Automobiles had designs before the 1920s, and observers likened annual model changes to changing fashions in clothes from the very beginning. What the American automobile industry lacked before the mid-1920s was a deliberately prominent and permanent commitment to selling automobiles by making regularly planned changes to styling. But a pause in sales momentum in 1924 pushed the industry to consider the possibilities more seriously. And now two technological developments made it possible for manufacturers to do more with the appearance of the automobile: the low-cost production of all-steel, closed-body automobiles and the development of fast-drying lacquer-based paints. Dodge introduced the first all-steel, closed car in 1922, a change made possible by advances in sheet and rolled steelmaking technology pioneered by the American Rolling Mills Company. Consumers loved the new closed cars because they were all-weather, all-season vehicles, and by the end of the decade nearly all the cars sold in America were closed cars.[8]

The closed car meant that more steel had to be made and more steel disposed of at the end of a car's life, but it also provided more surface area to be shaped and colored. Henry Ford (reputedly) said that buyers of his automobiles could have any color they wanted so long as it was black. That was because before the mid-1920s, the use of any other color paint meant a drastic slowdown in production—an extra two to four weeks per car depending on the temperature and humidity conditions that affected paint drying—with a commensurate increase in production cost. This step was unthinkable for the makers of low-priced vehicles such as Ford.[9]

Du Pont solved this problem with pyroxylin-based lacquer paints to which the company gave the brand name "Duco." Given Du Pont's de facto control of General Motors, GM naturally became the first automobile manufacturer to use the new paints, introducing them with the "True Blue" 1924 Oakland that came to market in 1923. Once GM had introduced Duco to the cars in all its divisions, Du Pont then made it available to the rest of the industry for the 1925 model year. In less than three years, automakers and consumers had a thousand colors or shades to choose from in pyroxylin enamels. "Duco . . . ," Alfred P. Sloan, Jr., the executive chosen by the Du Ponts to run GM, later observed, "made possible the modern era of color and styling. Furthermore, its quick drying removed the most important remaining bottleneck in mass production, and made possible an enormously accelerated rate of production of car bodies."[10]

Throughout the early 1920s, industry analysts constantly called the carmakers' attention to the importance of body styling in selling automobiles. "It does not seem to be sufficiently recognized in the industry that a high degree of aesthetic perfection of the outside form is one of the more important features which make a car desirable to the prospective owner," *Automotive Industries* chided the automakers in one 1924 editorial. But the manufacturers did recognize its importance. Distinctive styling was a crucial factor in the sales of many recently successful models. Even Ford changed the body styles on the Model T over the years. But these efforts had been ad hoc, and success with consumers had been more or less serendipitous.[11]

In spring 1924, Harry Tipper, writing in *Automotive Industries*, spelled out clearly the full importance of styling. Noting that after the saturation point had been reached demand for automobiles would only grow with the increase in real consumer purchasing power and that replacement purchases would soon predominate, he argued that two factors would govern the rate of these purchases: the average functional life of automobiles and the degree to which new styling encouraged people to buy new cars before the end of their old ones' useful lives. "The style element . . . not only reduces the value of previous cars," he noted, "but increases the amount of new buying because of the social affect [sic] of the style. The value of the style element is increasing and is bound to increase further." Tipper saw an opportunity for automakers. "The extent to which style is used as an inducement to buying," he argued, "is under control of the manufacturer to a large degree and should be carefully studied in order that its use may provide the proper incentive beyond the economic requirements of replacement. . . . It is along the line of women's fashions . . . that research will probably pay largest profits."[12]

For Alfred P. Sloan articles like Tipper's merely stated the obvious. The problem was not in recognizing that a styling-centered annual model change had the potential to sell cars but in devising a method to execute such a strategy successfully on a recurring basis. "Managers knew that styling sold cars," historian Sally Clarke has noted, "but . . . they could not predict consumers' tastes—that is, they could not be confident which styles would sell." Guaranteed success was essential because the extraordinary plant retooling costs associated with major changes to automobile body styles that came after the introduction of closed cars would quickly bankrupt even the largest firms unless the styles offered met with consistent enthusiasm from consumers. So it was one thing for *Automotive Industries* to editorialize that "really good body design is well worth seeking and production consideration should not be allowed to becloud this fact," and quite another for a firm like GM to figure out how to do it year after year while earning healthy profits.[13]

The consummate corporate executive, General Motors
president Alfred P. Sloan, 1927. © *Hulton-Deutsch
Collection*/CORBIS.

But Sloan drew a bead on this problem, and that meant something. The
management approach that he developed during the 1920s placed a premium
on dispassionate, rational, fact-based problem solving while carefully building
consensus with his fellow executives. Above all, he was single-mindedly and
self-effacingly devoted to hard work, his corporation, and maximizing profits.
General Motors, Sloan reminded his colleagues, was in business not to make
cars but to make money. Other goals were potentially profit-sapping distractions.
Handed the fiscal and organizational chaos of the General Motors organization
in 1920, Sloan and his colleagues eventually made it the most profitable corporation that had ever existed and the very definition of mid-twentieth-century

U.S. corporate excellence. But in the early 1920s Ford's commanding market position and the looming saturation point lay between GM and future success. Sloan feared Ford, and with good reason. He did not believe that GM had the financial resources or the organizational talent to compete with Ford head-to-head on the basis of low-cost production and price. Sloan and GM did not duck a fight on this front, but they recognized that Ford could not be dislodged over night. Yet Sloan had to compete with Ford successfully on some basis if GM was to make headway in the low-priced market where half or more of all vehicle sales occurred. The question was a matter, not merely of choosing an area where GM had a competitive advantage over Ford, but of finding one that also resonated with consumers and encouraged them to buy GM cars. If technological breakthroughs were slowing, then styling necessarily became a more important alternative.[14]

Sloan, however, was not a risk-taker. When he and the GM sales committee discussed the annual model change in July 1925, American consumers were showing greater interest in automobile styling. But many in the industry were deeply reluctant to commit themselves to annual model changes in which styling changes were central. GM proved no exception. While committed to continuous improvements and willing to continue with the industry practice of introducing new models each fall, the GM executives refused to commit formally to a policy of introducing new models from their car divisions each year, much less to new models that placed a premium on styling changes. "General Motors in fact had annual models in the twenties, every year after 1923, and has had them ever since," Sloan later recalled, "but . . . we had not in 1925 formulated the concept in the way it is known today. When we did formulate it I cannot say. . . . Eventually the fact that we made yearly changes, and the recognition of the necessity of change, forced us into regularizing change. When change became regularized, some time in the 1930s, we began to speak of annual models."[15]

The risk associated with an ongoing commitment to a styling-focused annual model change left Sloan and GM no choice but to do the best they could with what they had. Sloan's "car for every purse and purpose" marketing strategy targeted the five GM makes (Cadillac, Buick, Oldsmobile, Oakland [later Pontiac], and Chevrolet) at different income segments of the market. GM gave Chevrolet the daunting task of challenging Ford's dominance in the huge low-price sector. To lead this effort the company turned to a fired Ford production executive, William Knudsen, and a Harvard-educated Yankee sales manager par excellence, Richard H. H. ("Dick") Grant. Knudsen and Grant struck with the revamped 1925 Model K Chevrolet just as the Model T began to falter in the public's mind. We "modernized its appearance so as to remove the inevitable stigma which rests on low priced articles that show it," Knudsen explained. The car came in a range of Duco colors, but looks were far from the whole story, for in contrast to the Model T, the Chevrolet also had such up-to-date features as a standard gearshift transmission, a foot accelerator, four-wheel hydraulic brakes, balloon tires, and shock absorbers. The Model K boosted Chevrolet sales 70 percent in 1925 to 512,000, still a far cry from the 1.6 million Model Ts that Ford sold that year but a surprise to many in the industry and a suggestion of better things to come.[16]

Despite growing pressure, Ford—more precisely, Henry Ford—refused to consider a commitment to regularly improving the features and changing the appearance of its models—that is, to cater to this growing customer desire—a stance that reflected fundamental differences between the Ford and GM approaches to making and selling automobiles. Both companies sought to turn the issue of obsolescence to competitive advantage. But to do so each looked in the opposite direction: GM to accepting it and trying to manage product change profitably, Ford to postponing the day when its cars became obsolete. The two companies agreed that technological obsolescence caused by fundamental improvements in the mechanical functioning of the automobile was good. Before 1928, GM brought more of these improvements to market in its cars than Ford. Both firms, at least in their public pronouncements, agreed that functional obsolescence caused by parts wearing out sooner rather than later was bad. Ford worked on this problem more diligently than GM, trying to ensure that its original parts lasted as long as possible and then building a huge replacement parts business that by the mid-1920s produced most of Ford's profits. In fact, Griffin's finding that Ford cars lasted a third longer than other cars was undoubtedly due to the fact that Ford worked successfully to make all of its parts last as long as possible while still keeping them widely available and affordable.[17]

As GM hesitated over committing to styling-focused annual model changes and Ford tried to help Model T owners extend the lives of their cars, mass production moved into its second decade, and the problem of the mass disposal of old automobiles became an important issue. "This is a side of the automobile industry which thus far has attracted little attention," E. C. Barringer wrote in 1926. "The speed of the assembly line, the marvels of mass production, the personal appeal of a new car, these overshadow all else." An industrial system that produced physical objects like the automobile functioned very much like the digestive system. The more that went in at one end, the more that eventually came out the other. Like excrement, Americans treated discarded automobiles as an offensive nuisance that they preferred to keep out of sight and out of mind, even though junked cars served valuable purposes. The basic problem was what to do with the physical materials that remained after the automobile reached the end of its functional life. Americans did not have to address this problem until the first wave of mass-produced automobiles began to wear out in the 1920s. As they did, people such as Morris Roseman turned their attention to solving it.[18]

Before World War I, Roseman, an illiterate Russian immigrant, was a regular junk man, scavenging the vacant lots and back alleys of Lancaster, Pennsylvania, with a horse and wagon for the discarded iron, bones, bottles, and rags that were a junk man's stock in trade. After losing money on several used cars that he had purchased in the hope of a quick turnaround and profit, Roseman decided that selling still serviceable parts from discarded automobiles provided a better business opportunity. He set up a junkyard where he brought wrecked and junked automobiles so that customers could pick them over for useful parts. Junk men claimed that 25–50 percent of the parts on a junked automobile had additional life left in them. Among the more important parts were tires, especially those of popular sizes.[19]

After customers stripped a junker of its merchantable parts, the junk man pulled apart the remainder of the vehicle—the "hulk"—to get at the metals in it. The most important of these were the ferrous metals—iron and steel. However, aluminum, copper, brass, and smaller quantities of other metals had value and were saved, too. The metals were cut into small pieces and then stockpiled for eventual sale and shipment to a scrap dealer. The scrap dealer, in turn, sold the used iron and steel to the steelmakers, who used it to make new steel. As one former dealer observed, "Steel never dies." Despite the negative association suggested by the words *junk* and *scrap*, high-quality steel required large amounts of scrap, so the used metal had real economic value. The junk men and scrap dealers provided an important environmental benefit. Every ton of scrap used to make steel meant that four tons of iron ore, coal, and limestone did not have to be used, conserving natural resources and reducing the severe pollution from coke ovens and blast furnaces at places like the Rouge. Metals were not the only materials saved. Glass, leather, and cushion felt and hair were also recovered and stockpiled for resale.[20]

For these reasons an American institution, the automobile junkyard, was born. By 1924, Roseman had built "Rosey's Junkyard" and "The House of a Million Parts" into a lucrative business and a personal fortune of nearly half a million dollars. Although junk men like Roseman provided an important economic and environmental service, the advent of the automobile graveyard was not an entirely welcome development. Junk men made most of their money from selling parts and therefore needed to attract retail customers, which meant being

Migrant auto wrecker, Corpus Christi, Texas, 1939. *Photo by Russell Lee. Library of Congress, Prints and Photographs Division, LC-USF33-012019-M5.*

near a sufficiently large retail market. Since a junkyard was essentially a parking lot, a prospective junk man needed a fair amount of space. The unsightliness of a junkyard along a well-traveled metropolitan thoroughfare constituted the junk man's best and cheapest form of advertising. Critics, on the other hand, saw the junkyard as a visual nuisance that lowered surrounding property values.[21]

Not all discarded automobiles reached a junkyard, though, which presented a more serious environmental problem. Owners simply left many cars to rust in backyards and vacant lots. Junk men who operated without a junkyard picked over these cars for parts and valuable metals, leaving the hulk behind. By the late 1920s, abandoned automobiles were a problem for American cities. They were eyesores. Critics also complained that they were breeding places for rats and mosquitoes. In response to a reader's query about the problem, in 1927 *American City* published the results of a survey of 125 municipalities conducted by the New York State Bureau of Municipal Information. The survey suggested that American cities took a varied approach to the problem. Some required the owners, if known, to dispose of their automobiles. Other cities scrapped the cars themselves and sold the metal to scrap dealers. Many communities used old cars as landfill—and along the Mississippi River as levee fill. The rest sent the abandoned cars to private or city dumps, although one city complained that old autos occupied 50 percent of its dump space. Some cars were pushed into abandoned quarries as fill. Contractors used still others in the foundations of buildings. A good number ended up working as stationary engines to run farm equipment. But most found their way eventually to a junkyard and wrecker and from there to scrap metal dealers.[22]

By the mid-1920s, automobile wrecking and scrapping was a big business. Observers estimated that by 1925 a million cars a year were going to the junkyards and the wreckers. Figuring just under a ton a scrap per automobile, they suggested that nine hundred thousand tons of scrap ferrous metal made its way back to the steel industry. By the end of the decade, nearly three million cars a year reached the end of their life. Thus, even in the early decades of the automobile industry, a significant degree of recycling took place.[23]

The two largest American automakers wrestled with obsolescence and deliberate styling changes, a matter that had implications for the number of junked automobiles, until consumers settled it for them. Sloan and GM, while they had serious reservations about the risk that an ongoing commitment to annual styling-based changes posed to their firm's profitability, had no moral qualms about the practice. For Henry Ford such obsolescence was anathema. "It is considered good manufacturing practise [sic], and not bad ethics, occasionally to change designs so that old models will become obsolete and new ones will have to be bought either because repair parts for the old cannot be had, or because the new model offers a new sales argument . . . to persuade a customer to scrap what he has and buy something new," Ford wrote in *My Life and Work*. "We have been told that . . . the object of business ought to be to get people to buy frequently. . . . Our principle of business is precisely to the contrary. . . . We want the man who buys one of our products never to have to buy another. We never make an improvement that renders any previous model obsolete."[24]

These views were ideals prized by Ford (and his ghostwriters), speaking in a sense to himself and the nineteenth-century world that produced him, and not the rigid policy of the Ford Motor Company. But his beliefs, especially his abhorrence of waste, were strong enough and his personal control over his company complete enough that they became the single biggest impediment in slowing his firm's decision to replace the Model T, which allowed GM—less encumbered by scruples that interfered with profit maximization—a further opportunity to gain market share at Ford's expense.[25]

In 1926, as the tired Model T faltered in the marketplace, Chevrolet sold 732,000 cars, and GM became the first corporation to achieve a billion dollars in sales in a single year. As Model T sales collapsed, Henry Ford finally ac-

knowledged reality and bowed to the entreaties of his subordinates. He agreed
to replace the T with a new car. In 1927 Ford temporarily abandoned the entire
low-priced field to Chevrolet, shutting down Model T production in early June
and desperately struggling for the much of the remainder of the year to bring
the new Model A to market. Chevrolet sold more than a million cars. Ford,
borrowing a page from the Chevrolet playbook about combining the latest technological improvements and up-to-date styling, had a sensational success with
the Model A in 1928, although sales still lagged Chevrolet as Ford struggled to
ramp up production to meet demand for the new car. In all, the transition to the
Model A cost Ford $250 million. When the two companies went head to head
in 1929, Ford came out solidly ahead. But Ford still had not committed to continuous improvements or stylistic changes. Chevrolet introduced its six-cylinder
Model A-killer in 1929, and in 1931 Chevy finally passed Ford in sales for the
first time in true head-to-head competition. When consumers learned that they
could count on Chevrolet to have new styling as well as the latest technological
improvements every year, Ford's domination of the American automobile market ended. It did not beat Chevrolet again until 1957.[26]

Advertising played a newly important role in helping GM and Chevrolet catch
and pass Ford. There had always been automobile advertising, lots of it, but it accompanied consumer enthusiasm rather than created it. Some of it, as historian
Steven Watts has argued for Ford's early Model T advertising, was sophisticated,
anticipating themes that would soon become more widespread in American advertising. But the fact that Ford sold half the automobiles in the United States
from 1908 to 1927 and used little national advertising from 1911 to 1917 and none
at all from 1917 to 1923 (although the dealers did a great deal of local newspaper
advertising) suggests that advertising was not crucial to convince Americans to
buy automobiles before the arrival of the saturation point and the beginnings of
product differentiation through styling as a deliberate marketing strategy after
1924.[27]

General Motors, in contrast, clearly used advertising to create a positive corporate identity and to establish Chevrolet as an appealing alternative to the
Model T. In doing so, GM helped fuel the great surge in advertising that took
place in the 1920s. Between 1914 and 1929, the dollar volume of advertising in
America quadrupled from $682 million a year to almost $3 billion. Like Ford,
the GM car divisions ran advertisements that showed their factories. But GM
also played its part in pushing the changes in advertising content delineated by
historian Roland Marchand in *Advertising the American Dream* (1985). By the
1920s advertisers began to downplay direct "reason-why" appeals that focused
on products in favor of attractive settings that focused on consumers benefiting

from using the product. These new ads invited viewers to imagine themselves in similarly improved circumstances through product ownership. The aim of the new ads was therapeutic, satisfying the psychological needs of consumers, providing for their happiness. The approach was subtle, but that was its strength, since such ads were less apt to generate direct negative responses from consumers. By 1924, GM bought more magazine advertising than any other company in America, and it retained its position as America's largest advertiser for decades. Yet GM had no special advantage when it came to the ads, save perhaps persistence, and there was no magic relation between advertising and greater sales. In fact, Chevrolet's sales manager Grant told the International Advertising Association in 1928 that it would be refreshing to hear from advertising men an acknowledgment that they had merely shared in the success of the automobile rather than "made" the industry. Consumers no doubt validated what they saw in ads with what they saw in the real world. "The ordinary family likes to have the luxuries the smart people have, and the influential younger generation especially loves the 'right' thing," observed journalist and historian of American advertising Frank Presbrey in 1929. "The [theme] did not originate in automobile advertising particularly, but its skillful employment there has had a good deal to do with the tremendous growth of the industry."[28]

In his book *American Prosperity* (1928), Paul Mazur argued: "It is the degree to which the factor of obsolescence has been developed as an art or science of increasing consumer demand, and not the mere existence of obsolescence, that distinguishes the past few years." But as GM's quandary suggested, Mazur claimed too much for 1928. Businesses had not yet developed either the art or the science necessary to make planned style-centered changes consistently successful. What had changed was a growing acceptance on the part of many businesspeople of the necessity for style-driven obsolescence to bolster profits. If necessity had not yet made it a virtue, the value of the practice from a business standpoint was openly discussed and embraced. There was an opportunity here for any automaker willing to put aside the moral qualms and work patiently to find either a method or at least a market position in which such changes were regularly profitable. "The biggest business in America is founded on style," Ray W. Sherman argued in 1926, "—the desire of the people for things they don't really need—and in the automobile business there is today opening an epoch of style profits that will run into amazing millions." Sherman's prediction proved conservative.[29]

Stylistic obsolescence and its management in the interest of corporate profitability were impossible without the active encouragement and support of consumers. If people insisted on using things until they wore out, oblivious to changes in appearance in subsequent generations of the product, stylistic obsolescence as a conscious business strategy would not work. But producers already knew better. "Consumers were intensely interested in the new," Mazur observed. "America is a rabidly progressive nation in material things, and the possession of the latest this and the latest that is a favorite standard of individual progressiveness." As a consequence, Mazur asserted: "Obsolescence is what the sophisticated, well-fed, well-equipped consumer practically forced upon production in return for his or her favor."[30]

But consumers wanted the new, not just from regular stylistic obsolescence in a given model; they also wanted a greater variety of offerings to choose from. At every price segment Americans after the mid-1920s had a choice of cars that varied in contour and color. The opportunities for self-expression through choice expanded dramatically. At the lower end of the market especially, the chief thrust of expression changed from the simple fact of automobile ownership to the type of car that one owned—with the prospect of redefining one's self-image over and over again with each new purchase.

"Luckily for us," Sloan recalled, "during the first part of the 1920s, and especially in the years 1924 to 1926, certain changes took place in the nature of the automobile market which transformed it into something different. . . . I say luckily for us because as a challenger to . . . Ford, we were favored by change. We had no stake in the old ways . . . change meant opportunity. We were glad to bend our efforts to go with it and make the most of it." Sloan and GM did not initiate these changes. Nor did they simply "go with the flow." Rather they began the process of patiently learning how to influence these changes to their advantage. In time they became quite successful in doing so, especially in the realm of styling. The key to making styling-focused annual models regularly successful was gaining and keeping styling leadership. This could be done only by acknowledging, as *Automotive Industries* put it, that "a really fine body design is an artist's job" and by finding and beginning long-term relationships with good automobile stylists and integrating them into the process of designing new cars. In 1925, GM had not yet done this. Nor had Sloan decided to do it. But an opportunity soon presented itself, and Lawrence P. Fisher, general manager of GM's Cadillac division, seized it. For a good stylist he and General Motors turned to Hollywood.[31]

In the course of visits to Cadillac dealerships around the country in 1925, Fisher met Harley J. Earl, the chief designer in Los Angeles dealer Don Lee's custom body shop. After college, Earl had developed a business creating new bodies for Model Ts. Before long he had made a name for himself by designing custom car bodies for Hollywood movie stars and moguls such as "Fatty" Arbuckle, Mary

Lawrence P. Fisher, Harley Earl (on right), and the 1927 LaSalle. *General Motors Corp. Used with permission, GM Media Archives.*

Pickford, Tom Mix, and Cecil B. DeMille. He also experimented with automobile body paints and colors. Earl bragged to Fisher that he could make a Chevy look like a Cadillac. Here was a car stylist for General Motors.[32]

Fisher brought Earl to Detroit in January 1926 on a one-time consultant's contract to design a car for Cadillac, and in March 1927 the LaSalle made its debut. It was a sensation. The LaSalle was the first mass-produced car whose success was due to deliberate and predominant attention to styling. Sloan moved quickly to capitalize on this success. On 23 June 1927 he proposed to the General Motors executive committee the creation of a special department "to study the question of art and color combinations in General Motors products." Sloan suggested a staff of fifty for this new department, ten designers and the rest support staff. The committee approved his plan, and Sloan asked Earl to become director of the new art and color section. Earl headed the group—which became the styling section in 1937—for thirty-two years until he retired from GM in 1959, designing fifty million cars along the way.[33]

GM *eventually* achieved styling leadership in the American automobile industry because it was the first large automaker to make a permanent commitment to styling and because it did not waver in this commitment over time. A substantial market share was almost certainly a prerequisite for establishing long-term leadership, but more important, GM's largest competitors, Ford and Chrysler, were slower to make a permanent commitment to styling and then wavered in their commitments until after World War II. But GM's product differentiation strategy worked best during boom times when there were prosperous consumers eager to buy new cars on a regular basis. And almost as soon as GM committed to making styling-driven obsolescence work for it over time, those consumers suddenly became fewer in number.[34]

The severity of the Great Depression blindsided American consumers and automakers alike. Between 1929 and 1933, the real gross national product (GNP) of the United States fell by 30 percent. By March 1933, a quarter of the workforce was unemployed. Unemployment did not fall below 15 percent again until 1940. There was no national unemployment insurance program, and there were few programs at any other level. Eighty percent of American families had no savings when the Depression began. And many of those who did have savings lost all or a part of them in the five thousand bank failures between 1929 and early 1933. The American auto industry was decimated. New car sales fell 75 percent from 4.5 million in 1929 to 1.1 million in 1932. Sales revenue dropped even further. Despite a substantial recovery in the late 1930s, the industry did not sell four million cars a year again until 1949. The idea of consumers and producers exploring the mutually beneficial possibilities in styling and stylistic obsolescence suddenly seemed laughable.[35]

Although no one made the argument at the time, the Great Depression and the steep drop in new car production offered the natural world a reprieve, not just because consumers did not buy new cars sooner but because mass production lightened up on those parts of the environment touched by three of the four phases of the automobile lifecycle. Raw materials extraction, manufacturing, and the disposal of discarded (but not recycled) automobiles no longer had the impact that they had during the 1920s. Unfortunately, hard times also interrupted an important experiment to find high-volume economies in automobile recycling.

Perhaps the most interesting development to come out of the old car disposal issues in the 1920s was Ford's decision to create an automobile disassembly line at the Rouge. When the line began on 5 February 1930, it marked a milestone in the company's grudging acceptance of the reality of obsolescence. The disassembly line was the American automobile industry's most important step into postconsumer recycling, but for Ford it was also another step in backward integration. Ford's managers and engineers saw the same opportunities in the afterlife of automobiles that they saw when they looked at the industries that provided raw materials to the automobile industry—many, small, inefficient operators taking middleman profits between the used-car dealers and the steel mill or metal refiner. The question was whether a large firm such as Ford that purchased scrap to make steel for its own automobiles could find volume economies by cutting out the middlemen and dismantling cars on its own.[36]

Before the demise of the Model T and at the urging of Henry Ford himself, the company briefly toyed with the idea of taking back and refurbishing worn-out Fords. Now, it settled for taking apart both its own and other manufacturers' vehicles for scrap. In this undertaking Ford did not assume an obligation to take back or repurchase its automobiles at the end of their useful lives in order to reclaim and recycle materials, although some at the time argued that Ford should make this commitment. Ford claimed that the disassembly line served three ends: "It rids highways of motor menaces that are dangerous both to life and traffic, it helps to free the landscape from unsightly junk piles, and converts into usefulness material that would otherwise go to waste."[37]

The company's real interest was in reclaiming the material. Ford needed large quantities of scrap for steelmaking at the Rouge, the capacity for which the company was expanding. The disassembly line reduced Ford's need for market purchases. The company hoped through large-scale operation to achieve labor and operating efficiencies that could not be realized by local junkyards and scrap dealers, as well as to eliminate the profits paid to these parties as auto scrap made its way through the usual commercial chain from the used-car lot back to the steelmaker. But just how important cost savings were to the company is an open question.

Ford began its experiment by agreeing to buy "junkers" for twenty dollars a piece (less a small shipping fee) from its Detroit area dealers up to the number of new cars the dealer purchased from the factory, provided the junker had a battery and tires. The trailers that delivered new cars to the dealers brought the junkers back to the Rouge for disassembly. Ford plant engineers set up the disassembly operation in the Open Hearth Building so that once the vehicles had been stripped of everything that was not steel or iron, the hulks could be fed directly into one of the ten open-hearth furnaces as a scrap charge for making new steel. The open hearths used at the time required a charge mix that was 55 percent scrap and 45 percent pig iron.[38]

Men on the Ford disassembly line first drained the junked cars of their gasoline, oil, and grease—all saved and recycled. The cars then moved along two

Ford disassembly line at the Rouge, 1931. *From the Collections of The Henry Ford, P.833.56148.3.*

conveyors—one for Ford vehicles and one for other makes—where the workers stripped them of their tires, batteries, headlight lenses and bulbs, spark plugs, floorboards, glass, leather, cloth, upholstery stuffing, radiators, and nonferrous metal parts. The Rouge salvage sales department reconditioned and then sold at a discount from the regular parts list serviceable Model T parts, such as tires, tubes, spark plugs, headlights, steering gears, glass, carburetors, and generators. The department sold the better tires as used tires. Ford sheared the worn tires into sections and then sold them to rubber recyclers. Ford maintenance men sized and cut the larger sections of glass for use as windowpanes around the Rouge. They sent the broken glass to the Rouge glass plant for remelting. They used reclaimed leather for aprons, upholstery for handpads, and floorboards for crate tops. They baled and sold the cotton and hair from the upholstery stuffing. They carefully separated the nonferrous metals, such as the aluminum in hubcaps, the copper in ignition wire, the brass in oil cups, and the bronze in bushings, by metal type and sold it for scrap. At the end of these lines a twenty-two-ton press crushed flat the remaining wood, iron, and steel in the body and chassis, as a third moving line conveyed it to an open-hearth furnace for re-melting. The furnace simply consumed any remaining wood. The entire process took about half an hour for each junker. This first Rouge disassembly line employed 120 men and dismantled 375 cars in two eight-hour shifts. "It was," Ford production chief Charles Sorenson later recalled, "a very spectacular job." By summer 1930, the company had demolished eighteen thousand junkers at the Rouge. The company was careful to call the experiment practical rather than profitable.[39]

Despite making changes to increase capacity, Ford's local dealers could not supply enough junkers to keep the disassembly line going. It proved too costly to ship junkers from around the country to Dearborn. Moreover, scrap prices remained low. In April 1931 the *Wall Street Journal*, noting that Ford did not track costs closely on the operation, estimated that Ford obtained $5.50 to $8 in scrap steel and $5 to $7 in other recyclable materials from each junked car, while incurring a labor cost of 60 cents per car. The paper reported that the company paid $15 to $20 per junker. Therefore, Ford did no better than break even and sometimes lost up to $9.50 per car it disassembled.[40]

Ford ran the line only intermittently from 1932 until World War II, when the company abandoned it. "We did not come out ahead on the operation," Charles Sorenson conceded. Plant engineer Frank Hadas, who worked on setting up the line was blunter: the company "spent a million bucks there knowing that it wouldn't work." Ford made significant strides toward increasing the disassembly volume, but it was not able to generate enough savings there in relation to the cost of the junkers and the low value of the scrap to make the disassembly operation profitable.[41]

Since the company actually employed reengineered production machinery that had been used to manufacture Model Ts in the first place to help take junked Fords apart, Ford's plant engineers must have thought about what they could do to design and manufacture their automobiles on the front end to make them easier and less costly to take apart on the back end. Late-twentieth-century industrial ecologists called this approach "design for recycling." Ford's production engineers must have looked the need for such design squarely in the face while they worked to lower costs on the disassembly line. Thus, standing on the threshold of what might have been one of the great industrial and environmental breakthroughs of the twentieth century, but lacking sufficient incentives from Depression era markets or further prodding from Henry Ford in the face of losses, they did not pursue the opportunity.

In contrast to Alfred P. Sloan, who was lionized for orienting General Motors to give consumers what they wanted, historians have depicted Henry Ford as a stubborn fool. This assessment is too harsh. Historians have not only given Sloan and GM too much credit, they have overlooked other important measures of economic performance, including a broader conception of consumer welfare and regard for the environment. Henry Ford's actions during the 1920s and early 1930s repeatedly pointed to the potential power that businesses had to do more for consumer welfare and the environment.

But Ford's fate at the hands of consumers and General Motors also showed the limitation of pursuing these goals unilaterally in a competitive marketplace. They could not be achieved unless all producers made a commitment to achieving them. In making the effort Henry Ford may have been a fool, but he had a larger vision that was not entirely self-regarding. And he had the courage to act on it. Success depended on GM and the other automakers joining him. If they did not pursue these goals together, they could not be achieved at all, which meant either giving them up or inviting government to force consumers and producers to achieve them. Ford wanted neither of these outcomes.

Americans in the early 1930s had bigger problems than defining themselves through the latest automobile styling, let alone achieving additional goals like improving consumer welfare or minimizing environmental impacts. One of the biggest was simply holding on to their cars. Americans might not have had jobs or much money, but observers marveled at how tenaciously they clung to their automobiles, even at the expense of items that seemed far more necessary or useful. Americans, Will Rogers famously joked, became the first people in the world to go to the poorhouse in an automobile. Martha L. Olney has argued that the strength of this attachment contributed to the Depression. As the economy moved into a standard business downturn in 1929 and 1930, American consumers responded by dramatically cutting back expenditures in other areas so that they could continue to make their auto payments and not risk losing their cars or the equity they had invested in them. The result was a sharp 10 percent drop in consumer demand in 1930. This jolt, together with the more significant mistakes that the Federal Reserve made with monetary policy, sent the economy into a three-year tailspin.[42]

While the Depression hurt new car sales, it barely touched the relationship Americans had with their cars. Passenger vehicle registrations, which peaked at 23.1 million in 1929, declined by only 11 percent to 1933. By 1940, registrations were at 27.5 million, an increase of nearly 20 percent across the Depression decade. Nor did Americans cut back on their driving. Vehicle miles traveled increased 53 percent, and gasoline consumption 60 percent over the same period. Americans managed to do this mainly by holding on to the cars that they had. Consequently, the average life of an automobile increased from seven years when Griffin did his study to fourteen years in 1949.[43]

A man with a car still had something to smile about, 1939. *Photo by Dorothea Lange. Library of Congress, Prints and Photographs Division, LC-USF34-19645-C.*

Olney focused on the financial stake Americans had in their vehicles, but the dogged way people held on to their cars reveals the degree of psychological equity that Americans had invested in them. Pioneering sociologists Robert S. and Helen Merrell Lynd were deeply impressed by the strength of this attachment when they returned to their study of Muncie, Indiana. "If the word 'auto' was writ large across [Muncie's] life in 1925," they wrote, "this was even more apparent in 1935, despite six years of depression." The Lynds recognized that this attachment represented a profound psychological need. "Both businessmen and workingmen," they wrote, "appear to be clinging largely to tried sources of security rather than venturing out into the untried." But the need was especially acute for workingmen. "If the automobile is by now a habit with the business class . . . , it represents far more than this to the working class," the Lynds observed, "for to the latter it gives the status which his job increasingly denies, and, more than any other possession or facility to which he has access, it symbolizes living, having a good time, the thing that keeps you working." Car ownership, they concluded, was depression-proof.[44]

After 1930 Americans lived in hope for the return of good times, the day when economic circumstances again permitted them to use the latest automobile styling to say more about themselves and their relationship to the larger community. They nurtured this hope across a decade of depression and half a decade of war until a new economic horizon beckoned them to resume their love affair with the automobile and to reassert their symbolic claims to respect from their neighbors and themselves. But the Great Depression had terrorized many Americans. When they embraced consumer goods like the automobile with renewed gusto after the war, Americans exacted a fearsome, if unintended, revenge on the environment in the 1950s and 1960s. And the impact was worse because the market that consumers and automakers first created in the 1920s around deliberate styling-driven obsolescence then came of age.

6

CADILLACS AND COMMUNITY

"We always led a kind of a common life," he told interviewers. "We never had any luxuries, but we were never real hungry." In the years just after World War II, as millions of Americans realized long-deferred material dreams, his family straddled the line between the haves and the have-nots. He "would hear us worrying about our debts, being out of work and sickness," his mother later recalled, "and he'd say, 'Don't you worry none. . . . When I grow up, I'm going to buy you a fine house and pay everything you owe at the grocery store and get two Cadillacs—one for you and Daddy, and one for me.'"[1]

After one of America's most stunning ascents to fame and fortune, twenty-year-old Elvis Presley realized his dream. He bought a pink 1955 Cadillac Fleetwood. When he replaced it with a 1956 Cadillac Eldorado convertible, he gave the pink Cadillac to his mother, even though she couldn't drive. Within a year, Elvis owned four Cadillacs. He also purchased a ranch house for himself and his parents in a nice Memphis neighborhood. "What do you do with four Caddys," an interviewer asked him. Elvis was at a loss for words: "Well I . . . (laughs) I don't know. I haven't got any use for four, I just, ah . . . ." He then made a joke and changed the subject. But in a letter that he wrote to *Rod Builder and Customizer* magazine, Elvis explained that buying the Cadillacs was just something he needed to get off his chest.

Two years later, during his mother's last illness, he brought the pink Cadillac to the hospital to buoy her spirits and parked it so that she could see it from her window. A fan later told Elvis that she had visited Graceland and been shown the pink Cadillac by his mother. "What other boy would love his parents so much?" Gladys Presley had told the fan. For Elvis and Gladys Presley there was no greater symbol of accomplishment and thus no greater testament to a young man's love for his mother.

The King with his Cadillac and ranch house, 1956. *Philip Harrington, photographer, LOOK Magazine Collection, Library of Congress, Prints and Photographs Division, LC-L9-56-6706-M, #23.*

The Presleys' preference for Cadillacs was not anomalous. "In some part of his being," Thomas Hine writes in *Populuxe*, his homage to the mid-1950s American consumer marketplace, "every American wanted a Cadillac." The Presleys' choice reflected a much broader consensus about the meaning of the automobile—and the Cadillac in particular—shared by Americans from coast to coast in the early 1950s. The intensity of this consensus made 1955 and the first postwar decade a special moment in the relationship between Americans and their automobiles.[2]

And what a year it was. In 1955 Americans bought a record 7.9 million cars, 19 percent more than the previous best sales year in 1950. The Big Three—General Motors, Ford, and Chrysler—sold 95 percent of these automobiles. In fact, American automakers sold 99.2 percent of all passenger vehicles, the high-water mark of Detroit's domination of the American automobile market. The year began in a buying frenzy over 1955 models, vehicles notable mainly for the $1.2 billion the automakers spent on new styling. Like Elvis's pink Cadillac, these cars came in two-tone pastel colors with tail fins and lots of chrome. Most of them boasted V-8 engines with more horsepower than American cars had ever seen. Twenty million people turned out to see the new Chevrolets. "The people," one overwhelmed dealer told *Newsweek* early that year, "are like a pack of wild animals." Of their $398 billion gross national product, Americans spent $65 billion on cars in 1955. In one year automobile debt jumped 35 percent, from $10 billion to $13.5 billion. General Motors became the first corporation in history to earn $1 billion in a single year.[3]

The year that Elvis celebrated his success by buying a Cadillac also culminated a decade that was the golden age of the American automobile. After fifteen years of depression and war, including a three-and-a-half-year hiatus during World War II in which it was impossible for most people to buy a new car, Americans after the war suddenly had the means to satisfy what author David Halberstam called simply a "desperate hunger for products." Moreover, government officials, corporate spokesmen, and social commentators alike shared the conviction that the nation's welfare depended on consumers acting on this hunger if another depression was to be prevented. For many the first stop after the war was the dealer's showroom. A *Fortune* magazine survey published in December 1943 indicated that Americans longed to buy a new automobile before anything else. Not surprisingly, passenger car registrations doubled between 1945 and 1955 from 25.8 to 52.1 million. For much of the decade the Big Three strained to meet surging consumer demand, confident that every car that rolled off the assembly line had a waiting buyer.[4]

Prosperity, together with the affinity that Americans in the first half of the decade had for cars from the Big Three, breathed new life into—indeed, it was the long-wished-for fulfillment of—General Motor's "car-for-every-purse-and-purpose" marketing strategy. From Chevrolet through Cadillac, an American family with the requisite means could choose the automobile status marker of its choice. When President Dwight D. Eisenhower in 1953 tapped GM's postwar president Charles E. Wilson to be secretary of defense, Wilson reputedly said that what was good for GM was good for America. Wilson actually said just the opposite. That he was misquoted in this manner showed, not just a readiness to see corporate arrogance, but the utter plausibility that the interests of the nation and the corporation were the same. In the early to mid-1950s American automakers and consumers found a lot to be satisfied with in one another.[5]

No car epitomized this mutually satisfactory relationship more than the Cadillac, the postwar beau ideal of American automakers and consumers alike. General Motors actively cultivated the Cadillac mystique. After the war, the company embellished the line's historic reputation for engineering, quality, and comfort through special attention to styling and advertising. GM made Cadillac the styling flagship for its entire line. Here chrome and tailfins first insinuated

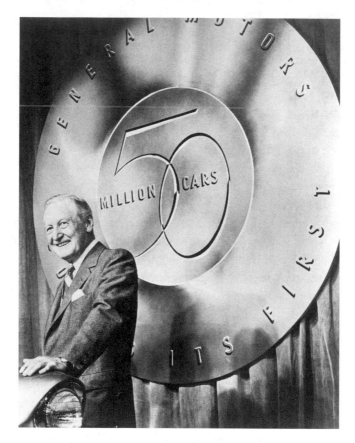

As good as it got. GM president Harlow H. Curtice with a
1955 Chevrolet Bel Air Sport Coupe, GM's fifty-millionth
car, November 1954. *General Motors Corp. Used with
permission, GM Media Archives.*

their way into the styling of American automobiles, as Harley Earl consciously
added diminishing degrees of Cadillac class to GM's entire product line. The
men who ran Cadillac in the 1950s took this cachet to the bank, expanding production to sell Cadillacs to the high end of the medium-priced market as well,
so that by 1955 there were eight hundred thousand of them on the road. In the
short run this approach was a profitable strategy for GM.[6]

During the early 1950s, the appeal of Cadillac was so strong that it became the
model for other automakers as well. Both Chevrolet and Ford explicitly copied
Cadillac, trying to appropriate some of the make's mystique for their own cars.
In Chevrolet's case most of the copying came at the urging of Ed Cole, one

of postwar Detroit's best young automotive engineers. In May 1952, Cole, who had been chief engineer at Cadillac, moved to Chevrolet with the understanding that he could design a completely new V-8 engine for the division. His son David E. Cole remembered the Chevy V-8 design team meeting at the Cole home on weekends and working on cars in the driveway. Ed Cole also pushed the stylists to give Chevrolet a new look to go with the new engine. With the V-8-powered, completely restyled 1955 Chevrolets, Cole created cars that Halberstam called "the virtual automotive equivalent of a Cadillac."[7]

The Ford Motor Company faced an especially severe marketing challenge after the war, one part of the larger managerial crisis that threatened the company's very existence. Henry Ford had suffered a series of strokes, and Edsel Ford died in 1943, leaving the company in the hands of managers better described as henchmen. Alarmed, the Ford women—Henry's wife, Clara (Bryant) Ford, and Edsel's widow, Eleanor (Clay) Ford—acted, installing Henry Ford II as president. A year earlier, Ford asked Elmo Roper and Company to conduct a confidential survey to find out what Americans thought about American automobile companies and about Ford and its cars in particular. From the survey Ford learned that it had an image problem. Although the public still thought highly of the elder Henry Ford and his firm, it held a lower opinion of its cars. American consumers preferred GM cars over Fords by better than two to one, largely because they believed that GM was "the company with the best styling" and the firm that was "most alert to the adoption of mechanical improvements." These results were the fruit of GM's persistent twenty-year commitment to meet consumer desires. The Roper report strongly urged Ford to give greater priority to styling. Ford had heard this message before, but the new postwar management team recruited by Henry Ford II (many of them younger men who looked to their former employer GM as their model) now took the message to heart. The result was the 1949 Ford, styled by industrial design consultant George W. Walker and introduced in the summer of 1948 at a cost of seventy-two million dollars for retooling and machinery and ten million dollars for promotion. Efforts to overcome the lingering vestiges of the "tin lizzie" image dominated Ford's marketing efforts throughout the late 1940s and 1950s. As one of Ford's admen from the J. Walter Thompson Company put it, the basic postwar Ford advertising strategy that resulted was to "compare with Cadillac to compete with Chevrolet."[8]

Among consumers, the Cadillac mystique took on a life of its own. A survey of ten thousand Americans who purchased cars in 1955 by *Popular Mechanics* found that, if money were no object, nearly 40 percent would have purchased Cadillacs. The next most popular make in the poll barely broke double digits. Similar polls put the Cadillac figure at 50 percent. A study conducted in 1953 by

Ford comparing with Cadillac to compete with Chevrolet.
*Ford Advertisements, JWT Domestic Advertisements Collection,
John W. Hartman Center for Sales, Advertising and Marketing
History, Rare Book, Manuscript and Special Collections Library,
Duke University, permission by Ford Motor Company.*

the Chicago consumer research firm Social Research, Inc. (SRI), found that no other make had as prominent or unambiguous a personality. In the minds of consumers, according to the study, Cadillac stood for "high economic status, ostentation, the mark of the man who has arrived." In 1955 *Fortune* editor William H. Whyte called this the "Cadillac Phenomenon." "Never before has one material object," he observed, "become so much the focus for so many of the aspirations that propel the American ego."[9]

Jim Adams, head of the ad agency MacManus, John and Adams, handled Cadillac advertising in the 1950s. A Cadillac owner himself, Adams was an old-school adman who didn't need a Ph.D. in psychology to recognize that Americans craved symbols of success. He urged GM to market the Cadillac to the man on his way up as a "symbol of achievement." Sophisticates in the advertising world labeled Adams's approach as "inexpressibly vulgar." But, as Whyte recognized, Cadillac's marketers had struck "a shrewd balance between the exclusive and the democratic." The Cadillac might be a symbol of achievement attainable only by a minority who were successful, but the door was wide open to all. A Cadillac said, in effect, that here was a man who through fair competition had emerged above the crowd, whose wealth, success, and station in life deserved respect.[10]

As might be expected, the Cadillac's appeal resonated strongly among Americans working to transcend commonplace or historically marginal backgrounds. "Newly prosperous and even low-income owners (a growing number) wish for attention," Guy Shipler, Jr., wrote in *Popular Science*, "[and] this car gives them the status in their own minds, which they may not otherwise be able to get." The Presleys—originally poor, white, rural Southerners—understood the symbolism of the Cadillac perfectly. Their ability to tell the world that they had arrived through Cadillac ownership was one of the most satisfying experiences of their lives. Whyte also called attention to Cadillac ownership among Jewish Americans and African Americans. He noted that *Ebony* endorsed the car to its readership as a symbol of economic and social achievement. "The fact is that basically a Cadillac is an instrument of *aggression* [emphasis added], a solid and substantial symbol for many a Negro that he is as good as any white man," the magazine editorialized in 1949, "To be able to buy the most expensive car made in America is as graphic a demonstration of that equality as can be found. . . . It is the acme of dignity and stature in the white man's world."[11]

The Cadillac's increasingly broad appeal did not go unnoticed, however. Although the make constituted just 2 percent of automobile registrations, other Americans found such associations as "most luxurious, flashiest, resented by some, increasingly appeals to new rich of deprived origins" actively repellant. Not surprisingly, Americans born to wealth were not Cadillac owners. Chastened by progressive income taxes and a half-century of Veblenite mockery of conspicuous consumption, old money had distanced itself from the sweaty upward striving of other Americans by adopting understatement or, as the SRI study had it, "conspicuous reserve," as its guiding principle of consumption. For established wealth in the 1950s the Cadillac was utterly déclassé. This consumer stance toward the Cadillac—and American automobiles generally—later be-

came more widely fashionable. But in the early to mid-1950s Cadillac was the
sun around which the entire American automobile market orbited.[12]

There are many reasons why a product dominates its market, but Cadillac's
symbolic dominance of the American automobile market in the early 1950s owed
something to Americans' experiences with the postwar economy and world. The
peace and prosperity of the 1950s were a sharp break with the immediate past.
But from another perspective the first postwar decade presented only a new and
different phase in a prolonged period of crisis and dislocation. Americans who
bought the Big Three's cars in the early and mid-1950s had lived through the
Depression in the 1930s and World War II, but now they found themselves in
the midst of an unwelcome Cold War involving conflict in Korea, a nuclear
arms race with the Soviet Union, and the prospect of a third world war. They had
experienced a generation of adversity that unsettled and, in many cases, severely
disrupted their lives. Hard times, they knew, could be just around the next corner. These postwar Americans were searching, as historian Elaine Tyler May has
argued, for "security in an insecure world."[13]

Policymakers worried, too, that economic adversity might be just around the
corner, and as historian Lizabeth Cohen has shown, they urged the consumption of "more, newer and better" products on Americans as a civic duty as well
as a private pleasure. The developing Cold War also provided a geopolitical
rationale for mass consumption—proof not just of the productive superiority
of the American economy but evidence of a growing economic egalitarianism
that would make the United States a "classless society" where people enjoyed a
material standard of living far above that of the Soviet Union. Thus, a lingering
sense of communal crisis still informed the context in which American consumers in the late 1940s and early 1950s made their individual and family buying decisions.[14]

The postwar economic boom presented opportunities that encouraged Americans to follow jobs to the more prosperous parts of the country. The boom also
drew many from older urban areas to new suburban developments. On the job,
in the neighborhood, at church or school, many Americans found themselves
prodded by the times into situations where they were newcomers and strangers.
These new experiences and the odd sense of finding oneself a stranger in one's
own land fostered anxiety. There was, Hine writes, an "insecurity and loneliness
that was part of their pioneering lives." Many Americans needed assurance that
they were not falling behind or failing to adjust, that they were coping successfully with a period of change and stress in their lives. In these new circumstances
traditional sources of authority no longer offered reassurance. People "looked
to their [new] neighbors for signals on how to behave," Hine argues. This com-

munication took place largely through the medium of consumer goods. Reflexively—and by the 1950s it was more a reflex than in the 1920s—Americans reached for those consumer goods that said most clearly, "Hey, I'm doing pretty good, aren't I?"[15]

Not only was the Cadillac the premier consumer good of the immediate postwar era, it epitomized nearly all the important developments in the American automobile market. In this Cadillac-centered market American cars became more expensive across the board. Although second-tier automakers like Crosley, Kaiser-Frazer, and Studebaker pushed alternatives that departed in various ways from the norm, the models offered by the Big Three generally became more powerful, as well as longer, lower, and heavier with greater size and embellishment—indeed, more luxurious. After World War II the average wholesale price of a car increased significantly for the first time since the Model T was introduced in 1908. When the new management team that Ford imported to mentor Henry Ford II decided to stress "value" rather than low price, a recognition that many Americans rejected automobiles advertised as low-priced for fear of the public loss of face that ownership of such vehicles involved, the price structure of the American market lost its low-end anchor and began to drift upward, opening a niche that low-cost foreign automakers later exploited. After World War II consumers no longer got more for less. They got more for more. Given that they largely ignored alternatives that offered less, consumers clearly wanted more, too.[16]

One area where consumers got more was under the hood. In the late 1930s the oil industry pioneered commercially viable processes to "crack" the hydrocarbon molecules in petroleum into high-octane gasoline. Engineers used substances that facilitated the necessary chemical reactions—catalysts—to accomplish the conversion rather than high temperatures. Catalytic cracking when combined with tetraethyl lead made for even higher octane gasoline, which, in turn, presented postwar automotive engineers like Cadillac's Ed Cole the opportunity to design much more powerful high-compression ratio V-8 engines. The long-term trend had been toward greater horsepower, but the increases with the new V-8s in the 1950s marked a sharp upturn. According to calculations made by *Automotive Industries*, the average horsepower of American passenger car engines increased from 54 horsepower in 1930 to 212 horsepower in 1958, a fourfold increase in just over twenty-five years. When the Big Three really began to compete on the basis of horsepower in the early 1950s a "horsepower race" broke out, and average engine power increased by more than 100 percent from 1950 to 1958. In fact, the average horsepower of an American car jumped nearly 25 percent in just one year after Ed Cole moved to Chevrolet and put V-8s in the

1955 Chevrolets. Automakers installed V-8s in 80 percent of the vehicles they made in 1955, and the engine remained standard in American cars for the next twenty years.[17]

For Hine, the styling of cars from the Big Three in the 1950s was the quintessence of "populuxe," mass-produced designs that conveyed luxury. The trend during most of the decade was toward greater size and embellishment. The long-term direction of styling in the industry had been toward longer and lower vehicles. Harley Earl had pushed this aesthetic on the industry with success since the late 1920s. After the war, with its divisions selling half the cars in America, General Motors successfully reassumed styling leadership in the American automobile market, a position that gave it extraordinary power to influence consumer tastes, and Earl became, in Halberstam's words, "the standard-bearer of the new age of affluence and abundance."[18]

Postwar American automobiles also saw a substantial increase in nonfunctional trim and body styling, of which the greater use of chrome and the advent of tailfins were the most notable flourishes. Insiders called it "gorp." As Earl told *Saturday Evening Post* readers in 1954, he introduced tailfins on the 1948 Cadillac as a mark of distinction to set the make apart from other automobiles. Unlike size and the use of chrome, tailfins were a marker with absolutely no previous associations with status. They might very well have failed in their intended purpose—in fact, they almost did—but because Earl put them on the Cadillac, the highest-status automobile at the time, they succeeded. GM then gradually and deliberately added fins to the vehicles made by its other divisions to lend them a touch of Cadillac class.[19]

The automakers initiated these postwar developments. Before 1948 no one had heard of tailfins, let alone clamored for them on a car or associated them with status. Yet by 1955 tailfins were not just a mark of distinction but incontestable proof that the major American automakers had real power to shape consumer tastes in automobiles in the interest of their own bottom lines. This power had limits, however. Automakers could choose what kinds of cars to offer, but consumers still chose which ones to buy. Despite important developments in market research and the arrival of a new advertising medium in television, the automakers and their admen failed to make any fundamental breakthroughs in the power of their promotions during the 1950s. When Earl described the annual styling changes that he orchestrated for GM as "dynamic obsolescence," he was not just deflecting responsibility away from the automakers' efforts to plan this obsolescence but acknowledging the truth that the automakers did not have absolute control over consumers' tastes. The second-tier domestic automakers, and, as we'll see in the next chapter, a growing number of foreign automakers

Tailfin, 1960 Cadillac Fleetwood. *Philip Harrington,*
photographer, LOOK Magazine Collection,
Library of Congress, Prints and Photographs Division,
LC-L917-59-8372, #30.

presented Americans with alternatives, so consumers could, and sometimes did, balk at what the major U.S. automakers offered. But, judging from behavior, the inescapable conclusion is that consumers liked what the Big Three offered in the early and middle 1950s.[20]

The major trends that characterized the American automobile market between 1945 and 1955 worsened the automobile's environmental impact on

America. The greater weight and horsepower of American automobiles meant that drivers consumed more gasoline and produced more lead-laden, smog-causing, climate-changing emissions simply to travel the same distance. The use of new materials in the manufacturing process caused new forms of industrial pollution. Ford, for example, discharged its chromium-plating baths at its Monroe, Michigan, plant directly to the Raisin River, where this toxic heavy metal flowed a short distance to Lake Erie. Larger cars consumed more natural resources and resulted in more materials to dispose of at the end of these cars' lives, a day that came sooner than it had in the 1930s and 1940s because greater consumer purchasing power, interest in new styling, and declining product quality all encouraged consumers to replace their cars sooner. Even trends like rising prices that in other circumstances might have lessened the impact by dampening sales failed to do so.[21]

Making bigger cars and turning out more of them meant a greater impact on the natural world from automobile manufacturing especially. Nowhere was this more evident than at Ford's River Rouge complex. With its iron and steelmaking operations and its extraordinary array and concentration of other production activities, the Rouge epitomized the impact of auto making on the environment like no other manufacturing complex that ever existed. Although Ford began to decentralize these operations after the war, more than half of the company's U.S. employees still worked at the Rouge as the postwar period began. While comparing itself with Cadillac to compete with Chevrolet, Ford burned six tons of coal to produce every car that it made, and the six hundred smokestacks at the Rouge released untold millions of tons of carbon dioxide to the atmosphere, as well as acid-rain producing sulfur dioxide, heavy metals, and other particulate matter that rained down on surrounding communities.[22]

The plant's impact on the Rouge River was as bad or worse. "I have seen some foul rivers," Bill Wolf wrote in 1948 in a series of articles surveying water pollution problems around the United States for *Sports Afield*, "but will stake the River Rouge from the Ford plant down to where the Rouge enters and contaminates the Detroit [River] against any stream for absolute, concentrated filth." When boats parted the thick layer of oil on its surface, it was clear that the river flowed, not rouge, but bright orange. One Michigan water quality official drew a bucket of water from the Rouge only to find that within ninety minutes the acids had eaten away the bucket's bottom. Nearly forty years later, the river was still one of the most polluted in the world and discussed as a possible Superfund site. As a Ford public relations official candidly admitted in 1953, the Rouge "may be the world's biggest proving ground for the control of air and stream pollution."[23]

Yet as bad as industrial pollution was at the Rouge in the late 1940s, the auto-makers and their suppliers could not and did not pollute with impunity—even when consumers were happy with the companies and their cars. There were as yet very few laws that actually constrained, let alone punished, corporate pol-luters, but when industrial pollution got bad it affected someone's pursuit of "the good life" and people complained. After World War II the complainers went public and the automakers disliked the negative publicity that ensued, even when still mostly local. In Ford's case, the complainers were often its own customers and employees who lived near the Rouge.[24]

Serious efforts to tackle pollution at the Rouge began immediately after World War II when Michigan sportsmen and journalists, Dearborn residents and smoke inspectors, and American and Canadian public health officials all pres-sured Ford to curtail air and water pollution at the complex. The first people to sound the alarm about pollution were Michigan's sportsmen operating through the Michigan United Conservation Clubs (MUCC), an umbrella organization that represented sportsmen who were organized into local clubs and the largest state affiliate of the National Wildlife Federation. When industrial pollution from places like the Rouge killed fish and harmed waterfowl, Michigan's hunters and fishermen complained. They quickly found a champion in the *Detroit Free Press*'s conservation and outdoor columnist, Jack A. Van Coevering.[25]

Between 1945 and 1949, Van Coevering used his column to lead a statewide crusade against water pollution. Using data collected by the Michigan Stream Control Commission, the hitherto quiescent state agency established in 1929 to monitor water quality, Van Coevering published by name industrial and munici-pal sources of water pollution, which included GM and Ford plants, the Rouge among them. His crusade garnered support from women's clubs, farmers, and junior chambers of commerce around the state in addition to sportsmen, con-servationists, and the tourist industry.[26]

Van Coevering and Michigan's sportsmen were not alone in sounding the alarm about industrial water pollution. In June 1945 G. H. Ferguson, chief of the Public Health Engineering Division of the Canadian Department of Na-tional Health and Welfare, contacted his counterparts in the U.S. Public Health Service to propose a joint study of pollution in the Detroit River. The upsurge in industrial activity in the Detroit–Windsor, Ontario, area during the war raised concerns in both countries about industrial pollution and the threat that it posed to sources of municipal drinking water. Managers of municipal water systems that took their water directly from the Detroit River below Detroit and the Rouge (like those in Wyandotte, Michigan) constantly scrambled to deal with sudden slugs of partially treated sewage from Detroit's inadequate sewage treatment facility and cutting oil and pickling acid from plants like the Rouge. In response, the U.S. and Canadian governments agreed to undertake a study of pollution in the Detroit River under the auspices of the International Joint Commission (IJC), the bilateral body created under the Boundary Waters Treaty of 1909 to address issues involving the shared water borders of the two nations.[27]

The IJC circulated an interim report in 1948 that demonstrated the remarkable extent to which the Rouge alone polluted the Detroit River. The Rouge was the only industrial plant of the sixty-two examined in the Detroit–Windsor area that contributed pollution in all seven of the categories used in the report: phenols, cyanides, ammonium compounds, oils and greases, suspended solids, biochemical oxygen demand, and miscellaneous (that is, acid and iron). According to measurements made in December 1947 by Public Health Service chemists as part of the IJC study, Ford released 5,400 pounds of ammonia, 3,300 pounds of cyanide, and 600 pounds of phenols into the River Rouge every day. These figures represented 47 percent of the ammonia, 89 percent of the cyanides, and nearly 10 percent of the phenols discharged by all the industrial polluters in the region.[28]

The efforts of the sportsmen, Van Coevering, and the IJC led to a slight strengthening of Michigan's water pollution law in early 1949 that brought the state favorable publicity as an early national leader in efforts to address industrial water pollution. For his efforts to improve stream quality for fishermen, Van Coevering received the Izaak Walton League's first national conservation award in 1948. From this point forward the myriad impacts of the Rouge on the Rouge River were cataloged, and Ford's progress in reducing them monitored by the newly established Michigan Water Resources Commission.[29]

In *Cleaning Up the Great Lakes*, historian Terence Kehoe argues that "cooperative pragmatism" characterized the relationship between the Great Lakes state water pollution control agencies and the industries and municipalities they regulated in the quarter century before 1970. With this approach, state water pollution regulators (many of whom had trained as sanitation engineers) worked quietly with corporate or municipal plant engineers to address water pollution problems without recourse to formal legal action or public pressure, an approach well illustrated by the relationship between the Michigan Water Resources Commissioners and Ford. Pollution problems were successfully addressed in this fashion, but in a manner where Ford controlled the pace and amount spent in order to minimize the potential negative impact on its earnings and reputation. This "go-slow" approach was comfortable for the corporate managers and tolerable for regulators, so long as the regulators did not come under serious public pressure from sportsmen and the general public. Cooperation did not involve a conspiracy to do little about pollution; rather it entailed a tacit understanding that in return for Ford demonstrating continuing progress in the right direction, the regulators would not press the company too hard or cause Ford public embarrassment.[30]

The revival of production at the Rouge during the war after slack times in the 1930s also brought a discernible increase in air pollution. Air pollution from the Rouge became an issue in 1946 when housewives in the surrounding Dearborn communities, tired of having their drying wash soiled on clotheslines by soot, began to complain. A local businessman, John Parkhurst, soon took up their cause. Their complaints came two years before the infamous incident of November 1948, in Donora, Pennsylvania, where seventeen people died over several days when heavy fog prevented industrial air pollution in that valley town from rising and dispersing. But it took the Donora incident to convince the Dearborn City Council to pass a smoke control ordinance and appoint a smoke inspector to enforce it. Dearborn's first smoke inspector, Henry J. Hartleb, a former Ford employee, soon found himself filling the shoes of Don Quixote, confronting what may have been the greatest fixed source of air pollution in the world.[31]

The Rouge Production Foundry where V-8 engines for Ford cars were cast proved to be the biggest air pollution problem. As Ford's own in-plant newspaper, *Rouge News*, explained: "Huge quantities of wind—15,000 cubic feet per minute—are blown into the cupolas. This causes fine coke and iron rust particles to blow out into the atmosphere. The dust fell on adjoining buildings, roads, automobiles and personnel." And, the article might have added, quoting Harry Hanson, on Dearborn generally. Foundry air pollution dogged Ford as its biggest and most persistent air pollution problem over the next twenty-five years. But the company also had serious problems with the hundred-plus coke ovens, the two (soon to be three) blast furnaces, the steel mill complex, and the great eight-stacked Power House No. 1.[32]

In Michigan water pollution from plants like the Rouge initially received more attention than air pollution because sportsmen played an especially important role in fighting the problem. Hunters and fishermen worried companies like Ford a good deal. They were customers, and in Michigan, they were numerous. They also showed a willingness to embarrass the automakers publicly. When they complained in dealer showrooms about industrial pollution from the Rouge, the dealers told Ford it hurt sales. Many sportsmen worked for Ford, too, in both hourly positions and management. Conservationists who did not hunt could be derided as "bird and bunny boys," but hunters and fishermen could not be marginalized so easily. Protestors and the plant engineers often

Hunter with buck, gun, and car, Huntingdon County, Pennsylvania, 1937.
Photo by Arthur Rothstein. Library of Congress, Prints and Photographs Division,
LC-USF34-26101-D.

shared this common identity and basis for respect, one reason why water pollution got more attention.[33]

"Rapid growth in outdoor recreation in the 1950s," historians Samuel P. and Barbara D. Hays have written, "extended into the wider field of the protection of natural environments, then became infused with attempts to cope with air and water pollution and still later with toxic chemical pollutants." The postwar relationship between the automakers and Michigan sportsmen certainly demonstrated this dynamic. The automakers surely recognized the irony here—that the most important piece of equipment sportsmen took with them on a Saturday outing was not a rod or a gun but an automobile. Whether they saw the further irony, that in buying industrial peace with the United Auto Workers in the landmark "Treaty of Detroit" contract with GM in 1950, they created even more sportsmen to police their environmental behavior, is doubtful. But without question the auto industry's own product and labor relations policy helped create an army of environmental stewards who naturally wanted cars and a clean environment in which to hunt and fish.[34]

Like the people of Dearborn, who bore the brunt of the air pollution from the Rouge, local residents also had power to constrain neighborhood industrial polluters when sufficiently angered by the impact of air pollution on their home life. Prosperous postwar Americans embraced even more fervently the cherished ideal of a safe and healthy home. "Millions of urban Americans," Samuel and Barbara Hays have written of this generation, "desired to live on the fringe of the city where life was less congested, the air cleaner, noise reduced, and there was less concentrated waste from manifold human activities." Any industrial firm that crossed purposes with the American homeowner's conception of "home, sweet home" risked messing with a force more powerful than itself.[35]

The quality of home life embodied in the suburban ideal remained an important concern even when economics forced many to settle for single-family detached homes in the vicinity of plants like the Rouge. If a man's home was his castle, he could not stand by and allow a neighbor to defile it with impunity. Having a safe, clean home, in an era when many married women were fiercely proud homemakers, was, if anything, even more important for women. But as the early air pollution control efforts in Dearborn showed, protesting housewives found it harder to find a basis for mutual respect with male automakers, plant engineers, and city officials. A common masculine interest seemed to be a precondition for generating the additional leverage that sportsmen had when it came to water pollution. Tellingly, Dearborn officials disparaged John Parkhurst, who took up the housewives' cause, as an "old woman." And after four years of tilting at windmills Henry Hartleb took a job with the Edward C. Levy Construction Company.[36]

While sportsmen, public health officials, and homeowners confronted industrial pollution from the Rouge and other plants in Michigan, Southern Californians discovered an automobile-related air pollution problem of their own: smog. For the first half of the twentieth century, many people viewed Southern California, and the Los Angeles metropolitan area especially, as the Eden at the end of the road of making it in America, a place to move after saving some money from manufacturing and selling things elsewhere. The abundant sunshine, limited rainfall, and year-round mild-to-warm temperatures of its Mediterranean climate made the region a gorgeous and healthy place to live.[37]

They also made it a driver's paradise. Because of an extensive electric street railway system that had encouraged widely dispersed settlement, people instantly appreciated the usefulness of the automobile. Elsewhere in the United States, the adoption of the automobile preceded the development of good roads. By contrast, in Southern California, the distances between settlements, the greater than average wealth of the region's early inhabitants, and the mild, dry climate had all combined to encourage road building and discourage road deterioration. As a result, the first automobile owners found good roads waiting for them, which made the region ideal for those twin pleasures of early automobiling, speeding and touring. From the beginning Southern Californians had more automobiles per person than any other place in the world, and driving fast with the top down, the sun and the wind in your hair, somewhere between the mountains and the sea became a quintessential Southern California experience.[38]

Southern California's civic leadership was putting the finishing touches on a consensus to build the world's greatest metropolitan freeway system, an engineering marvel that would triumphantly reconcile speed and safety, when it got blindsided by smog. Like people who lived near the Rouge, Angelenos noticed an increase in air pollution with the upturn in economic activity during World War II. People later recognized that there had been episodes of air pollution in Los Angeles in the late 1930s and early 1940s, but the problem as a matter of significant public concern can be dated to the last ten days of July 1943, when the city suffered its first severe episode. The most striking characteristics of this air pollution were a brownish blue haze that greatly reduced visibility and a stinging sensation in the eyes that caused tearing. Although people also experienced a variety of other physical symptoms, the smog did not seem to pose an immediate health threat unless one already had respiratory problems.[39]

Citizens of Los Angeles reacted to their smog problem much more vigorously than Dearborn's citizens responded to air pollution around the Rouge. People didn't go to Detroit for the weather or even, for that matter, clean air. But many of those who moved to Southern California had moved there for the climate, including many health-seekers who settled there in the early decades of the region's development for the beneficial effects of clean air. More than any other region in the United States, dirty air struck at Southern California's very identity and posed a direct threat to one of the central tenets of the region's marketing. In Los Angeles upper-class elites, who fully appreciated this aspect of their region's appeal, led the antismog fight. Stephen W. Royce, owner-manager of the Huntington Hotel in Pasadena, became an antismog crusader because he worried about smog's impact on the region's tourist trade. In September 1944, the president of the Los Angeles County Chamber of Commerce joked that he was getting letters and phone calls from businessmen complaining that they had moved their businesses to Southern California on the basis of promises the chamber had made about the climate and were now claiming breach of contract.[40]

But a tarnished climate was no joking matter, as the members of the chamber of commerce well knew. Powerful real estate interests well represented in the chamber were not about to see decades of favorable publicity about the region's healthful climate and the prospects for even greater postwar commercial, industrial, and residential development undermined by this unwelcome but surely solvable problem. As Los Angeles County Supervisor John Anson Ford plaintively put it in 1946, "We must all keep at this postwar evil for the sake of beloved sunny California."[41]

With remarkable unanimity and dispatch, the region's business and civic leaders moved directly into problem-solving mode. Air pollution, everyone knew, was caused by factories, and since the region had seen a great deal of recent wartime industrialization, the city's leaders assumed that the source must be a new factory of some sort. Los Angeles County soon formed a Smoke and Fumes Commission to ferret out these industrial polluters. Scientific research, in this initial view, was not necessary to solve the problem. Someone was putting this stuff in the air, and it was a matter of finding the culprit and making him stop. Smog would be quickly eradicated — or so the region's leaders and residents confidently expected.

In 1947, after legwork by the Citizens Smog Advisory Committee, Pasadena state assemblyman Albert I. Stewart and Los Angeles County counsel Harold W. Kennedy drafted and introduced legislation in the state legislature that gave California counties the power to create air pollution control districts and to pass ordinances to control the problem. Crafted specifically for Los Angeles County and with near universal support of the county's business and civic establishment, the California Air Pollution Control Act passed unanimously and was signed into law by Governor Earl Warren on 10 June 1947. The Los Angeles County Board of Supervisors used the new power to create the Los Angeles Air Pollution Control District (APCD). The county quickly established the largest, best-financed, and most technologically sophisticated air pollution control organization in the nation — and probably the world.[42]

The automobile had always been a suspect in the Los Angeles smog mystery, just not a prime suspect. The first head of the APCD, Dr. Louis C. McCabe, who earned his reputation combating industrial pollution in Saint Louis, explicitly played down the possibility that the automobile had a significant role in smog production, arguing that studies of concentrated automobile exhaust in tunnels and mines demonstrated that even the maximum concentrations experienced there were considered safe. "I do not mean to imply that improperly operated and obsolete motor vehicles should be allowed to pollute the atmosphere," he commented, "they should not, but neither should folklore be encouraged that will place the onus of metropolitan area atmospheric pollution on the automobile, without proof." Air pollution officials had little enthusiasm for the prospect of regulating automobiles. "You must realize . . . ," an APCD press briefing document stated the following year, "that an attempt to control the emissions from more than one million vehicles would require a vast police force. As soon as the major sources are under control, more attention will be given the minor problems." Oddly, even though the APCD worked aggressively and successfully to bring the major sources of industrial pollution under control, the smog problem did not improve.[43]

A deeper understanding of the problem emerged because of a special effort made by Arnold O. Beckman. A former Caltech scientist turned scientific instruments entrepreneur, Beckman was a prominent civic leader who chaired the Los Angeles Chamber of Commerce scientific advisory committee. Concerned about smog-caused damage to local crops and plants, he asked Arie J. Haagen-Smit, a Dutch-born, internationally renowned plant biochemist at Caltech, to join the committee in September 1947. Beckman and Haagen-Smit both became APCD consultants, and under a contract with the APCD, Haagen-Smit soon began to conduct smog experiments at Caltech. In February 1949 he announced that organic peroxide from the incomplete combustion of fuels was responsible for smog-caused tearing.[44]

The Western Oil and Gas Association responded by hiring chemist Abe Zuram, who announced that Haagen-Smit didn't know what he was talking about. This development surprised and angered Haagen-Smit, who was not accustomed to having his professional competence questioned, let alone as a matter of public debate. In May 1950, at Beckman's urging, he took a leave of absence from the university to work as a full-time consultant for the APCD on the smog problem. On 17 November 1950 Haagen-Smit gave a public demonstration lecture at Caltech during which he made smog. He announced that Los Angeles smog was caused by a chemical reaction involving unburned hydrocarbons and nitrogen oxides in the atmosphere that produced ozone (O_3). The two chief sources of unburned hydrocarbons in the atmosphere, he said, were oil refineries and automobiles.[45]

Southern California's unique climate and geography played key roles in the explanation that Haagen-Smit advanced. The two crucial factors were sunlight and the concentration of the pollutants. Dissipation of pollutants in the atmosphere was the traditional solution to air pollution and the rationale for building tall smokestacks. In most places and at most times the atmosphere becomes cooler with increasing altitude. Because the upper atmosphere is cooler, warm air rises. In most places this natural updraft greatly aids in dissipating or at least widely dispersing air pollution, particularly when assisted by the horizontal movement of air—wind.

In Southern California, however, ocean breezes from the west blow a layer of cooler ocean air in underneath a layer of warmer air that comes over the mountains from the higher-altitude deserts that surround the region. The effect is called a temperature inversion—cool air below a layer of warmer air. This warm layer of air functions as an invisible ceiling that, in effect, prevents contaminated air from rising and dissipating. This temperature inversion exists in Los Angeles on upward of 340 days a year. In fact, on some days the temperature at the top of Los Angeles City Hall might be eight degrees (F) warmer than on the ground 464 feet below, creating an inversion layer that trapped contaminants and held them within only a few hundred feet of the ground.[46]

Southern California is surrounded by mountains and has very light average wind speeds. Although the breezes are strong enough to bring in a cool layer of air from the ocean, they are not strong enough to dissipate the air pollution by blowing the pollutants horizontally up and over the surrounding mountain ranges. Southern California is thus like a bowl with the mountains as its sides and the inversion layer as its lid. Hydrocarbons and nitrogen oxides from automobile exhaust are then concentrated and exposed to sunlight, which causes the chemical reactions that produce ozone, or photochemical smog. "If this area were hidden continuously by heavy clouds," Haagen-Smit noted at the time, "we never should have heard of smog." What he discovered was that the climate and geography of the Los Angeles basin made it one of the worst places in the world to have a high concentration of automobiles. Unfortunately, the region already had the world's highest concentration.[47]

Given people's assumptions about the causes of air pollution, the reaction to Haagen-Smit's findings could be predicted. His work confirmed what many had suspected. Industry was to blame, and now they knew the real culprit: the oil industry. The reaction of the oilmen was also predictable: they stepped up their efforts to discredit Haagen-Smit's explanation through research they funded at the Stanford Research Institute (SRI). Rather than retreat to his plant research at Caltech, Haagen-Smit stuck by his findings and deepened both his smog research and his public engagement with the issue.

The finger-pointing disturbed Southern California's civic leaders, who worried that the credibility of the scientific research that now clearly seemed needed to understand and address the smog was being unduly clouded by special interests. Beckman again stepped forward, convening a group of more than seventy-five Los Angeles businessmen, industrialists, civic leaders, and scientists who met at

Los Angeles City Hall with smog trapped by temperature inversion layer. *Photos by Los Angeles Air Pollution Control District. Courtesy National Archives.*

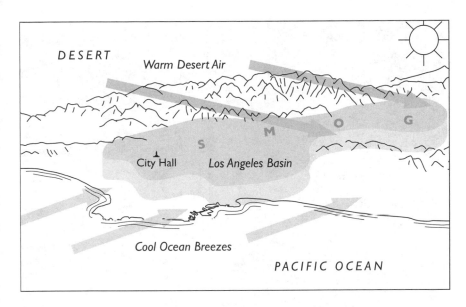

DESERT

Warm Desert Air

City Hall   Los Angeles Basin

Cool Ocean Breezes

PACIFIC OCEAN

Southern California smog. *Map by Bill Nelson.*

the Ambassador Hotel on 2 November 1953 and established the Air Pollution Foundation, a private, nonprofit research corporation to monitor, coordinate, and fund research to address the problem. The goal was both the production of disinterested research and its dissemination to business, government, and the public. "This should be a program carried out . . . without fear or favor," said Asa V. Call, chair of the organizational meeting and president of the Pacific Mutual Life Insurance Company. "If we can come up with an answer and somebody gets hurt, that's too bad." Of course that was easier for an insurance executive to say than someone from the oil or auto industry.[48]

Despite some initial problems replicating Haagen-Smit's findings, research by both SRI and the Air Pollution Foundation confirmed the Haagen-Smit thesis: the petroleum industry and the automobile bore responsibility for Los Angeles's photochemical smog. By the end of 1956, the foundation had concluded that the automobile was the principal cause of smog. As the foundation's director W. L. Faith wrote the following year, "In Los Angeles, automobile exhaust is now the principal source of air pollutants that lead to smog. All of the factors known to contribute to smog, including composition, amount, distribution, and reactions of automobile exhaust, support this conclusion." The research suggested that the automobile accounted for about 60 percent of the contaminants. By the end of the 1950s, scientists, regulators, and local civic and business leaders no longer had any doubts. The automobile caused smog. But who would address

Arie J. Haagen-Smit. *Photo by James McClanahan. Courtesy of the Archives,*
California Institute of Technology.

the problem, three million car owners in Los Angeles County or the automobile
manufacturers in Detroit—the consumer, the producer, or some combination
of the two?[49]

The last thing Southern Californians wanted to hear was that using their
automobiles caused the problem. Many people never accepted this explanation.
Faced with changing their behavior—by driving less—or installing emission control technology on their cars to deal with the chemical causes of smog, Southern
Californians turned their eyes toward Detroit. Although local officials had the
power and several times came close to imposing temporary emergency driving
bans during several bad smog attacks in the mid-1950s, reducing the amount of
Angelenos' driving as a permanent long-term solution was never seriously discussed. The car and not the driver had to be the basis for solving the problem.
Moreover, Angelenos expected freeways, the solution to traffic congestion, to
cut smog, since freeways would reduce stop-and-go driving.[50]

Southern Californians bought 5 percent of the cars made in the United States,
so they had some leverage with Detroit. The manufacturers just needed to make

and sell cars that were pollution-free. No Angeleno pushed this view with more vigor than Kenneth Hahn, who in 1953 was just beginning his forty-year tenure as an elected Los Angeles County supervisor. An outspoken smog-fighter from his days on the Los Angeles City Council, Hahn embraced the Haagen-Smit thesis without waiting for confirmation from the Air Pollution Foundation, taking the issue directly to the automakers. On 19 February 1953 he wrote the first of what proved to be decades of prodding letters to auto industry executives asking each what his company was doing to develop emission control devices. Ford responded by denying any connection between its cars and smog. GM indicated that it continually sought solutions to the problem of exhaust.[51]

Despite the response from company spokesmen, the APCD started passing data from smog studies on automobiles to their engineering counterparts in Detroit soon after Haagen-Smit's discovery. The automakers apparently expected that "cooperative pragmatism" would also characterize their relationship with the Los Angeles regulators. Hahn's initial letter to GM, in fact, had been passed to its industrial waste-control committee, the same people that dealt with the Michigan Water Resources Commission. This assumption proved mistaken because business and civic leaders in Southern California were much more serious about eliminating smog quickly than their counterparts in Michigan were about ending water or air pollution. Smog struck at the sunshine and automobiles heart of the Southern California dream. A lot of people had a lot of money and a lot of themselves invested in that dream.[52]

The automakers decided to tackle the problem cooperatively within the industry, which sounded like a good idea at the time. On 4 December 1953 the Automobile Manufacturers Association (AMA) established the Vehicle Combustion Products Committee to be the industry's official link to the Southern Californians. The committee set up task groups to study the problem and several promising lines of solution. At the urging of Chrysler's representative, Charles M. Heinen, the automakers agreed to share subsequent emission control technology developments with one another by cross-licensing. Scientists and engineers representing their firms on the committee were given the responsibility for keeping the Californians and one another informed of progress in these efforts. All actions were to be approved by the AMA's board of directors, which included the heads of all the automakers. The public rationale was that working together would expedite a solution to the problem.[53]

A considerable amount of interchange between the companies did occur both through the AMA committee and through the Society of Automotive Engineers (SAE). But the committee also created a climate of interfirm deferential courtesy that permitted the automakers to dawdle without risk that one of them

California meets Detroit, the beginning of a long relationship. LA County
supervisors John Anson Ford and Kenneth Hahn (far right) greet GM's John M.
Campbell and the AMA Vehicle Combustion Products Committee, January 1954.
*Courtesy of University of Southern California, on behalf of the
USC Specialized Libraries and Archival Collections.*

would suddenly introduce a solution that tilted 5 percent of the market in its
direction. As an AMA creation, the committee also placed a premium on public
relations. These two proclivities led the committee to agree that no individual
company would announce a solution on its own, that all announcements would
be made industrywide through the AMA, and that all solutions would be installed at the same time. This approach came back to punish the automakers.
Despite a visit by the committee to Los Angeles in January 1954 and a return visit
by Hahn and APCD scientists to Detroit the following year, industry spokesmen
adopted a pose of cautious concern, arguing that the connection between automobiles and smog was not fully proven, that the automobile's contribution to
the problem was far less than suggested, or that the problem was confined to the
unique geographic and climatic conditions in Southern California.[54]

This strategy met its nemesis in Hahn. When he went to Detroit in 1955, several engineers informally told him that they had the technology to install a fifteen-dollar emission control device and could do so in relatively short order if they had sufficient backing from senior management. A few days later, an upbeat Hahn told the *Los Angeles Times's* Washington, DC, bureau that GM engineers had shown him a carburetor device that cut emissions. Hahn's optimism was not misplaced. It was already clear that there were two basic ways that engineers could reduce emissions—by modifying the engine's induction system or by attaching a treatment device to the exhaust system.[55]

Hahn also knew that the famous oil industry chemist Eugene J. Houdry was working on the issue. A French World War I tank commander and battlefield hero (Croix de Guerre), auto racing enthusiast, and 1922 visitor to the Indianapolis 500 and Ford's Highland Park plant, Houdry was one of the twentieth century's most important chemist-inventors. In April 1927 he discovered a chemical catalysis process to "crack" petroleum and yield high-octane gasoline, a development that, had it been available three years earlier, might have derailed GM's efforts to add tetraethyl lead to gasoline. Working in the United States with the Standard Oil Company of New York (later Mobil) and the Pew family, who did not want to put tetraethyl lead into their Sun Oil Company (later Sunoco) gasoline, he commercialized a catalytic cracking process for petroleum that cost far less than thermal cracking and significantly boosted gasoline octane.[56]

The announcement of the process in 1938 stunned the oil industry. Catalytically cracked gasoline meant low-cost, high-octane premium gasoline. It meant that adding tetraethyl lead to gasoline was no longer necessary, or, better yet, with the lead left in, the possibility of even higher-octane gasoline, which paved the way for the post–World War II high-compression, high-horsepower engines that Ed Cole and General Motors brought to market. Houdry's catalytic cracking process also led to the low-cost, large-volume production of high-octane aviation fuel and synthetic rubber, two developments that played important roles in the Allied victory during World War II.

Convinced that unburned hydrocarbons released to the atmosphere from automobiles and oil refineries were responsible for a steep rise in the number of lung cancer cases, Houdry decided to see whether catalysts could be found to reduce automobile air pollution. In February 1948, nearly three years before Haagen-Smit's public smog demonstration, he retired from his oil industry work to begin experimenting with potential catalysts to convert the hydrocarbons and carbon monoxide in auto exhaust into carbon dioxide and water, the natural by-products of complete combustion, a line of research that had been pursued by others sporadically since 1912. The following year he started a new company, Oxy-Catalyst, in suburban Philadelphia to pursue commercial applications. He

Eugene J. Houdry with his catalytic converter. *Courtesy*
Chemical Heritage Foundation Image Archives.

filed his first patent for a catalytic converter in August 1949. Early in 1950, he
settled on a platinum-impregnated aluminum oxide wash coat bonded to a ceramic support structure as the best catalyst device to oxidize hydrocarbons and
carbon monoxide, an approach with important future implications.[57]

After Haagen-Smit's announcement Houdry began to work directly with
Gordon P. Larson, Louis McCabe's successor as head of the Los Angeles APCD.
By 1954, Houdry and his company had developed a specialized catalytic converter, called the OCM Muffler, used on indoor forklift trucks. Concluding that
Houdry's device was the only one available for immediate use, the APCD tested
it for several weeks and found that it oxidized 80 percent of the hydrocarbons in
car exhaust. On the basis of those encouraging results, the Air Pollution Foundation in November 1954 contracted with Southwest Research Institute to road

test the OCM Muffler. Southwest attached the Houdry devices to six vehicles and ran them for ten thousand miles each under conditions approximating typical driving in the Los Angeles area. Southwest found that if the device was used with unleaded gasoline, it eliminated 80 percent or more of the hydrocarbons at each thousand-mile interval test through ten thousand miles. However, if the device was used with leaded gasoline, the elimination efficiency dropped to only 22 percent by four thousand miles. Since unleaded gasoline was not widely available, owners would need to defoul or replace the converter several times a year to keep it functioning at high levels. Southwest therefore concluded that it was not suitable for adoption by the public.[58]

Developments in Southern California and elsewhere did not go unnoticed in Detroit. Despite public statements from the AMA that soft-pedaled the Los Angeles smog problem, the automakers were more concerned than Hahn believed. At GM Research, William G. Agnew remembered that his boss John M. Campbell came back to Michigan from his 1954 visit to California with the Vehicle Combustion Products Committee quite concerned and told his staff "we'd better do something about it."[59]

Scientists at GM Research and the Exhaust System Task Group of the AMA's Vehicle Combustion Products Committee, chaired by GM's George J. Nebel, examined several early catalytic mufflers. By October 1954, GM had already reached the same conclusion that Southwest soon reached: since the catalysts required the use of precious metals, rapidly deteriorated from leaded gasoline, and needed to be replaced frequently, the approach was too expensive for commercial use. GM's concerns about the availability and cost of platinum were valid, but the key sticking point was lead in gasoline. GM could not have been disinterested in this matter. It still owned 50 percent of Ethyl Corporation and made millions each year from the presence of lead in nearly all the gasoline sold in the United States.[60]

Nonetheless, a catalytic converter along Houdry's lines looked like the most promising solution. His invention meant that a highly effective antismog device was at hand in the mid-1950s, provided that the oil industry removed lead from gasoline and the auto industry redesigned its engines to run on slightly lower-octane gasoline. When the Vehicle Combustion Products Committee visited Los Angeles in January 1954, its members promised a device by the 1958 model year. If the industry and others sometimes sounded optimistic, it was due to the hope that a less expensive base metal catalyst that cut emissions from leaded gasoline for an extended period would be found. Unfortunately, to the growing frustration of Southern Californians, Detroit produced no such device.[61]

Developments in Dearborn and Southern California remind us that the consumer and business agendas of postwar Americans could simultaneously worsen the automobile's environmental impact and ultimately constrain it. In Michigan people living close to the sources of pollution, especially those who were not wealthy enough to retreat to pollution-free neighborhoods, reacted to the industrial pollution caused by the automakers and their suppliers. When pollution impinged on their recreational agendas and diminished hard-won improvements in their standard of living, people protested and thereby began to set the outer limits of the permissible. The power given to the early environmental regulators, limited as it still was, owed everything to the sportsmen and local residents who complained about the problems, prodded legislators to pass or strengthen laws, and pressured regulators to enforce them. These groups limited the degree to which automakers like Ford could pollute the environment from their factories.[62]

The early response to air pollution in Southern California was different. Smog there touched a broad spectrum of the community, even those who lived and worked far from industrial activity. Homeowners felt its effects, but so did businesspeople with a stake in Southern California's reputation for sunshine and health. Not surprisingly, the region's traditional civic elite quickly mobilized a powerful response that created the world's foremost air pollution control agency. With this backing, the APCD cleaned up industrial sources of air pollution in short order. The contrast with Henry J. Hartleb, Dearborn's lonely smoke control inspector, could not have been starker. Yet when it became clear that the automobile rather than local industry caused most of Southern California's smog and the Californians took on Detroit, they made little initial headway, much to their frustration.

Craving "community" as well as respect within that community during an unsettling period of prosperity, insecurity, and change, postwar American consumers like the Presleys pursued agendas that again made the automobile— especially the Cadillac and Cadillac wannabes from the Big Three—the central consumer icon in American life. They needed a Detroit that provided emphatic, unambiguous ways to announce that they were no longer marginal Americans but instead possessed the goods to command community respect. "In a society where their own and others' positions had been shifting so rapidly," Whyte observed of the Cadillac, "they needed a fix—a visible symbol that would affirm, to themselves as much as others, where they had got to, and how it would be from here on in." E. Franklin Frazier has described middle-class African Americans in the early 1950s as "exaggerated Americans," but the same could be said of many Americans as they sought to create community from common consumer goods.

The strength of this need brought conformity to the market. With their postwar material achievements, Americans ministered to psychic wounds caused by two decades of privation and insecurity, but they also did so with an "outright, thoroughly vulgar joy in being able to live so well . . . ," Hine observes, "the sense that a new car [was] an achievement worth celebrating."[63]

When Americans felt the need for more community, they increasingly turned to large publicly used material goods like the automobile to create it. This proclivity contributed heavily to making the years between 1945 and 1955 — and the two decades that followed — the worst period of the automobile's impact on the environment. Most Americans not only failed to perceive a need for concern, but their affinity for greater speed, size, styling, and expense in cars from the Big Three encouraged trends that worsened the environmental impact of automobiles. Yet consumer demand for chrome and tail-finned behemoths from the Big Three (and disdain for alternatives that departed from the industry norm) did not grant the automakers a license to pollute with impunity; even in the absence of significant government regulation, the environmental impact from making and using automobiles eventually collided with other strongly held individual, family, and business agendas. Thanks to these agendas, the pressures to mitigate the automobile's environmental impacts soon grew.

7

DISENCHANTED WITH DETROIT

The romance did not last. American automobile sales fell in 1956 and 1957. Slumping auto sales triggered a nationwide recession in 1957–58 that further reduced sales. In 1958, the industry sold just 4.3 million passenger vehicles, only 54 percent of the number sold three years earlier. By that spring, Americans were having a national conversation about what was amiss in the automobile market. Mystified industry executives called it a "buyers' strike." At the time, automakers viewed the sales downturn as an unforeseen bump in the road. In some respects it was. But it was also something more.[1]

After 1955, a growing number of consumers began to seek alternative automobiles that departed from the prevailing American standard set by the Big Three, especially small European imports. When imports and smaller cars from second-tier domestic manufacturers like American Motors suddenly took 10 percent of the market, the press, casting about for an explanation, gave critics of American automobiles a public hearing. These critics gave specific direction to the disenchantment that became the Buyers' Strike of 1958. Detroit's reaction to these consumers and critics—introducing small cars of their own—showed that a comparatively small percentage of consumers, when they had grown numerous enough to be noticed, had some power to force the automakers to meet their desires more closely, even when the makers were not enthusiastic about doing so. Small cars also happened to be easier on the environment, a reminder that consumers could change the impact of the automobile on the environment directly by making alternative product choices. Except for those in Southern California concerned about smog, few Americans in the 1950s made this connection. But the possibility that consumer choice might have been exercised to this end, a prospect raised by the Big Three's response to the Buyers' Strike of

1958, is worth pondering before examining the government regulation that followed.

For Americans in the 1950s the "alternative über alles" was the Volkswagen "Beetle." Against a backdrop of postwar ostentation in American automobile styling, the Beetle's novel look stood out. Nothing intrigued Americans more than that its design remained unchanged year after year. A Volkswagen television commercial from the early 1960s captured the essence of this appeal. "If you owned an eight-year-old Volkswagen and had it washed now and then," the voiceover suggested, "who'd know it was an eight-year-old Volkswagen — except you?" Volkswagen even poked fun at the industry's annual model announcements by declaring periodically that their car's appearance would not change. Small, light, underpowered, and unchanged in appearance, the Beetle was a rolling mockery of the U.S. auto industry. From the Big Three's perspective, American consumers should have laughed the car out of the country. Instead, they reacted positively. Volkswagen sales remained small in the 1950s, reaching just 150,000 and a 2 percent market share in 1959. Its best years lay ahead. But as the first import to establish a niche in the American market, the car received media attention far out of proportion to its sales. For some it became the method of choice to needle Detroit. The last thing that any American automaker in the 1950s believed was that the Beetle would become the best-selling automobile of all time.[2]

The secret of Volkswagen's success was not advertising but the chemistry between consumers and the car itself. The company did no national advertising during the car's first decade on the U.S. market (1949–1959) and little dealer advertising. Sales took off due to word of mouth among foreign car enthusiasts and favorable articles in the press. The car's strong practical characteristics appealed to consumers. It had the lowest purchase price in America ($1,495). It got thirty miles or more per gallon and cost little to maintain. It did not depreciate as rapidly as American cars because new models looked just the same as the old models. "The main reason why the Volkswagen has been doing well," Edgar Snow observed, "is because it offers a degree of power no small car in the same field possesses." Although the Beetle was no sports car, there was in the mid-1950s still some conflation between small foreign car and foreign sports car in the American consumer's mind that worked to Volkswagen's early benefit.[3]

"Our first customers," New York City area distributor Arthur Stanton recalled, "were faddists." These buyers, foreign car saleswoman Dorothy M. Mattimore observed, were "bugs for knowing every dimension and specification relative

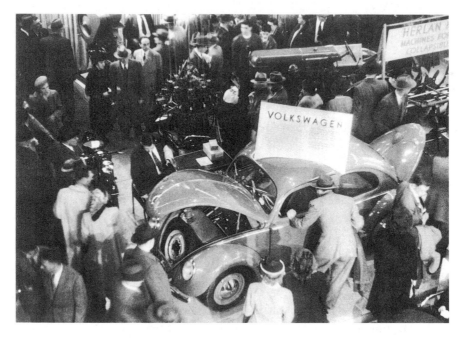

Anti-status symbol status symbol. The Volkswagen Beetle at an industrial exposition
at the Museum of Science and Industry, New York, 1950. *Library of Congress,
Prints and Photographs Division, LC-USZ62-108184.*

to each foreign make." They quickly formed a community of early adopters.
"A spirit of camaraderie . . . seems to infect Volkswagen owners," an early VW
dealer, William Dretzin, told a reporter. "They blink their lights to one another
and honk their horns on passing." Commentators called the Volkswagen "an
honest car" that "doesn't pretend to be anything it is not."[4]

Americans were smitten with the Beetle, but it took considerable work, luck,
and irony to reach that point. The German "people's car" had been designed in
the 1930s by Ferdinand Porsche to remarkably consumer-oriented specifications
provided by Adolf Hitler. Both men had been deeply impressed by Henry Ford's
achievements with the Model T and were committed to providing a German
car for the German masses as an integral part of the Nazi program. The Nazi
leader wanted a car that was fast (more than sixty miles per hour), economical
(thirty-three miles per gallon), family-sized (four to five passengers), and affordable ($250) so that the German people could revel on his regime's new highway
system, the *Autobahn*. In the course of planning the Volkswagen plant, Porsche
visited the Rouge, where he met his hero Henry Ford.[5]

The war frustrated these plans, but the vehicle Porsche designed was the only car that Volkswagen had to sell when British postwar occupation authorities asked Heinz Nordhoff to revive the company. Nordhoff had been an executive with General Motors's German subsidiary, Adam Opel. In the 1930s, he, too, made the pilgrimage to Michigan to visit the parent company. Though not a Nazi like Porsche, Nordhoff continued to work for Opel during the war. Because of this, GM refused to rehire him, and American occupation authorities certified him for manual labor only. Nordhoff moved north to the British occupation zone, supporting his family as a mechanic in Hamburg before the British sought him out in 1947 to run Volkswagen. As a GM-Opel executive, Nordhoff had scorned the small car Volkswagen project, but he now decided to make the best of his new circumstances. There was still room in the automotive world, he decided, for Henry Ford's ideal—a small, well-made, affordable car whose appearance did not change from year to year. There remained nothing to do but make a success of it, which he did, by selling the car not just to Germans but to the rest of the world.[6]

If 1955 was the golden year for the Big Three, it was also the breakthrough year for Volkswagen in the United States. Sales more than quadrupled, as Volkswagen pulled away from British sports car makers Jaguar and MG to sell 28,907 cars, 56 percent of all the imports sold in the United States. The next year Volkswagen sold 50,011 cars, or 51 percent of the imports. By summer 1957, Volkswagen had a six- to nine-month backlog of American orders that it was unable to meet, while the Big Three's unsold dealer inventory approached a million cars. Used Volkswagens sold for as much or more than new ones because they looked like the new ones and buyers didn't have to wait for delivery. Volkswagen's success and its difficulties in meeting consumer demand signaled opportunity for other import makers and sparked a general boom in import sales that lasted for three and a half years. Volkswagen's sales climbed, but its share of total imports fell to about 25 percent as Americans sampled a range of imports in 1958 and 1959.[7]

Soon headlines appeared in the national press with titles like "Who Buys Those Small Foreign Cars?" Detroit's spokesmen soft-pedaled the development, dismissing the buyers of small or foreign cars as "gray flannel nonconformists." "There will always be 5 per cent of the car market," one Detroit executive scoffed, "that will be made up of individualists . . . who cannot permit themselves to choose a car built by one of the Big Three." But consumer surveys revealed disturbing information. Foreign car buyers tended to be professionals and white-collar managers, the very masters of "making it" in America. They were middle-income to wealthy, highly educated suburbanites in metropolitan areas in the Northeast and on the West Coast, found in especially large numbers around Los Angeles and New York City. Nearly half bought these makes as first cars. In Los Angeles County, 16.5 percent of the new cars registered in 1958 were imports. "Every three minutes, day and night," wrote Bill Carroll of *Automotive News*, "some Californian buys a new imported car." More of these buyers were trend-setting sophisticates than Detroit was willing to admit.[8]

Small foreign cars had a reverse snob appeal Detroit could do little about. A small, inexpensive import, in sharp contrast to a low-priced domestic car, suggested that its owner was a discerning sophisticate rather than a poor man. "It's like Bing Crosby wearing a sweat shirt to a party," one Los Angeles Renault dealer observed. "Everybody knows he has a tuxedo if he wants to wear it." "Owning a Volkswagen . . . ," Huston Horn wrote in *Sports Illustrated*, "does not stamp you as *nouveau riche*, and it does not suggest you are strapped for funds either. Among other things, it says you are above striving—and that maybe you have a swimming pool in the backyard anyhow." Rather than sweaty material striving, the Beetle suggested conspicuous distance and understatement, arch humor, and detachment. The press called consumers who bought them "an avant-garde of American car owners."[9]

Many who soon followed the first Volkswagen owners recognized the opportunity to put some distance between themselves and their fellow citizens by decamping from the Cadillac-dominated community of chrome and tailfins. "The car that presented itself as the antidote to conspicuous consumption," *Ad Week* critic Bob Garfield later observed, "was itself the badge product for those who fancied themselves a cut above, or at least invulnerable to, the tacky blandishments of the hidden persuaders." Volkswagen owners inverted the status hierarchy of American automobiles, replacing large, powerful, expensive makes like Cadillac at the top with the Volkswagen and effectively changing the rules for automobiles and status on Detroit and their fellow citizens in midgame—or at least starting a game of their own.[10]

Criticism of the Big Three's cars grew as their sales slumped and those of Volkswagen and the other small imports grew. Critics had been taking the industry to task for years, but from World War I to the mid-1950s consumers could not care less. Automobiles were America's glamour product. Americans loved cars, and with good reason. The average price of an American-made automobile fell sharply after 1908, continued to fall for several decades, and then remained low until 1945, while the automakers increased horsepower, added technical improvements, and provided attractive and varied styling. For most consumers the biggest problem was not what the automakers were offering but simply finding enough money to buy the model they wanted. After World War II, prices soared; the automakers piled on horsepower, turned to chrome and tailfins, and searched for more effective ways to convince consumers that they were actually getting more "value" for their dollar. Although consumers could afford these prices and many no doubt delighted in paying them, these developments opened the auto-makers to criticism. They also made the industry, which accounted for 20 percent of American GNP by the mid-1950s, a lightning rod for growing anxieties about postwar affluence.

Criticism did not precipitate Detroit's Buyers' Strike of 1958, but it eventually influenced consumers once it reached the mainstream media. The news magazines *Time* and *Newsweek* devoted regular coverage to events in the auto industry during the 1950s, but apart from occasional articles that mentioned criticism of American automobiles or the auto industry, they mostly gushed about Detroit. The magazines did report that each year's models were getting longer, lower, more powerful, and more expensive, but they did not cast these developments in a negative light. Criticism was confined to highbrow magazines such as the *Nation*, the *New Republic*, the *Atlantic Monthly*, and *Harper's*. Critical articles that appeared in these magazines between 1955 and 1958 provided the agenda and vocabulary for the very public criticism of 1958, but it is hard to imagine that when legendary designer Raymond Loewy complained that the Big Three's cars had become "jukeboxes on wheels" in a talk to the Society of Automotive Engineers and in an *Atlantic Monthly* piece that he had much direct impact on American consumers. Yet buyers began to desert the Big Three (or, at least, stay out of the market) in 1956 and 1957.[11]

Whether the recession of 1957–1958 was cause, effect, or both, it depressed auto sales. So, too, did the reporting in the popular newsmagazines, which in its search for causes for the slump finally turned critical in early 1958. As a disappointing 1957 became the marketplace disaster of 1958, previously isolated criticism became part of a larger story that the news magazines told about Detroit being out of touch with American consumers. The criticism peaked with a five-page cover story in *Time* on 12 May 1958. "Detroit," the magazine noted, "has become the center of a vast family argument." Even President Eisenhower weighed in. "I personally think our people are being just a little bit disenchanted by a few items that have been chucked down their throats, and they are getting tired of them," *Newsweek* reported the president as saying. In case anyone found the president's remarks cryptic, the magazine helpfully told its readers that he meant Detroit's automobiles. The news magazines and Eisenhower gave previously scattered criticism wider currency and far greater credibility. Suddenly, riding around in a big, overpowered, chrome-laden, be-finned automobile from the Big Three was like having a sign on one's forehead that said "chump."[12]

In 1958 a full-blown battle for American automobile consumers developed between American automakers and their critics. For the first time large numbers of Americans read that their love affair with the cars offered them by the Big Three might not be in their own best interests. Never before had they heard or read so much about their thralldom to the automobile and the problematic nature of their affinity for consumer goods in general. At specific issue were the kinds of cars they purchased and how these cars were sold. For a brief moment in late 1957 and early 1958 the critics gained the upper hand in this struggle and the behavior of a substantial portion of American consumers changed in the directions urged by the critics. This development raised two larger issues: whether the automobile would remain the central consumer icon in American life, and whether consumers would use their purchasing power to force the automakers to change their product in directions that better met their desires. On this score, the critics in 1958, as well as those who followed them over the next fifteen years, were hopeful. Even the automakers worried, at least briefly.

John Keats's best-seller *The Insolent Chariots*, published in September 1958 shortly after criticisms of the industry peaked and Detroit's sales reached their nadir, provided the most comprehensive and colorful criticism of the American automobile industry. Although his book did not create the crisis, it said much about what had caused it. American automobiles, he wrote with a candor calculated to offend everyone, were "overblown, overpriced monstrosities built by oafs for thieves to sell to mental defectives." He also found the cars lacking in quality, practicality, economy, and safety.[13]

Keats focused his criticism on the obvious developments in the American automobile market of the 1950s. One concern, horsepower, had been removed (at least temporarily) by the time that Keats's book was published. Detroit found it impossible to market increasing horsepower without arousing the corresponding concern about automobile safety and accidents. In June 1957 the automakers finally decided that the best solution to the criticism was to stop touting horsepower, speed, and racing in their advertising, although by then much of the potential horsepower in their postwar engines had been realized.[14]

Keats thus addressed himself to the remaining concerns. "Automobile prices," he wrote, "seem to be set more by greed, whim and opportunity than by cost, supply and demand." He especially loathed the Big Three's styling. "It is to General Motors' credit that it discovered, or sensed," he wrote with sarcasm, "that the public's desperate, but unknown, unvoiced need was not for a completely safe automobile but for tail fins." "In styling," Keats noted, "the word is still careful sloth and General Motors is much admired in Detroit for selecting new styles which go just so far, but not too far ahead in any one year." "Automotive progress in America today boils down to this," Keats wrote. "The basic steel shell is bent a little bit this way, this year, and is bent slightly that way next year . . . no significant changes take place, except in price, and the change there is certainly significant." He lamented the passing of the Model T's form-follows-function aesthetic and could see only foolish irrationality in the choices of most consumers.[15]

Keats also sharply criticized how Detroit marketed its cars. In 1955 the Big Three constituted three of the four biggest corporate advertising spenders in America. But Keats worried less about what the automakers spent on advertising. His real concern was the growing use of motivation research (MR) by marketers and ad agencies, the subject also of Vance Packard's 1957 number-one best-seller *The Hidden Persuaders*. In a tone of moral outrage, Packard argued that the use of MR constituted an invasion of privacy and an unconscionable manipulation of consumers by devious marketers and admen intent on padding their own bottom lines regardless of consumer welfare.[16]

MR's promoters argued that scholarly research into the motivations behind consumer behavior had much to offer marketers and admen interested in selling people retail products. Academics called the attention of business people to the deeper, often hidden, meanings and emotional associations of goods. The basic idea about consumer goods and symbolic communication was not new, but the experts told retailers that they could sell their products more effectively by identifying these meanings and by crafting appeals that were congruent with them. "M.R. differs from orthodox market research," *Fortune* told its readers in 1956, "in seeking the psychological and sociological reasons for consumer behavior, and refusing to take at face value the conscious, often rationalized, replies of those interviewed." The idea appealed to postwar marketers and admen who were already pouring considerable time and money into the search for more "scientific" and effective ways to sell products. In what was partly an accident of timing, the American automobile industry got caught in the crossfire between motivation researchers and their critics at what was already a bad moment for the industry. Worse, Packard, Keats, and others singled out the automakers as the primary example of the manipulative use of motivation research.[17]

The rationale behind MR, that consumers purchased products by acting on unconscious motives, has been borne out by psychological research. However, making concrete connections between specific cars and specific motives, given the interpretive leeway involved, was speculative, to say the least. The automakers commissioned motivation research. But a close look at the decision-making processes in the 1950s at companies such as Ford or the creative process at advertising agencies such as J. Walter Thompson or Chevrolet's ad agency, Campbell-Ewald, suggests that MR had little impact on their marketing and

advertising practices. Consumer research that encompassed more than just motivation research did grow in importance, but it remained peripheral to the far
more pressing everyday demands of selling cars and grinding out new ads to help
sell them. MR certainly did not drive decisions about styling, engineering, or
marketing among the automakers.[18]

In the 1950s the automakers successfully made the transition from advertising
principally through mass circulation magazines to network television, but they
achieved no attendant breakthrough in the persuasive power of their advertising.
Indeed, the content of automobile advertising changed relatively little during this
period. As author David Halberstam has argued, only when Campbell-Ewald's
Kensinger Jones teamed with Gerry Schnitzer in 1957 to make television com-
mercials for the 1958 Chevrolets did the industry start to grasp the possibilities
in the new medium. More fundamental changes in the nature of advertising, as
Thomas Frank has shown in *The Conquest of Cool*, came in the 1960s.[19]

Unfortunately, Packard's *Hidden Persuaders* and Keats's *Insolent Chariots*
clearly linked the auto industry to the use of motivation research. Packard's book
singled out the industry for criticism on this point, while Keats's book merely
featured MR in a longer list of obnoxious practices. Packard called attention
to the same research that had noted the Cadillac's unambiguous symbolism, a
study done for the *Chicago Tribune* in the early 1950s by Social Research, Inc.,
called "Automobiles, What They Mean to Americans." "The automobile tells
who we are and what we think we want to be . . . ," wrote the *Tribune*'s Pierre
Martineau, the foremost proponent of MR in the 1950s, "It is a portable symbol
of our personality and our position . . . the clearest way we have of telling people
of our exact position." Quoting the study, Packard told his readers that one of the
most important functions of the automobile was "to express aggression."[20]

In November 1957, after Packard's book had been at the top of best-seller lists
for three months, *Reader's Digest* ran a condensed four-page article by Packard
that mentioned the automobile twice and featured an illustration of a fedora-
clad male consumer ogling the ghost-image of a buxom young woman in a strap-
less evening dress standing next to a convertible in a showroom window—an
illustration of MR-maven Ernest Dichter's famous "Mistress vs. Wife" study
where he urged Chrysler dealers to put convertibles in their showroom windows
if they wanted to sell more sedans because while men fantasize about having
mistresses (convertibles), they usually go home with their wives (sedans).[21]

Keats, who also shared Packard's sense of moral outrage, argued that motivation research was just another way that the automakers and their ad agencies foisted their dubious products on a gullible public. "The preposterous pre-
sumptions of motivation research are equaled only by Detroit's willingness to

accept them," he wrote. "Detroit believes that the cars people buy reflect the personalities of the people who buy them, because motivation research thinks so. Thus, automobiles are advertised not as machines, but as extensions of one's id." Whether Detroit merited this criticism—and the automakers certainly bore responsibility for the intention and the attempt, if not the successful commission, of the alleged crime—the industry's reputation suffered.[22]

Much of the public criticism directed at American cars, automakers, and consumers in 1958 came in the form of ridicule, which made it especially effective in changing consumer behavior. In *Insolent Chariots* Keats (ably assisted by illustrator Robert Osborn) dished it out in spades. "Ah, the Midwest! Land of simple plowmen!" he wrote. "No doubt it was the Midwesterners' immemorial custom of attending agricultural fairs which led them to think in terms of the biggest pumpkins, and thus to believe that if it's the biggest, it's the best." The problems with Detroit's automobiles, in Keats's judgment, were due to "the rustic caution and provincial limitations of the Midwestern mind." Like many crude stereotypes, this one contained enough truth to make it believable. More than 90 percent of the cars sold in America came from three companies located in southeastern Michigan. The companies designed cars similar in appearance and performance. To many outsiders the industry looked insular—and even more so for being the largest and most important in the United States. "The men who ran [GM]," author David Halberstam later wrote, "were square and proud of it, instinctively suspicious of all that was different and foreign. They were American, and above all else Americans knew cars." Keats was not the only critic to indulge in ridicule once it became open season on Detroit.[23]

Keats did not spare consumers, either. They clearly liked the cars that they bought from Detroit, so for Keats they were fair game, although he admitted to being mystified by their tastes. Rather than pursue a question that led to the crux of the relationship between consumers and the automobile's social and environmental problems, he opted instead for condescension. In his judgment, American consumers were "mental defectives" for not demanding cars from Detroit that were "cheap, safe, practical, simple, sturdy and economical." "The hidden persuaders' efforts," he wrote, "are largely wasted on [the typical American consumer] because he buys what he must, and he has almost no depths to probe, anyway."[24]

When it came to the cars themselves Keats naturally directed much of his scorn at the Cadillac. But deriding the Cadillac, a car in which millions of Americans—owners and nonowners alike—had a sizeable symbolic investment, was apparently less fun for the public than ridiculing the automakers as a group. Consequently, much of the wider public mockery directed at the cars themselves was aimed, not at the established Cadillac, but at a new make that arrived with considerable hype at precisely the wrong moment. The Edsel was not a single car but a new line of automobiles that Ford, again modeling itself on GM, introduced on 4 September 1957 to add another division targeted at the medium-priced market. As a Lehman Brothers report noted in 1956, "Ford [was attempting] to exploit two rather dependable tendencies of auto buyers: to trade up and to stick to the same family of cars."[25]

The publicity buildup for the Edsels far exceeded any in American automotive history since the introduction of the Model A thirty years earlier. Ford spent ten million dollars on advertising alone. Curiosity was intense and expectations were high as the first Edsels hit dealer showrooms. Nearly 2.9 million Americans visited Edsel dealers on the day the new line was introduced. Ford's public relations staff orchestrated favorable initial press reviews. A national survey indicated that 84 million Americans age twelve or older could identify the Edsel as a new car on the day after its introduction. Ford, expecting to sell two hundred thousand in the first year, waited for sales to take off.[26]

Americans looked, but they did not buy. Even a four-hundred-thousand-dollar Edsel television extravaganza, viewed by fifty-three million Americans in October 1957 that preempted Ed Sullivan and featured Bing Crosby, Frank Sinatra, Louis Armstrong, and Rosemary Clooney, could not salvage the new line. Although Ford carried the cars for another two years to save face, the Edsel was dead on arrival. Ford sold only 109,000 of the cars before discontinuing the line in November 1959. The company's loss was $350 million, one of the great marketing disasters in American corporate history.

The Edsel quickly became the chief scapegoat for Detroit's excesses. The press, which did an about-face when the cars failed to sell, pinned the failure on the two most prominent public criticisms of Detroit that critics had served up, motivation research and styling. Packard's book was the number-one best-seller when the Edsel was introduced. Keats's chapter on the Edsel dwelt on the failings of MR and on how Ford missed an opportunity as a consequence of mindlessly aping GM. Critics made fun of the Edsel's vertical radiator grille, then a departure from the industry norm. "I may see it as a classic eleventh-century Norman shield," Arthur W. Baum commented in the *Saturday Evening Post* early on, "you may see it as a horse collar." As Edsel became a byword for failure, critics took full liberties with the grille, one likening it to a horse's vagina. Others said the Edsel looked like an Olds sucking a lemon or a Mercury pushing a toilet seat.[27]

But the real styling problem was that the Edsels were not different enough. The completely different automobile that Ford conceived when it designed the

Edsels from consumer feedback in 1955 was not the completely different automobile that consumers wanted by fall 1957. With too much of the same old styling, the Edsels arrived just as Americans decided that maybe Detroit was not offering the kinds of cars they really wanted. Between design and introduction, consumer tastes changed. Motivation research, conducted after the styling was set and only to provide a theme to promote the line ("the smart car for the younger executive or professional family on its way up") had nothing to do with the car's styling. "If Edsel had introduced an American-made, European-oriented automobile," a dealer told *Automotive News*, "I am certain they would have had the entire small-car market almost to themselves this year."[28]

The closest students of the Edsel's demise offered two additional reasons. C. Gayle Warnock, public relations director for the Edsel Division, argued that the Edsels were priced too high. Expecting a radically new car that justified the higher price tag or a respectable full-sized automobile at a substantially lower price, American consumers were offered neither, a classic case of over-promise and under-deliver. With growing fears of inflation and the economy entering a recession, Americans paused, and that proved fatal, not just because the benefit of the extraordinary buildup and the initial favorable reviews was lost, but because a far more damaging problem emerged—quality. "Within a few weeks after the Edsel was introduced," John Brooks wrote in a careful 1960 postmortem in the *New Yorker*, "its pratfalls were the talk of the land." A former Edsel executive told him that only about half the first Edsels worked properly.[29]

Fall 1957 was a bad time for an American corporation to have quality problems with an important consumer product. On 4 October the Soviet Union launched *Sputnik*, the first manmade satellite to be placed in Earth's orbit. With the Cold War shifting to the high frontier and America apparently behind, suddenly it seemed obvious that too much of the country's engineering talent went into chrome, tailfins, and lousy cars and not enough into rockets and space technology. "Fear of rockets, the atom bomb, automation and a whole world of dehumanized, vast and uncontrollable phenomena finds expression in a continued trend to smaller cars," the Center for Research in Marketing reported in December. "The Edsel," *Consumer Reports* concluded in its scathing January 1958 review, "has no important basic advantages over other brands." "Too much of the effort apparently went into making a piece of novelty merchandise," the magazine's reviewer suggested sarcastically, "too little into creating a superior vehicle." Bill Carroll, writing for *Automotive News* that same month, called quality "a gaping wound in Edsel's armor." Asked to develop a public relations campaign to transform the hapless line into a winner, Kenyon and Eckhardt told Ford several months later that "it is difficult to persuade many people to consider Edsel seriously because of the fear of their neighbors' and friends' comments." The Edsel had become America's chump-mobile.[30]

Americans often tune out criticism but never ridicule. Some people derive a certain ego satisfaction from being the object of public criticism, but no one enjoys being laughed at in public. Consequently, people make a real effort not to get caught on the wrong side of public ridicule. Public ridicule led to disassociation from Detroit's standard fare. The more opinion-sensitive automobile consumers reacted to the criticism by adopting a pose of ironic detachment toward their cars, adding to the trend already evident in Volkswagen sales. For a significant group of trendsetting consumers on the East and West coasts "American Iron," as they now disdainfully began to call Detroit's cars, became "uncool." The American automobile industry never entirely recovered from this moment.[31]

A substantial change in consumer preferences took place between 1955 and 1959 that was not orchestrated by the Big Three or in their financial interest. This change was well under way before criticism of the Big Three's cars and marketing practices was amplified by the mass media in 1958, which then fueled these changes. Chicago motivation researcher Louis Cheskin claimed to have measured it in his surveys. "As recently as last year [1957]," he observed, "our tests showed that people reacted favorably to elaborate ornamentation, gaudy color combinations, and elaborate chrome trim on cars and other steel products. . . . Recent studies show that people are reacting unfavorably to such ostentatious ornamentation. . . . Now the consumer is in revolt."[32]

The automakers were keenly aware of, even somewhat alarmed by, the criticisms and the sudden new direction in consumer tastes. Most of the industry's leaders had risen to their positions while auto making basked in the glow of being America's glamour industry, so national criticism and ridicule was new and discomforting. They were uncertain about how to proceed. One part of their collective mind was inclined to be dismissive. "Detroit decided," as Keats put it, "that the criticism was nothing but a lot of nittering and nattering emanating from a few aesthetes and intellectuals from the effete East—from the kind of people who drove Volkswagens and read highbrow magazines just to show off." This view had truth in it. But unfortunately, as the automakers well knew, such people were the traditional style leaders. Now they were disassociating themselves from the chrome behemoths of the Big Three in favor of alternatives like Volkswagen, and other consumers were following their lead. That process had to be stopped.[33]

But the automakers also knew not to confuse critics with consumers. Although many critics adopted the rhetorical stance of speaking for consumers, the industry knew that the critics were not giving voice to universally shared dissatisfactions with American-made automobiles. Consequently, the companies were not inclined to engage in much deep soul-searching and opted instead for their standard, low-cost first-line response—public relations. The Big Three sought to counter criticisms as effectively as they could in order to stem consumer disenchantment and stave off any threat of government regulation or antitrust action while it pondered any next steps.

The goad that forced the Big Three to go a little farther came from the 1958 marketplace in the form of American Motors. By all rights, the downturn should have delivered the deathblow to this perennially unprofitable automaker. Instead, the company became the major domestic beneficiary of consumers' new interest in alternatives. Through the second half of the decade, the company's outspoken maverick president, George W. Romney, relentlessly touted the virtues of the small car and criticized the cars of the Big Three as "gas-guzzling dinosaurs." Nash-Kelvinator chairman George Mason had converted his protégé Romney to the merits of the small car in the early 1950s. But in that Cadillac-centered market their efforts to sell the smaller (100-inch wheelbase) Nash Rambler failed. Romney, who succeeded Mason when Mason died shortly after Nash merged with Hudson to form American Motors in 1954, took over a company well on its way to becoming one more independent that had failed to find a profitable niche in the shadow of the Big Three. Even Romney had reservations about small cars in the American market of the mid-1950s. The problem, he candidly told the *Wall Street Journal* in December 1954, was that the public was so wedded to big cars that smaller cars were stigmatized. It "looked as if the buyer couldn't afford a more expensive car."[34]

But with American Motors ninety million dollars in debt, its creditors threatening to pull the plug, and consumers abandoning the company, Romney felt he had no other choice than to take the gamble that everyone told him would be corporate suicide: recommitting to selling a smaller car. Like Nordhoff at Volkswagen, he really had no choice. Dismissed by the industry ("We pay no attention to Romney," Ford chairman Ernest R. Breech said at the time) and lacking support within American Motors, Romney, too, found inspiration in Henry Ford's ideal of offering the American consumer a useful car as a form of public service. So he forged ahead, trusting that the market would reward this commitment. Although he also embarked on a crash cost-cutting program that included reducing his own salary by 40 percent at the same time, Romney decided to bet the company on the 108-inch-wheelbase Rambler Six and a revamped version of the Nash Rambler, finally brought to market in late 1957 as the 100-inch-wheelbase 1958 Rambler American.[35]

This decision was an act of pure faith on Romney's part. Small imports might be selling, but the recent history of small American cars offered no encouragement. After the war, a number of the smaller American automakers—Crosley, Kaiser-Frazer, and Nash-Kelvinator, all desperate for a market niche not dominated by the Big Three, tried the small car market and failed. Even Chrysler made a timid stab at this market by shortening the length of its 1954 automobiles a bit, only to see its market share decline from 20 percent to 13.1 percent in a single year. Consequently, as late as 1954, the *Wall Street Journal* was writing the epitaph for the small car in America.[36]

But Romney's intuition and timing were perfect. Larger than the 94-inch-wheelbase Volkswagen but smaller and plainer than the other American cars, the new American Motors Ramblers hit the market just as criticism of the Big Three's cars peaked. Romney has "identified himself with this 'hate American cars' movement, and is reaping the benefits right now," a rival noted sourly. In 1959 Rambler sales shot past Buick, Oldsmobile, and Pontiac and nearly topped Plymouth for third place, trailing only Chevrolet and Ford.[37]

While Detroit's sales slumped after 1955, sales of foreign cars soared, eventually reaching 614,000 in 1959, a more than tenfold increase and an 11 percent share of the American market. Seventy-five to 80 percent of these imports were smaller economy cars. At the same time, sales of American Motors's smaller, plainer Rambler jumped from 37,779 in 1954 to 401,422 in 1959. When these cars sold well through the recession of 1957–1958, the conclusion was inescapable. A significant number of Americans had become small car buyers. With nearly a million vehicle sales between them in 1959 (14 percent of the U.S. passenger car market), the Ramblers and the small foreign cars like Volkswagen and the Renault Dauphine demonstrated beyond question that a new market temporarily beyond the Big Three's control had emerged, a development all the more remarkable because GM and Ford had closely studied the possibilities of just such a market for years.[38]

Ultimately, the industry's executives decided that their slumping sales and the criticism could be reduced to one issue: many Americans now wanted a smaller, plainer car from the Big Three. The growing market share of small imports led by Volkswagen and the obvious success of Romney's small car strategy at last forced the Big Three to introduce the 1960 "compacts." Romney coined this euphemism for small cars with a wheelbase under 110 inches, and the industry adopted it now that it was listening to him. "We're going to run them out," bragged one Big Three executive to the *Wall Street Journal*, as the Ford Falcon, Chevrolet Corvair, and Chrysler (soon to be Plymouth) Valiant all came to market in fall 1959. But the Big Three did not aim squarely at Volkswagen. They instead offered

Open season on Detroit. *Copyright © 1958 by Bill Mauldin.*
Reprinted courtesy of the William Mauldin Estate.
Papers of Bill Mauldin, Library of Congress,
Prints and Photographs Division.

cars slightly larger and slightly more expensive. The Detroit compacts sold for around $2,000 (versus $1,650 for Volkswagen). The Falcon (109.5 inches) and the Valiant (106.5 inches) also had longer wheelbases than the Volkswagen (94.5 inches). But Americans responded positively, especially to Ford's Falcon, buying more than six hundred thousand of them in 1960. By 1961, compacts accounted for 35 percent of automobile production in the United States.[39]

As it transpired, Detroit's commitment to this market went no further than stemming the inroads of the imports, which the compacts did for several years. Paced by more than 150,000 in Volkswagen sales, sales of imported passenger vehicles peaked at 614,000 in 1959 before declining to 339,000 in 1962, with market observers reporting that the novelty or snob appeal of the imports had faded. But Detroit's victory was temporary. Volkswagen, almost as insurance against the entrance of the Big Three into the small car market, began national advertising in 1959, fortuitously choosing a young ad agency, Doyle Dane Bernbach (DDB), to handle its Beetle account. After revolutionizing the U.S. automobile market by opening it to non-U.S. automakers, Volkswagen and DDB revolutionized American advertising, crafting honest, humorous ads inspired by its honest, humorous car. Detroit's compacts flattened Volkswagen's fellow foreign competitors, but not Volkswagen. After falling to 4.8 percent of sales in 1962, the market share for small foreign passenger cars—with Volkswagen still in the vanguard— steadily expanded again, reaching 15 percent of the American market in 1970, when Volkswagen's U.S. sales exceeded half a million.[40]

Apart from entering the small car market, Detroit otherwise ignored the criticisms that gained public currency in the late 1950s. The long-term consequences of this choice proved disastrous. As the Edsel story suggested, the automakers ignored one element of criticism that, though not a significant focus of media criticism at the time, was borne out in persistent feedback from consumers: dissatisfaction with the quality of American-made cars. Poor quality, combined with rising prices, rendered suspect Detroit's claims to be providing consumers with "value." More than any other issue, poor quality ensured that criticism of the American automobile industry remained an ongoing part of public discourse, which in turn prepared the ground for regulation and for a full generation of repeated public embarrassments and painful marketplace adjustments, as the Japanese automakers (who were just testing the U.S. market in the late 1950s) systematically exploited this competitive weakness. Quality was also the natural ally of prestige. When Detroit did not aggressively defend its reputation for selling the best cars in America, it opened the door to the possibility that some day the American market might revolve around makes that were not made in Detroit and that its own cars might occupy the downscale niches in the new order.[41]

The Buyers' Strike of 1958 marked a watershed in America's relationship with the automobile. The criticisms that gained public currency in 1957 and 1958 prepared the way for a generation of industry critics, including critics that later called attention to the environmental harm caused by the automobile. American automakers had been exposed—or at least publicly ridiculed—as clumsy, clownish, Midwestern manipulators pedaling shoddy, expensive goods to pad their bottom lines. As David Halberstam has suggested, GM more or less accepted rather than repudiated this approach—with negative consequences for the future of the industry—when it appointed the dour accountant Frederic

Donner as president in 1958. The mystique that the automakers and their cars epitomized American corporate excellence, carefully cultivated over the preceding generation by cosmopolitan executives like Alfred P. Sloan (who ran GM from New York City), crumbled.[42]

Thereafter, a succession of critics dogged the industry, and they found a sizeable, sympathetic audience for their criticisms. Each critic in turn ensured that the problems of the automobile and the seeming incompetence and perfidy of industry executives were never far from the public's attention. The critics of the 1950s thus scored a long-term victory by winning a permanent place at the ear of American automobile consumers and making public criticism an ongoing part of the discourse on automobiles in America. After 1958 there could be very few Americans who did not understand and share dissatisfactions with American automobiles, at least to some extent. The criticism that came to a head in 1958 took the edge off America's romance with the automobile and ended the broad—at times near-complete—public tolerance for many of its problems. If there was ever a golden age of the automobile in America, it ended then.

The Buyers' Strike of 1958 revealed a less than all-powerful Detroit. Automobile critics with the help of the media successfully influenced what Americans thought about the Big Three's cars just as consumers opened themselves to alternatives. For a brief period in fall 1957 and spring 1958, Detroit lost control of the imagery associated with its cars and its grip on the American market. Automakers, through the millions that they spent on promotions and advertising each year, soon reasserted their dominance in the dialogue with consumers about the imagery associated with automobiles. But consumers in the late 1950s showed that they still had the power to force reluctant automakers to meet their desires. Although consumers could do so without the assistance of critics, together the two made potent allies for change. Bringing them together, however, required the mainstream press, an institution dependent for its existence on the advertising revenues provided by the Big Three and, not surprisingly, one that reflexively withheld criticism until a major problem demanded an explanation. When the press brought the critics' message to the public and the public acted on it as consumers, considerable pressure could be brought to bear on the automakers to change their cars. This same pressure could be brought to bear on legislators, too, who could force the automakers to make changes through regulation. The alignment of critics, press, and public behind change, if it could be created, had potentially positive implications for the automobile's relationship with the environment as America moved into the 1960s, although history thus far offered scant cause for optimism.

8

IF WE CAN PUT A MAN ON THE MOON . . .

Americans in the late 1950s still viewed Los Angeles smog and industrial pollution at plants like the Rouge as local problems, so John Keats and other critics perhaps understandably did not make environmental concerns part of the larger public criticism they directed at American automakers. But they also ignored an obvious larger trend that caused more harm to the environment from automobiles than any other in the 1950s or 1960s—indeed, in the entire second half of the twentieth century. In 1956 the number of registered passenger vehicles in the United States surpassed the number of households. Although 28 percent of families still did not own cars for various reasons, including poverty, this moment effectively marked the one-car per household saturation point and the beginning of a new challenge for Detroit. Ford designer Gordon M. Buehrig summed it up for the American Society of Body Engineers in 1954: "If every family in the country has an automobile, how do you prevent a saturated market? The answer is obvious—you sell every family a second car."[1]

In 1949 only 3 percent of American families had two or more vehicles. In hindsight, it seems inevitable that postwar economic and social developments encouraged multiple car ownership. The growing real wealth and purchasing power of the working and middle classes, the further suburbanization of American living, shopping, and work arrangements, the unwillingness of Americans to patronize or invest in mass transit systems, the increase in average family size and the eventual maturity of the Baby Boomers, and the continuing departure of women from the home for the workplace all could be expected to stimulate the purchase of additional vehicles per household.[2]

Yet those who studied consumers most closely in the years immediately after the war—like Ford and its ad agency, J. Walter Thompson—found little evidence that Americans wanted second cars. On the contrary, until the early 1950s, many families still clamored to buy their first postwar car, which for a sizeable number was the first car that they ever owned. Other consumers used the prosperity of the 1950s to do what Americans normally did with prosperity—trade up to more prestigious makes like Buick and Cadillac. Consequently, automakers such as Ford had to convince consumers that life with a second car was more attractive than life without one.[3]

The decision to go after the second car market was an act of opportunism motivated by Ford's difficult postwar predicament. The company's executives realized that the greater purchasing power that most Americans had recently acquired presented automobile manufacturers an opportunity to sell many more families a second car, but they also knew from the 1944 Roper survey that there was a two-to-one chance that a second car purchase would result in a sale for GM. If the second car idea caught on, it would increase the size of the total market for all the automakers, but it would not necessarily increase the total Ford market share at the expense of archrival GM. The company needed to convince Americans to purchase a second car with greater assurance that the car would be a Ford.

The J. Walter Thompson admen offered a solution, one that turned the company's weakness to advantage: selling station wagons and convertibles as second cars. Both were market niches then dominated by Ford. Beginning with its 1950 advertising, Ford touted the idea of owning two Ford automobiles with the theme: "Two fine cars for the price of one." The two cars shown in the ads were, of course, both Fords, but one was always either a station wagon or a convertible. Every ad suggested that trading in a higher-priced car (read GM) might well cover the down payment on two Fords—cars with the styling, V-8 power, comfort, and all-around value of higher-priced cars. A smiling husband, wife, and children interacted with the cars in some way. The text invariably highlighted the convenience or "freedom" associated with having a second car and invoked a then largely bogus sense of peer pressure by mentioning the hundreds of thousands of families that were already "2-Ford families." Suburban living, family, convenience, and "keeping up with the Joneses" provided the recurring motifs in the second car advertising that Ford ran over the balance of the decade.[4]

Women were a deliberate focus of Ford's second car strategy, as Thompson explained in an April 1956 ad entitled "*Stranded* in Suburbia" that it ran to tout its own services in advertising periodicals. "When the male population leaves Suburbia each workday morning—millions of housewives are left virtually prisoners in their own homes," the ad's text read, "For, while many harried wives drive Dad to the station, 11 million others stand and watch Dad go, taking with him their link to the outside world—the family car. In millions of other homes, however,

4 bedrooms, 3 baths... 2 FORDS

Selling Americans a second car. Ford billboard, 1957. *Ford Advertisements, JWT Domestic Advertisements Collection, John W. Hartman Center for Sales, Advertising and Marketing History, Rare Book, Manuscript and Special Collections Library, Duke University, permission by Ford Motor Company.*

Mother is not cut off from civilization when Dad departs. For the family has a second car in the garage."[5]

The Ford-Thompson strategy of using the station wagon to build the second car market worked because it was congruent with the changing consumer agendas of Americans in the 1950s. The station wagon became the iconic vehicle of the 1950s and the Baby Boom generally. Sales soared from 100,000 to 890,000, or 15 percent of the passenger car market, between 1949 and 1957. Between a third and half of station wagon sales were to families as a second car. By 1959, 15 percent of American families had a second car, a number that grew to 31 percent over the next decade.[6]

While Ford worked to convince Americans that they needed a second car, General Motors had been working to destroy what remained of urban mass transit systems around the United States, which made having a car ever more necessary for people living in urban areas. In order to sell more of its buses GM decided in 1936 to create National City Lines to acquire, motorize, and then resell electric transit systems around the country. Over the next thirteen years, GM converted more than one hundred systems in forty-five cities to its buses. The most famous example of GM's "creative destruction" occurred in Los Angeles. By 1955, 88 percent of the nation's electric streetcar network was gone. In 1974 U.S. Senate general counsel Bradford C. Snell caused a sensation when he told this story to the Senate Judiciary Committee. "The effect of General Motors' diversification program," he wrote, "was threefold: the substitution of buses for

Stranded
in Suburbia...

When suburban fathers go off to work—
as suburban fathers must—their absence
is felt in more ways than one

When the male population empties out of Suburbia
each workday morning—millions of housewives are left
virtually prisoners in their own homes.

For, while many harried wives drive Dad to the station, 11 million others stand and watch Dad go, taking
with him their link with the outside world—the family car.

In millions of *other* households, however, Mother is
not cut off from civilization when Dad departs. For the
family has a *second* car in the garage.

How did it get there?

How did millions of American families get the idea
they could afford a second car?

It didn't just occur to them. Few American families
automatically think of themselves as two-car families.
The notion had to be *planted* there.

And there was nothing haphazard about the planting.

Ford anticipated a growing <u>need</u> in the postwar migra-
tion to the suburbs. By taking advantage of this need,
Ford hastened the development of the "two-car family."

Result? Well over a quarter-of-a-million families in

In that *left-behind* feeling, Ford finds
a billion-dollar opportunity

the United States are "Two-Ford Families." More than a
million other two-car families include a Ford.

A vast new market has been opened for the future—
well ahead of its time.

• • •

The expanding *needs* of "Suburbia" extend to virtually *every* product. Indeed, people everywhere have gained
the *ability* to live better. *Creative marketing* can cut down
the time it would normally take these people to raise their
living standards to the level they can now afford.

J. WALTER THOMPSON COMPANY

New York, Chicago, Detroit, San Francisco, Los Angeles, Washington, D. C., Miami,
Montreal, Toronto, Mexico City, Buenos Aires, Montevideo, São Paulo, Rio de
Janeiro, Santiago (Chile), London, Paris, Antwerp, Frankfurt, Milan, Johannesburg,
Cape Town, Bombay, Calcutta, New Delhi, Sydney, Melbourne, Tokyo, Manila

FORTUNE May 1956 45

Creating discontent. J. Walter Thompson bragging about
helping Ford sell Americans second cars. *JWT Advertisements,
JWT Domestic Advertisements Collection, John W. Hartman
Center for Sales, Advertising and Marketing History,
Rare Book, Manuscript and Special Collections Library,
Duke University, permission by JWT.*

passenger trains, streetcars and trolley buses; monopolization of bus production;
and diversion of riders to automobiles." His tale rapidly passed into American
folklore as the principal reason for America's automobile dependence. There is
no question that GM worked hard and successfully to destroy electric streetcar
systems, but this perspective needs to be tempered by the recognition that the
streetcar systems were unpopular with the riding public and in long-term decline before GM helped the process along. If not yet terminally ill, the streetcars

were in serious trouble, a point later strongly made by historian Scott L. Bottles in *Los Angeles and the Automobile* (1987). Urban mass transit worked for millions, especially as routine transportation between home and workplace, but it simply could not compete with the multifaceted allure of the automobile. As an early automobile advertisement had aptly put it, "Why be part of the ten-cent common herd?"[7]

While recognizing that Ford and General Motors aggressively pushed their product, it is also important to remember that in the new consumer agendas of the 1950s the essential "goodness" of automobile ownership went unquestioned. Automobile ownership was still the blessing that both best symbolized and facilitated much of what Americans considered the "good life." In the generation after 1945 Americans began to define full personhood as requiring ready individual access to an automobile. Naturally, women at home in the suburbs, women who worked outside the home, and the children of the wealthiest nation in history must all have cars. This assumption, which Americans embraced during the 1950s without quite realizing it, lay at the core of the consumer agendas of nearly all Americans. The automakers began to realize that new vistas of market segmentation by taste, interest, and lifestyle were possible if they were selling cars to individuals and not families. Even the automobile's critics shared the assumption that all Americans were entitled to automobiles. The problem in their view was, not this expanding sense of entitlement, but the kind of automobiles that Detroit offered and Americans purchased.[8]

The wisdom of this assumption soon came under scrutiny. Critics recognized that the growth in the total number of cars on the road and the fact that Americans drove these cars more miles than in the past multiplied the automobile's environmental impact. Passenger car registrations grew by 139 percent from 25.8 million in 1945 to 61.7 million in 1960. Over the same period the number of annual vehicle miles traveled by passenger cars increased by 194 percent from 200 billion to 587 billion, as Americans used their cars, the growing Interstate Highway System that Congress authorized in 1956, and suburbanization to spread out. Car design and driving trends came together in the amount of gasoline consumed, which increased by 210 percent from 13.3 billion gallons in 1945 to 41.2 billion gallons in 1960, as the fuel economy of the typical American passenger car fell 5 percent from 15.0 miles per gallon to 14.3 miles per gallon.[9]

As Americans entered the 1960s, these trends made the environmental impact from all four phases of the automobile life cycle far more evident. So, too, did journalists in the mainstream media who articulated and amplified these concerns. A growing national perception that the nation was failing to cope with the disposal of junked automobiles, industrial pollution, and smog, among other issues, coalesced in the late 1960s and early 1970s to create the modern environmental movement. When this national concern reached a Congress controlled by the Democratic Party that was predisposed to use the powers of the federal government to address national problems, one of the most consequential episodes of reform in American history resulted.

More cars on the road eventually meant more cars reaching the end of their lives. The bigger cars that Detroit sold in the 1950s also meant more car to be recycled. Deliberate stylistic obsolescence, increased driving, and quality problems encouraged Americans to get rid of their cars sooner. Even the shift in consumer tastes and the criticism that made Detroit's standard fare seem so unsatisfactory in 1958, an otherwise positive development in the long run, actually accelerated the process of obsolescence in the short run by encouraging Americans to replace their large vehicles with compacts. Unfortunately, these developments coincided with a change in the steel and scrap industries to produce one of America's great environmental blights of the 1960s.

The postwar afterlife of American automobiles remained in the hands of the two parties who had come to the business in the late 1910s and 1920s, some eight thousand to thirty thousand junkyard owners and four thousand scrap dealers. Not everything that the junkyard owners did was environmentally friendly. They usually burned the automobile hulks to remove as much of the remaining non-metallic materials as possible, a source of air pollution. Still, the work of junkmen and scrap dealers remained the most significant step to reduce the environmental impact associated with the automobile's lifecycle before the 1960s. For their efforts they received little positive recognition and instead were dealt a series of blows by steelmakers, automakers, and government regulators that threatened their livelihoods and the process of automobile recycling.[10]

The postwar scrap metal market peaked in 1956, when dealers sold forty-one million tons of scrap to domestic and foreign steelmakers for three billion dollars. The market then collapsed, as the steelmakers began to change their furnace technology from open-hearth furnaces to basic oxygen furnaces. Where open-hearth furnaces took a charge that was about 50 percent scrap metal, the new basic oxygen furnaces used just 20–25 percent scrap. This change alone effectively halved the steel industry's demand for scrap metal. Indeed, the price of a one-ton No. 2 bale of automobile scrap fell from $42.86 in 1956 to $19.84 in 1963, its lowest level since World War II.[11]

The crisis in the scrap metal industry soon became obvious to Americans, as the changed economics of automobile recycling rippled all the way back through the recycling chain to the consumer. Scrap from obsolete equipment like auto-

Abandoned automobile, Bronx, New York, 1964.
Photo by Phil Stanziola. Library of Congress,
Prints and Photographs Division.

mobiles constituted about 50 percent of all scrap purchased by the steel industry, but due to impurities, it was the least desirable. When steelmakers began
substantially to reduce their overall demand for scrap, the market for purchased
obsolescent scrap practically vanished; scrap dealers stopped buying hulks from
automobile junkyards; and junkyard operators, in turn, stopped buying junkers
from dealers and consumers. When the junkmen stopped paying and towing,
consumers lost an important incentive for channeling their vehicles into the
recycling system at the end of their cars' useful lives. Overnight, the number of
abandoned cars in the United States exploded, creating problems that plagued
both urban and rural America. In New York City, the number of abandoned automobiles quintupled between 1960 and 1963 and then quintupled again between

1964 and 1969, growing from twenty-five hundred in 1960 to seventy thousand in 1969. In Detroit, the number grew from two thousand in 1964 to thirteen thousand in 1966. The problem overwhelmed the responsible municipal agencies. By 1966, an estimated thirty million abandoned cars littered the landscape in America. Complaints swamped local officials. Abandoned cars provided a prominent visual index to what seemed to be the deteriorating quality of urban life.[12]

The cars removed from the streets, vacant lots, and fields had to be taken someplace, and that place was mostly to automobile graveyards owned by the auto wreckers. If the car had been "stripped" of its valuable remaining parts while abandoned, it was worthless to the junkman. All he could do was add it to his growing inventory of automobile hulks, which now often sprawled across acres. According to one estimate, the cars abandoned in the United States in 1967 covered thirty thousand acres—forty-seven square miles. Complaints soon focused on the automobile graveyards as well, whose owners were stuck with the growing number of junked cars.[13]

Americans viewed abandoned automobiles and automobile graveyards, not as environmental or economic problems, but as visual blights. The problem became much more obvious to Americans, not just because the number of junkers rapidly accumulated, but because the postwar suburbanization of the United States made possible by the automobile and the growing Interstate Highway System also brought people to live, work, and shop in areas outside cities where the junkyards had been banished in the 1920s. The junkyards clashed with the visual aesthetic of the upscale suburban lifestyles to which postwar Americans aspired. Perhaps for some suburbanites they were also too obvious a reminder of the excesses of an affluent society.[14]

People tackled the problem in a manner consistent with their perception of it. They tried to make the visual nuisance disappear. In this spirit the federal government took a stab at the problem in 1965 with the Highway Beautification Act, legislation that was also motivated by the proliferation of billboards. The bill was a special interest for Lady Bird Johnson. Lady Bird, who spent a good deal of time in automobiles traveling around the country before her husband became president, had noticed and deplored the overflowing automobile junkyards. Under the new law, all 17,500 junkyards in the United States located within a thousand feet of interstate right-of-way were to be screened by 1 July 1970 or removed. The federal government provided the states with 75 percent of the financing needed for screening or fencing junkyards. But the law was weak, and changes were slow in coming. In 1979 the U.S. Department of Transportation reported nearly eleven thousand junkyards that were not in compliance.[15]

Junkers piling up, 1957. *Library of Congress,*
Prints and Photographs Division.

The American scrap metal industry had its own problem—the survival of
its industry. After 1956 the challenge was how to produce smaller-sized, readily
meltable, comparatively pure scrap that could be used in the new basic oxygen
furnaces of the steel industry without recourse to intensive (and increasingly expensive unionized) labor that drove up the cost of scrap and made it uncompeti-
tive with virgin iron ore. Israel Proler and his three brothers tackled this problem
in 1958. The Prolers ran the Proler Steel Corporation in Houston. Like other
scrap dealers, they produced and sold baled scrap to steelmakers. In fact, the
Prolers had purchased the huge baler that Ford used at the Rouge to crush automobiles at the end of the disassembly line. Following an approach pioneered by
New Mexico scrap dealer Alton Newell in the late 1930s, they built a very large,

powerful machine that literally pounded automobile hulks into smithereens, the Proler automobile shredder, or Prolerizer.[16]

The Prolerizer was a machine the size of a small factory, three stories high and occupying an area slightly larger than a football field. Operators fed junked automobiles, either baled or flattened, into the machine by a conveyor belt, and a battery of rotating eight-hundred-pound hammers moving at a speed of four hundred feet per second, or about half the speed of a moving bullet, pounded the hulks to pieces. Within two minutes, a hulk was reduced to chunks of fist-sized material that were then run past magnets that separated the ferrous scrap from the rest of the automobile debris. At the end of this process, the Prolers had what they wanted: low-cost chunks of iron and steel that could be sold to the steel industry as a 90 percent-pure, readily meltable raw material for furnaces. This shredded automobile scrap marked a substantial improvement over the older No. 2 bundle of automotive scrap that was typically 65–80 percent ferrous metal.[17]

The Proler brothers built the first of their machines in Houston. By 1964, they also had similar machines operating in Chicago, Kansas City, and on Terminal Island off Long Beach, California. The operation on Terminal Island, a joint venture with metal importer-exporter Hugo Neu Corporation, cost two million dollars to build but could process fourteen hundred automobiles a day. Instead of $18.64 for a ton of baled scrap, a ton of Proler's fragmentized scrap commanded $33. Proler quickly had competition from the largest scrap metal firm in the United States, Luria Brothers of Philadelphia. Luria developed its own automobile shredder and in 1964 set up a shredding operation at Vernon, California. Although they did not immediately resuscitate the industry, the automobile shredders pioneered by Proler and Luria gave automobile scrap metal dealers a new lease on life — at least those companies large enough to afford the expensive new technology.[18]

The American steel industry did not initially embrace the new form of scrap, still finding it too expensive relative to virgin ore due to railroad rates that continued to penalize the transportation of scrap. Instead, Luria sold the output of its Vernon operation to its next-door neighbor, Bethlehem Steel, and Proler sold its Terminal Island scrap to Japanese steel mills, where the scrap could be sent by water rather than railroad. In fact, exports of fragmentized automobile scrap to both Japanese and European steelmakers during the 1960s permitted the American scrap metal industry to survive during this period. Ironically, much of this exported scrap returned to the United States in the form of Japanese and European automobiles.[19]

Automobile shredder. *Courtesy ISRI.*

Ford, the lone steelmaker among U.S. automakers, was not so reticent about trying the new scrap and agreed with Luria Brothers to purchase the entire output from two new shredder operations that Luria built outside Detroit and near Cleveland in order to provide scrap for Ford's Rouge and Cleveland foundries. The thirty-acre Taylor Township, Michigan, plant that opened in July 1967 cost three and a half million dollars and had the capacity to process a thousand cars a day or a quarter million junked automobiles a year. Luria promised Ford pellets that were 99 percent steel. By 1973, Ford purchased half a million tons of recycled automobile scrap a year. Scrap metal gave Ford the raw material to manufacture 93 percent of its ferrous castings and 28 percent of its steel.[20]

The prospects for the American scrap metal industry finally brightened at the end of the 1960s. In 1968 the industry dismantled more cars than were abandoned. At the same time, smaller steel mills—the so-called mini-mills—began to employ electric arc furnaces that could use 100 percent scrap charges. By the end of the 1970s, the automobile recycling industry in the United States had assumed the form that it retained for the balance of the century, dominated by large firms that owned the two hundred automobile shredder operations distributed around the major population centers of the country. About 90 percent of deregistered vehicles each year were recycled for their scrap metal content, and 60 percent of all deregistered vehicles were processed by the large auto shredders. Studies indicated that recyclers successfully recovered 85 percent of the ferrous metal available in deregistered cars. It was a long, bumpy road from the fateful intersection of postwar stylistic obsolescence and the collapse of the scrap metal market in the late 1950s to this new recycling regime, but the large scrap companies that pioneered automobile shredding offer an environmental success story—abandoned cars excepted—achieved without direct government regulation and largely outside public view.[21]

The booming economy and record auto sales of the 1960s also made life much harder for the men responsible for curtailing industrial pollution at automobile manufacturing sites like the Rouge. By 1965, sportsmen, journalists, Dearborn residents, and regulators concluded that Ford's efforts had failed to keep pace with the pollution generated by the company's growing postwar output. Even Ford conceded the point. Problems began again in summer 1959, when pollution shut down Sterling State Park, a popular picnic and swimming destination for Detroiters on the Michigan shores of Lake Erie about thirteen miles below the mouth of the Detroit River and thirty miles south of Detroit. Then in March and April 1960, oil in the lower Detroit River killed some to ten thousand to twelve thousand wild ducks. The next winter upwards of twenty thousand ducks were killed on the same stretch of water by a release from Detroit's sewage treatment plant. These events, which received wide publicity in the Detroit area, convinced the public and many elected officials that Michigan's Water Resources Commission (WRC) was not doing enough to combat water pollution, a view the commission's staff vigorously contested.[22]

Matters came to a head in late fall 1961 after Congress amended the federal Water Pollution Control Act. Thanks to the efforts of Dearborn's Democratic congressman John D. Dingell, the amendments created a mechanism that for the first time permitted the federal government to intervene in a pollution problem affecting the waters within a state. Over the objections of WRC staff, but with the support of Michigan's two Democratic senators, Philip A. Hart and Patrick Mc-Namara, Michigan's Democratic governor John B. Swainson requested federal assistance on 6 December 1961. The Public Health Service (PHS) then conducted a three-year study of water pollution in the Detroit River watershed, the first comprehensive study since the IJC-sponsored survey in the late 1940s.[23]

The PHS report, completed in June 1965, identified the Rouge as the largest source of water pollution in the Detroit area. Analyzing discharges from thirty-seven industrial plants in the region, the investigators found that the Rouge was either the first or second greatest source for ten of the eighteen pollutants measured. Despite nearly two decades of efforts by Ford engineers, problems persisted with the phenols, cyanides, and ammonia from the company's coke operations, iron dust from the blast furnaces, and acid from the pickling liquor used in the steel rolling mills. The report broke new ground in identifying Ford's substantial discharges of toxic heavy metals such as cadmium, chromium, nickel, copper, and zinc to the Detroit River, reflecting a growing concern with these substances. With the incidents of 1959–1961 in mind, the investigators focused far more attention on the threat that pollution posed to wildlife and human recreational pursuits than the IJC report of 1951, which had stressed the threat to human health from tainted drinking water.[24]

Ford refused to accept the report's findings and recommendations but carefully refrained from saying so publicly. For the better part of a year, the company balked at committing to a program and timetable to fix the problems that the WRC and PHS jointly demanded. But the company also faced overwhelming public pressure from the Michigan United Conservation Clubs, the League of Women Voters, and the Michigan State Chamber of Commerce. Sportsmen remained adamant. Thanks in part to the advocacy of recreation director Olga Madar, United Auto Workers (UAW) president Walter P. Reuther and Rouge UAW Local 600 both also urged Michigan's Republican governor George W. Romney to step up efforts against water pollution. The ever-maverick Romney, who had left American Motors on election to the governorship in 1962 and who had his sights on national office, needed no convincing when it came to pollution. Faced with the prospect of highly negative public controversy and possible prosecution in the midst of rising national concern about industrial pollution, Ford relented.[25]

Over the next four years, Ford's engineers dealt with a number of the regulators' concerns. The company converted its remaining two steel pickling lines from sulfuric to hydrochloric acid with the wastes returned to Dow Chemical for recycling. Engineers redesigned the Rouge glass production facilities to eliminate sand and polishing rouge from its liquid wastes in June 1967 and routed the coke oven phenol wastes to Detroit's municipal waste treatment plant in January 1969. Ford naturally touted these steps, but not without public contradiction. In December 1970 the WRC reported a 42 percent decrease in the volume of industrial waste discharges to the Detroit River since 1965. But water problems remained, especially from toxic heavy metals.[26]

Dearborn residents also renewed their opposition to air pollution at the Rouge in 1965 after Ford announced plans for a thirty-million-dollar foundry expansion. Residents in Dearborn's South End pressed Dearborn mayor Orville L. Hubbard, the Dearborn City Council, and Ford officials to do more about the plant's pollution. The South End was an ethnically diverse, white, blue-collar neighborhood of about five thousand. Most of the men worked for Ford at the Rouge or for nearby Great Lakes Steel. Henry Ford had recruited these people and their parents to work in his plants: 40 percent from southern and eastern European backgrounds, 40 percent Arab, the remainder mostly migrants from Appalachia. The only Dearborn neighborhood east of and thus downwind from the Rouge, the South End was literally sandwiched between the Ford complex on the west and a series of cemeteries on the east.[27]

Soot rained down on the neighborhood from the Rouge powerhouse, foundry, and steel mill. Hanging laundry out to dry was impossible, and people could not open their storm windows in the summer without coal soot piling up on their windowsills. The streets smelled like rotten eggs—hydrogen sulfide—when trucks hauled steaming slag from the Rouge through the neighborhood to the nearby Levy Slag Company. Fly ash from the Rouge pitted the finish of cars and ruined the paint on homes. The Salina School had to seal its windows and install a special air filtration system to keep out the soot. Children had to go to the doctor to have steel particles removed from their eyes. The anger was widespread—perhaps close to unanimous. "You couldn't keep your house clean," recalled Paul J. Hardwick, the minister at the Neighborhood Baptist Church, "You couldn't keep your kids healthy. . . . We couldn't send our kids to the playground." Residents launched a letter writing campaign to senior executives at Ford.[28]

Dearborn city councilman George Z. Hart, who excoriated Ford representatives before the city council in November and December 1965, took up the community's cause first. Hart called the Rouge "a monster—a necessary monster." "We need your tax money," he conceded, "but our people in the south end are paying a costly price in health." The new pastor of Saint Bernadette's Roman Catholic Church, Father Albert U. Lombardi, and John Parkhurst both ably backed Hart. Lombardi said the sky in his neighborhood in the morning looked like fallout from an atomic bomb explosion. Parkhurst, in a pointed dig, charged that Ford could have solved the air pollution problem at the Rouge if it had spent 1 percent of the money it had lost on the Edsel.[29]

Complaints from the community continued through 1966. Hardwick, with his wife, Linda Mae, and Vincent Bruno, who was president of the Southeast Dearborn Community Council as well as an employee in the Rouge foundry, joined Hart, Parkhurst, and Lombardi in their efforts. But the community council made little headway with either Ford or the city. "We talk to Ford's quite frequently about it (air pollution) without carrying a big stick and hitting them over the head," Orville Hubbard told the council at its annual dinner on 15 May 1967. "They are spending a lot of money on it." Ford provided Dearborn with more than half its tax revenues, and Hubbard was not inclined to provoke the source of the revenue that made it easier for him to do his job as mayor. Nor did Hubbard have much patience for constituents who did not see the wisdom in this order of things. "He was the kind of institution," Hardwick recalled, "that could teach old Richard Daley tricks for Chicago." "It's not a smart political thing to say," Hubbard finally conceded to the group, "but it's a hopeless situation." Either the Rouge had to go or the people had to abandon the neighborhood. The mayor urged his audience to accept the long-term solution that the city had already launched, an urban renewal project where the city slowly bought out and converted the neighborhood from a residential to an industrial section. Residents rejected that plan. As they educated themselves about pollution, they decided that Ford could do more and that the city could do more to force Ford to do more. People "were concerned [that] nothing [was] happening with the pollution that [was] drifting over the neighborhood," Paul Hardwick remembered. "They would get these letters back two or three months later from [Ford] public relations that said, 'Yea, we're working on it.' And after that happened two or three times . . . you know there's nothing happening."[30]

Frustrated, the Southeast Dearborn Community Council decided to enter a protest float in Dearborn's 1967 Memorial Day parade. The activists recreated a miniature version of the famous eight smokestacks of the Rouge Power House No. 1 on the roof of a van. They even used smoke bombs to make the stacks belch smoke. The van pulled a flatbed trailer on which children with gas masks hung laundry on clotheslines. The front of the van bore a sign: "Greeting. From the Valley of Death." An estimated forty thousand people turned out for the parade on 30 May 1967. The float made it about halfway through the parade, but when it was within a block or so of the reviewing stand occupied by Hubbard and Congressman Dingell, someone gave Hubbard the heads-up, and he ordered the police to remove the float from the parade. They did so, turning it on to a side street. While Bruno and Lombardi argued with city officials, a photographer from the Associated Press took a picture of the float. The next day that photograph ran in newspapers around the United States.[31]

Several weeks after this negative national publicity, Ford spokesmen announced that the company had spent or was planning to spend sixty-seven million dollars for air and water pollution control at the Rouge. Community leaders also received a phone call from the office of the number-three man at Ford, executive vice president Charles H. Patterson, inviting them to the executive offices at the Rouge for a luncheon presentation, discussion, and tour in early August 1967. Never before had so senior a Ford official faced the public to answer complaints about pollution from the Rouge. Ford also invited city officials and the local press to what it clearly intended as an orchestrated event to tamp down protest and tout a new round of planned pollution controls.[32]

The meeting did not go as Ford hoped. "We've seen no evidence of cleaner air in our area," Vince Bruno complained. "The only thing we have to go by is visible evidence," Father Lombardi explained. "You tell us about pollution control measures," Linda Mae Hardwick, then eight months pregnant, added, "but we can't see any difference. We still see the dirt, and smell it, and our children have to play in it. . . . We're embarrassed when relatives come to town." After lunch had been cleared from the marble-topped conference table, Lombardi reached into his jacket pocket and pulled out a tobacco tin. He opened it and dumped the contents, black soot, on the table. "My parishioner gave this to me. It is his job between one and five in the morning to turn on the fan that blows this out of the smokestack at the foundry."

Patterson lost his temper. "Don't tell me that we haven't done anything," he snapped. "Don't doubt our integrity. The company from Henry Ford II on down is dedicated to meeting its civic responsibility." But Patterson was forced to admit that Ford was struggling to control emissions from its smokestacks and that it had no comprehensive control program under the direction of a single person. Paul Hardwick urged this step. Lombardi, turning to Mort Sterling, director of the Wayne County Health Department's newly formed air pollution control bureau, said, "From now on, we want Mr. Sterling and the law to take over." People around the country were reaching much the same conclusion as Father Lombardi about their local industrial polluters.

Ford could see it coming. The August 1967 meeting and the public commitment to do more about air pollution were attempts to meet and defuse the mounting anger head-on. The company's stepped-up commitment was real. Ford acted on Paul Hardwick's suggestion by appointing Frank Kallin head of its worldwide pollution control efforts in May 1968 and announced plans to spend eighty million dollars worldwide on pollution over the next four years. Early in 1969 Ford's pollution control people reported to Dearborn officials that work to install new electrostatic dust collector systems on all seven boilers in Power House No. 1 had been completed but that they were still having problems with dust collection in the foundry. In March, Ford threw in the towel and announced that it was closing the production foundry at the Rouge in order to build a new state-of-the-art foundry in Flat Rock, Michigan. These steps by Ford to regain control over its pollution problems culminated in a major statement by Henry Ford II in December 1969 committing his company to a leadership role in the auto industry in combating pollution from both its plants and its vehicles.[33]

Father Lombardi also got his wish. In 1968 Wayne County created a centralized, countywide air pollution control division under the direction of Sterling to supersede municipal air pollution control activities like Dearborn's. Sterling was soon backed by stricter laws and regulation that came from county, state, and federal governments over the next forty months. Ford did not substantially reduce air pollution from the Rouge overnight. In fact, it took the better part of two decades and a lot of court time. But a corner had been turned, and by 1975, real progress had been made. "Those little kids on the back of that Memorial Day float," Hardwick noted, recalling how Ford's behavior seemed to change after the parade protest, "had way more power than they ever guessed."[34]

With the number of cars on their roads increasing, Southern Californians in the second half of the 1950s waited for their technological fix from Detroit with mounting frustration. Bad smog attacks in 1953, 1955, and 1958 created a sense of urgency about the problem. Despite the occasional optimistic report, 1958 came and went without an emission control device. For Kenneth Hahn, the Los Angeles Air Pollution Control District, and Angelenos generally, Detroit's behavior was exasperating. "I have gained the impression," Hahn noted sarcastically in 1958, borrowing language then current among the industry's critics, "that . . . air pollution control is not as important as new styling, grille design and more horsepower."[35]

Disappointing Hahn was a mistake. When the solution that seemed imminent in 1954 and 1955 did not soon arrive, he decided that the companies were holding back the technology needed to address Southern California's smog problem and the replies that he received to the pleading letters that he sent to the companies year after year, in which spokesmen explained the difficulty of solving the problem, were essentially lies. He sensed correctly that the automakers were not

Traffic on Pasadena Freeway, 1958. *Photo by*
Los Angeles County Air Pollution Control District.
Courtesy National Archives.

pushing hard enough to make ideas like Houdry's catalytic converter a reality. The idea that he was being lied to angered him. As a result, Detroit acquired a persistent foe.

On 4 December 1959, after several years of pressure from Los Angeles officials, California legislators passed the Motor Vehicle Pollution Control Act in an effort to increase pressure on the automakers to introduce a control device that treated the major source of smog-causing emissions, engine exhaust. The act established the California Motor Vehicle Pollution Control Board (MVPCB) and charged it with evaluating and certifying exhaust control devices for installation on California vehicles. California governor Edmund G. Brown, Sr., named Arie J. Haagen-Smit and twelve others to the first board. One year after the board had certified a minimum of two devices, all new cars sold in California would be required to have them. The legislation provided an incentive for inventors, since

Waiting for Detroit. Smog attack, Los Angeles, 1958.
Photo by Art Worden. HERALD EXAMINER COLLECTION /
Los Angeles Public Library.

it guaranteed them a piece of a statewide market estimated at seven hundred million dollars. Some thirty groups—nearly all of them nonautomotive manufacturers—soon filed applications with the board to submit devices for evaluation.[36]

Detroit was working on the problem. In January 1957 GM launched a joint engineering and testing program with Houdry's company Oxy-Catalyst to develop a lead-tolerant catalytic converter. GM's program was no public relations sop. It did want a catalytic converter that tolerated lead. In fact, the next year, Ethyl, still 50 percent owned by GM, asked U.S. Surgeon General Leroy Burney to bless a 33 percent increase in the amount of tetraethyl lead in gasoline to boost octane further and better accommodate the more powerful V-8 engines that the automakers had introduced during the 1950s. Burney, who lacked regulatory authority to block the request, nonetheless convened a committee that recommended granting the request, but only subject to agreement by the producers to cooperate in a new study of the health effects of atmospheric lead with the Public Health Service. On 1 October 1961 Oxy-Catalyst ran an ad in the *New York Times* touting the joint research and the fact that the Californians had chosen its catalytic device as the first for full field-testing. In March 1962 GM engineers and Houdry reported that they had made and tested a "functionally and structurally satisfactory" catalytic converter based on technology licensed from Oxy-Catalyst that met the California emissions standards but that deteriorated from lead poisoning well short of the minimum twelve-thousand-mile life (using leaded gasoline) required by California law. At the same time, Houdry announced a more expensive base-metal device that met the twelve-thousand-mile requirement. In June 1962 he was awarded a patent on his catalytic converter (without specifying the type of catalyst). Unfortunately, he died three weeks later with his device still under review by the Californians.[37]

However, GM apparently concluded, as it had nearly a decade earlier, that the fundamental problems with commercializing catalyst technology remained. With lead still in nearly all the gasoline sold in the United States, platinum devices would have to be replaced every three thousand to six thousand miles. Putting such devices on millions of cars and then replacing them this often would probably outstrip the available supply of platinum, making this already expensive metal far more costly, unless a major new supply could be tapped. GM continued to be constrained by its relationship with Ethyl. Although the company sold its stake to the Albemarle Paper Manufacturing Company on 30 November 1962, GM could not take an action like making cars that required unleaded fuel that would have destroyed the value of Ethyl to its new owners so soon after the sale.[38]

But perhaps the most important reason why GM held back on further development of the catalytic converter at this point was the dominance within GM Research of mechanical engineers. "The main engineering brainpower in GM was dead-set against catalysts," remembered GM's Richard Klimisch. "They were just bound and determined to do this mechanically. They were all mechanical engineers and the idea of putting this chemical thing on cars was just abhorrent to them. They didn't want any part of it." GM's decentralized divisional structure posed another significant internal barrier to moving forward. Even if the engineers at GM Research had been favorably disposed toward catalytic converters, they could not force a GM car division to put the device on one of its cars. Only a car division executive or a very senior corporate executive above him could make that decision. The clamor coming from California did not provide enough incentive for a division executive to run the risk of putting a new device on one of his division's cars first. And no champion stepped forward at the top of GM either.[39]

On 17 June 1964, the California MVPCB certified four exhaust control devices (three catalytic converters and one thermal afterburner) from third-party vendors only days after Automobile Manufacturers Association spokesman George A. Delaney had reiterated the industry's contention that no such devices could be expected from the Big Three before the 1967 model year (that is, fall 1966). The certification of the third-party devices was a bluff by the board to force a response from the automakers, since the MVPCB members and staff knew that the certified devices required replacement short of the twelve-thousand-mile stipulation.[40]

Faced with the prospect of installing devices made by third parties on their own cars in order to sell them in California, the automakers formally announced on 12 August 1964 that they could introduce exhaust controls of their own in time for the 1966 model year, ending all chance of the vendors selling their devices. Over the protests of the third-party device makers, automakers sought and got a change in the certification procedure from the California legislature to accommodate their suggested approach. The automakers did not offer a catalytic converter; rather they made modifications to the engine, including changes to the fuel induction systems, the combustion chamber, and engine timing. It was the mechanical engineers' preferred solution. As it transpired, this approach was not very effective either.[41]

The automakers' abrupt about-face incensed Hahn and Los Angeles APCD head S. Smith Griswold. If the companies had a solution, why wasn't it already installed on 1964 model vehicles or scheduled to appear on the 1965 models? Why had Detroit's offering been forthcoming only after competitors from outside the industry threatened to enter the field? The presumptive answers to these questions seemed to prove what Hahn had believed since 1955: the industry was holding back technology. Already fed up, Griswold ripped the industry in a June 1964 speech before the Air Pollution Control Association in Houston, Texas. He called for federal legislation to require the installation of emission control devices on all cars.[42]

Shown a copy of Griswold's Houston speech by a minor official at the Public Health Service, the not-yet-famous Ralph Nader called Griswold and discussed the problem. Griswold's claims struck Nader as outlining the basis of an antitrust action, not the traditional price-fixing collusion that harmed the public by keeping prices higher than they would have been if there had been competition, but a kind of collusive "product-fixing" that harmed the public's health or safety to the degree that the products fell short of what was technologically feasible. Nader's idea suggested an important new front in antitrust law with wide application, and the Antitrust Division at the Justice Department decided to look into it.[43]

On 19 January 1965 the Justice Department issued subpoenas to the four automakers and the AMA, seeking documents related to the industry's emission control efforts. The department soon sent Samuel Flatow to Los Angeles to investigate the matter. An eighteen-month investigation led to the impaneling of a Los Angeles federal grand jury on 21 June 1966 to hear the evidence. In December 1967 Flatow, in a sixty-four-page memorandum that summarized the evidence presented to the grand jury, asked his superior Donald F. Turner, head of the Antitrust Division, for permission to seek a criminal indictment against the automakers and the AMA.[44]

Documents from the AMA and the automakers as well as testimony before the grand jury showed plainly, among other things, that GM had held back introduction of the positive crankcase ventilation (PCV) valve (that rerouted "blow-by" emissions from the crankcase back to the engine intake manifold and the combustion process, capturing about 20 percent of the unburned hydrocarbons otherwise emitted) until the 1961 model year in California; that the industry had conspired to delay the valve's national introduction for an additional two years; and that the approaches that the Big Three announced in 1964 to counter the certification of third-party devices and meet the California emission standards had been developed in the case of Chrysler's Cleaner Air Package as early as 1960 and GM's ManAirOx-Manifold as early as March 1962 but by agreement were not to be introduced until the 1967 model year. Turner nonetheless rejected Flatow's request. Outraged, the foreman of the Los Angeles grand jury offered to indict the automakers without Justice Department permission, but Flatow counseled the grand jurors to accept the department's decision.[45]

Turner and others later argued that it was longstanding Justice Department policy to reserve criminal actions, always a rare step, only for cases with overwhelming evidence of price-fixing. Nader's "product-fixing" argument and Flatow's evidence failed to convince Turner. Flatow resigned from the department shortly thereafter and died in 1969, reportedly dejected that his hard work on the case had come to naught. Turner may also have decided not to proceed with a criminal indictment because he was considering a much more important antitrust case against GM—one that would have asked for the breakup of the company. In any case, Turner and Edwin Zimmerman, who succeeded him in 1968, sat on the case, mulling civil charges, while Hahn pestered them to bring an action.[46]

Democrats in Washington were not eager to take on the auto industry during the 1968 election year and waited until 10 January 1969, ten days before the end of the Johnson administration, before filing a civil suit against the AMA and the major automakers as a parting present for the incoming Republican administration of Richard M. Nixon. The suit asked for an injunction against an unlawful conspiracy rather than the maximum penalty of triple damages allowed under the Sherman Antitrust Act. Fending off vigorous efforts by California attorney general Thomas Lynch to obtain access to the grand jury evidence for use in a suit on behalf of the state against the automakers, the new Republican Justice Department under John Mitchell quickly began negotiations with the AMA's Washington attorney Lloyd N. Cutler to settle the case out of court. In the resulting consent decree, announced on 11 September 1969, the automakers — without admitting guilt — simply promised not to conspire to withhold emission control technology in the future. In return the Justice Department agreed to seal permanently all evidence taken in the case. The automakers wanted the evidence sealed, not just to spare them a public relations disaster, but also to ensure that California and other potential plaintiffs never had access to it.[47]

The consent decree provoked a firestorm of protest from members of Congress and people contemplating lawsuits of their own, which included state and local governments, environmental organizations, and individuals around the country. In October 1969, Federal District Court Judge Jesse W. Curtis, a Los Angeles Democrat, explaining that legal precedent prevented him from altering the terms arranged by the parties, allowed the consent decree to stand. In March 1970 the U.S. Supreme Court unanimously upheld the decree. Twenty-eight states eventually filed their own lawsuits, charging the automakers with conspiracy and asking that the automakers be forced to take steps to address emissions. In November 1973 Federal District Court Judge Manuel L. Real dismissed these suits, ruling that the states could not use the antitrust laws to compel the automakers to find a solution for air pollution.[48]

John D. Caplan, head of GM's Fuels and Lubricants Department and a chairman of the AMA's Vehicle Combustion Products Committee, provided what might stand as the final word on the case. Asked to comment before the Los Angeles grand jury on statements that Ralph Nader made in a chapter in his book *Unsafe at Any Speed* about the automakers' work on emission controls, Caplan told the grand jurors that "his criticism of the lack of recognition of the problem by the industry is easily refuted. Where we must give the 'devil his due' is in the area of implementation of our findings. Does such implementation occur only in response to legislative pressure and public criticism? Development of material to refute this criticism is difficult."[49]

Although the automakers avoided punishment and the potential for further serious embarrassment from a trial or release of the grand jury documents, the smog antitrust case made their approach to the emissions problem look less than urgent at a moment when the appearance of urgency as well as substantive action mattered a great deal to the public. It came just three years after GM's humiliating public confession that it had harassed Nader in retaliation for his assertion in *Unsafe at Any Speed* that GM had knowingly placed profit ahead of safety in designing and selling the Chevrolet Corvair. The consent decree further damaged the American automobile industry's already slipping public reputation at a highly inopportune time in American history. Public concern about automotive and industrial pollution was cresting, stoked by extensive media coverage of the oil spill off the California coast near Santa Barbara in January 1969, fires on Cleveland's Cuyahoga River during summer 1969, the continuing Los Angeles smog problem, and a bad smog summer in 1970 along the northeastern metropolitan corridor from Washington, DC, to Boston. To the sound of taps, in February 1970 students at San Jose State College buried a new Ford Maverick purchased with student donations in a symbolic show of contempt for the automobile.[50]

Throughout the 1960s, Democratic Senator Edmund S. Muskie of Maine, who chaired the Senate Public Works Committee's Environmental Pollution Subcommittee, and the subcommittee's staff director, Leon G. Billings, led efforts to strengthen federal pollution laws. Muskie was an avid sportsman, but he got into the pollution issue in the 1950s, when as governor of Maine, he had to confront stream pollution. In Washington Muskie worked diligently and largely out of the limelight for many years to make himself Congress's foremost expert on both water and air pollution.[51]

In 1965 Congress passed amendments to the Clean Air Act that made the 1966 California emission standards the 1968 federal standards. In November 1967 Congress passed the federal Air Quality Act and, over the strenuous objections of Dearborn congressman John Dingell, gave California the right to seek a waiver to set and enforce emissions standards that were tougher than federal standards if it chose to do so. In 1968 the federal government announced even tighter emissions standards for the 1970 model year. Each bill gradually increased federal involvement in what had traditionally been areas of state and local responsibility. Even so, nothing dramatic had been done at the federal level before late 1969. But Muskie epitomized the problem-solving confidence of the Americans who came home from the service after winning World War II. He and many of his colleagues believed that the federal government should focus the national will on solving technical problems. "If we can put a man on the moon," Billings remembered Muskie repeating over and over again, surely pollution problems could be solved, too.[52]

Concerned about the rising tide of antipollution sentiment, the magnitude of which soon became dramatically evident when an estimated twenty million Americans participated in the first Earth Day on 22 April 1970, Richard M. Nixon early in his administration moved to coopt the environmental issue from the congressional Democrats. On 1 January 1970, to launch the new decade, Nixon signed the National Environmental Policy Act, which created the White House Council on Environmental Quality (CEQ) as a companion to the National Security Council and the Council of Economic Advisors. "While the President may have had little personal interest in or enthusiasm for the subject," Nixon's first CEQ chairman Russell E. Train recalled, "he and his immediate advisors recognized the political significance of the environmental issue." Naturally, Nixon was not opposed to a cleaner environment, but he mainly wanted to deny the Democrats an issue that they could use to electoral advantage. He knew as well that Muskie might capitalize on the issue as the leading Democratic contender for the presidency in 1972. "Nixon gave us all pretty much a blank check as far as environmental initiatives were concerned," Train remembered. "He wanted a strong environmental record, but he didn't want to get involved."[53]

On 10 February 1970 Nixon sent a thirty-seven-point environmental program to Congress that featured several steps designed to address automotive air pollution. Touting the several rounds of tightened automobile emission standards that the federal government had authorized to date under the existing clean air legislation, Nixon asked for new legislation authorizing emissions testing on representative samples of each automaker's vehicles and the regulation of fuel composition. He ordered a federal research and development program into unconventional vehicles. He also asked for nationwide air quality standards, with the states to prepare within a year plans for meeting those standards. Muskie, too, seized the window of opportunity offered by the cresting of public concern about the environment by introducing a strong, comprehensive federal air pollution bill of his own on 4 March.[54]

Muskie's commitment to eliminating air pollution, Nixon's political interest in outmaneuvering Muskie on the environment, and overwhelming public pressure created strong bipartisan support for a significant strengthening of federal air pollution laws. The House passed an initially White House–crafted version of the toughened laws 375–1 on 10 June 1970. Stung by the publication in May 1970 of *Vanishing Air,* a report on air pollution legislation by a Ralph Nader study group that was highly critical of him, Muskie successfully urged his Senate colleagues to draft an even tougher bill. Senator Howard H. Baker of Tennessee, the ranking Republican on Muskie's subcommittee, played an especially important role in creating a bipartisan consensus behind a tougher air pollution law. Over the bitter but largely ignored opposition of the auto industry (especially from Ford president Lee Iacocca), the Senate on 22 September passed the Muskie bill 73–0. Although the public might have been surprised at the assessment, Capitol Hill insiders characterized the lobbying efforts of the Big Three, and GM in particular, as simply "inept." In a perceptive analysis of the Muskie bill, Train urged Nixon to endorse it publicly while working with the conference committee to modify the statutory language about granting extensions in meeting the proposed deadlines to permit more leeway and to eliminate language that allowed citizens groups to sue the federal government to force it to act to implement the law. In conference between the two houses the tougher Muskie bill largely prevailed with these provisions unchanged. Nixon signed the Clean Air Act amendments of 1970 into law on 31 December 1970 at a public ceremony at the White House. He did not invite Muskie.[55]

The lawmakers and their committee staff identified two shortcomings in earlier federal air pollution legislation. First, the laws lacked a definition of air pollution. Second, they precluded effective enforcement because consideration had to be given to whether abatement technology existed and its use was cost effective. At Muskie and Nixon's urging, the new law contained National Ambient Air Quality Standards. Muskie made sure that these standards were set solely on the basis of protecting public health, regardless of economic cost. At Baker's urging, the law contained "technology forcing regulations" that required the automakers to meet specified emission reduction standards, even if the technology did not yet exist. Another committee member, Democratic Senator Thomas F. Eagleton of Missouri, successfully argued that the standards would be worthless without concrete deadlines.[56]

Taken together, these provisions meant that the automakers had to spend whatever it took to invent whatever technology was required in order to meet the standards and safeguard the public health by the deadlines. Muskie expected the automakers and industrial corporations to say that they didn't know when or how to reduce emissions substantially, and federal bureaucrats, like the state and local pollution control officers before them, would have no choice but to accept their claims. His legislation aimed to prevent this ploy. In fact, the Clean Air Act

Freedom to breathe. *Courtesy National Archives.*

of 1970 was a law designed, not just to show that Congress cared, but to actually solve a problem—and in short order. It was not only the toughest environmental law ever passed in the United States, it may well have been the most antibusiness piece of legislation ever passed by Congress. Billings, who drafted much of it, called it "a very radical bill." Tom Jorling, another staffer who worked on the bill, predicted that the law "will have greater impact on society than any previous exercise of government power." It soon became the most controversial.[57]

By the mid-1960s, the American public had lost patience with limited state and local efforts because progress against pollution, much of it overwhelmed by increases in industrial output paced by the ever-growing number of automobiles, was not evident enough. This grass-roots pressure peaked in 1970 and 1971. The media played a powerful role in generating and stoking concern. A survey of editorials in five leading newspapers during this period found the environment to be the leading domestic issue of concern. In October 1972 Congress overrode a Nixon veto and passed the Clean Water Act, establishing the National Pollutant Discharge Elimination System (NPDES). The law required industrial plants that wished to discharge to water bodies to obtain permits from state water pollution control agencies like the Michigan Water Resources Commission that either certified that their discharges met federal water standards or committed the company to a specific program to reduce or eliminate the discharges by 1977.[58]

Enforcing the new laws fell to the federal Environmental Protection Agency (EPA), created in December 1970 from existing pollution control agencies within the federal government. With the 1970 Clean Air Act and 1972 Clean Water Act, Congress decisively rejected "cooperative pragmatism" in favor of "forcing regulations" that required the EPA and state regulators to make polluters to meet strict standards by specific deadlines. The EPA stood ready to intervene should state regulators and polluters prove slow in complying with the new laws. An automaker like Ford, which found itself in three of the most environmentally troublesome industries—autos, steel, and power generation—now faced far greater pressure, not only to clean up the tailpipe emissions of its cars, but also to clean up its manufacturing operations at the Rouge and elsewhere. "Never before in history," Train observed, "has a society moved so rapidly and so comprehensively to come to grips with such a complex set of problems." Federal regulators soon learned how hard it was to win compliance from corporate polluters. They also learned that there was a big difference between implementing regulations addressed to industry and those addressed to the nation's eighty-nine million automobile owners, who had come to view the ready use of a car as an individual entitlement that defined the American dream.[59]

9

THE ONE WHO GOT IT

The automakers knew that tighter emission standards were coming, not only for California but for the rest of the nation. Even by 1967, three years before passage of the 1970 Clean Air Act amendments, this prospect seemed clear. "If you were going to address the overall issue of air pollution," Tom Jorling recalled, "you had to address mobile sources." Between January 1965 and passage of the federal Air Quality Act in November 1967, both GM and Ford concluded that the engine controls that had placated the Californians for the 1966 models and that had been adopted as the new 1968 federal emission standards would not be enough to meet future requirements that now seemed imminent. The engineers at the automakers chafed at the fact that the engine modifications they had made to reduce emissions also brought unwelcome side-effects, suboptimal engine performance and a fuel economy penalty of 10–12 percent. Yet, as late as 1967, the industry devoted only about 4 percent of research and development (R&D) spending to emission controls. Tighter standards, if they came sooner rather than later, meant a catalytic converter or, more accurately, addressing the serious impediments necessary to bring it to market. After 1967, Ford and GM finally intensified their efforts in this direction. By 1970, the Big Three were spending real money on the problem, $157 million a year.[1]

Ford moved first. In 1964 James C. Gagliardi and Henry Nickels at Ford Powertrain Systems Research divvied up emissions responsibilities, with Nickels taking engine modifications and Gagliardi catalysts. "We did pretty much what we wanted to do," Gagliardi recalled. He read a paper about a platinum catalytic converter called the PTX that the Engelhard Minerals and Chemicals Corporation of Newark, New Jersey, had developed with 3M to remove carbon monoxide from the exhaust of gasoline- and propane-fueled vehicles being used in Swedish mines. Engelhard's Carl D. Keith had been influenced by Houdry's work. Gagliardi invited Keith and a younger colleague, John J. Mooney, to Dearborn to discuss the device.[2]

With guidance from Joseph T. Kummer, Ford's point man on scientific research on catalysts since the 1950s, Gagliardi built a team to work with Engelhard on a platinum-based catalytic converter using a one-piece, or monolithic, ceramic substrate. As part of this effort, E. Eugene Weaver and Harry Lassen discovered in 1968 that the conversion (that is, elimination) efficiency of a catalytic converter could approach 90 percent for hydrocarbons, carbon monoxide, and nitrogen oxide if the engine ran in a narrow band near the stoichiometric point in the air-fuel ratio (14.6:1). With such an engine, all three of the exhaust contributors to photochemical smog could be treated with one catalytic device. Weaver and Lassen wanted to patent their idea for this device, but Ford's legal staff were not encouraging about the prospects for winning protection, and there the matter rested, which proved unfortunate for Ford.[3]

These early Ford-Engelhard efforts culminated in a prototype 1971 Mercury Capri that four young Ford engineers who were also graduate students at Wayne State University entered in the August 1970 "Intercollegiate Clean Air Car Race." Forty-three student-built automobiles "raced" across the country from MIT to Caltech in a competition to prove that cars could be built that generated substantially lower emissions than those sold by the automakers. Engelhard provided its catalytic converter free to all the entrants with cars using internal combustion engines and unleaded gasoline. The Ford–Wayne State car produced the least emissions and "won" the race. At this point, Ford needed only a corporate executive to give the green light to move forward with catalytic converters. But Henry Ford II, who waxed hot and cold on environmental issues, was reluctant to champion the device. Arnold O. Beckman remembered visiting Ford at some point in the late 1960s to tell him that the industry had to substantially reduce emissions or face government regulation. Ford, who was drinking at the time, as Beckman recalled, resented the outside pressure. "From my level up," Gagliardi recalled, "nobody cared."[4]

GM did have an executive who understood emissions science and who decided to champion the catalytic converter—Ed Cole. When he was promoted to the GM presidency in 1967 *U.S. News and World Report* joked that Cole was one of the last auto executives who could fix his own car. Compared to his more conservative finance-oriented colleagues, he seemed to be a risk-taker, pushing projects that had one foot outside the box, like the rear-engine Corvair and the rotary engine. But the risks that he took were grounded in an engineer's insights. Cole had followed GM's antismog research from as early as 1954. He knew that the catalytic converter not only was the best available technology to cut hydrocarbon and carbon monoxide emissions substantially but that it also preserved the high-performance engines that he had done so much to put into GM cars. Speed and power with low emissions: Cole-the-engineer liked the challenge of achieving that goal.[5]

But it was Cole-the-executive that made the crucial difference. No Detroit executive played a more important role in forcing the industry to reduce smog-causing emissions. In 1967 GM had only two people still working on catalysts, engineers Harry Schwockert and Chuck Scheffler. Cole pressured GM Research to go outside the company and hire chemists who were experts in catalysis. In August 1967 Richard L. ("Dick") Klimisch joined GM Research from Du Pont as the company's first catalysis expert. His mechanical engineer colleagues quickly dubbed him "Captain Catalyst." Long the world's leading source of research on fuels and combustion, GM Research now made emissions a top priority, and research in this area grew to nearly half the company's R&D spending. In 1967 GM ran two 1968 model vehicles on unleaded gasoline for fifty thousand miles with a catalytic converter from Universal Oil Products (UOP) (one of the four certified by the Californians in 1964) that proved the effectiveness and durability of the technology.[6]

These tests gave Cole confidence that the catalytic converter was the way to go. But the research scientists and engineers at GM Research found that many of the once-burned third-party catalytic converter manufacturers, who remembered what the automakers had done to them in California in 1964, were cautious about working with them. Cole recognized that the catalyst makers would not take GM's interest in the catalytic converter seriously so long as they were dealing only with GM Research, a staff department, so in 1969 he created the Catalytic Converter Task Force to move the catalytic converter into production engineering. He charged an operating division, AC Spark Plug in Flint, Michigan, and a car division, Buick, with responsibility for making the technology a commercial reality.[7]

Cole now stepped up to the biggest challenge of all: leaded gasoline. GM played the major role in putting lead in gasoline in the 1920s, but, having sold its stake in Ethyl, it was no longer in a direct position to get it out. The oil industry had to remove the lead, at a loss of five to seven octane numbers and substantial costs in refinery modifications to make up the loss. Only one oil company, Amoco (Standard Oil of Indiana), offered unleaded gasoline in the United States at the time. The automakers needed to make the rest of the oil companies spend money to solve their emission problem. Papers to this effect presented at the January 1969 SAE conference caused substantial controversy. That same year researchers at Ford Scientific Laboratories informed Ford's management that lead would have to go in order to move forward with catalysts. But it was not Henry Ford II or Ford president Lee Iacocca who stepped up to the responsibility of persuading the oil companies to help the auto industry solve its emissions problem, it was Cole, whose Catalytic Converter Task Force had told him the same thing.[8]

Cole decided that he needed a second company inside the oil industry to produce unleaded gasoline in order to crack what might otherwise be a unified industry facade against taking the step. He turned to his friend and fellow engineer Fred Hartley, then president and CEO of Union Oil Company of California. Hartley was not a favorite among environmentalists in 1969. When his company's offshore oil wells fouled sixteen miles of Santa Barbara beaches he had been quoted, in error it turned out, as being "amazed at the publicity for the loss of a few birds." Hartley agreed to develop and market unleaded gasoline.[9]

On 14 January 1970 Cole went before the annual SAE convention in Detroit and stunned the auto and oil industries by predicting that the federal government would soon tell automakers to produce virtually pollution-free cars by 1980 and that it could be done but only if lead was removed from gasoline. "We have already demonstrated in our laboratories that these improvements are technically feasible . . . ," Cole told the convention. "It is my opinion that the gasoline internal combustion engine can be made essentially pollution-free." Because Cole wanted unleaded gasoline on the market well before the catalytic converter arrived, he called for regulation to remove lead. Eric O. Stork, a self-described "get it done" federal bureaucrat just then beginning his involvement with automobile emission issues, remembered attending a California Air Resources Board (ARB) meeting in Sacramento around this time. (The ARB replaced the Motor Vehicle Pollution Control Board in 1968.) A man sitting behind him stood up to speak. It was Cole. "If you will give me unleaded gas," Cole told ARB chairman Arie Haagen-Smit, "I'll put catalysts on my cars." When he presented his antipollution program to Congress on 10 February 1970, Richard Nixon requested federal regulatory authority to control gasoline additives, a step interpreted as a prelude to eliminating lead from gasoline.[10]

The oilmen went nuts. It was more expensive to make high-octane unleaded fuel than it was to achieve high octane numbers simply by adding lead. Conversion cost estimates for the change ranged from two billion to eight billion dollars. Cole personally lobbied the oilmen as well as tetraethyl lead suppliers, Ethyl and Du Pont. To cushion the loss of the octane points from removing lead and to give the oil companies some customers for the unleaded gasoline, Cole had GM reduce the compression ratio in the engines of all 1971 GM cars to 8.2:1 so that they would run on 91-octane gasoline—at a cost of performance and a

3–5 percent fuel economy penalty. Ford and Chrysler quickly followed suit. Oil companies soon began to announce that they would offer unleaded fuel. With cooperating oil companies, the threat of federal legislation, and rising public ire over pollution, the rest of the oil industry had no choice but to go along grudgingly, and a number of companies had 91-octane unleaded suitable for the 1971 cars on the market by the end of summer 1970.[11]

The leaded gasoline story was far from over, however. In December 1972 the Environmental Protection Agency, exercising authority granted by the Clean Air Act amendments of 1970, issued a proposed regulation requiring that unleaded gasoline be "generally available" at gas stations around the country by 1 July 1974. The EPA followed this regulation with a second one in November 1973, ordering a gradual multiyear phase-down of lead levels in gasoline in the United States. This step precipitated one of the young EPA's great and all-too-typical early fights, part of a larger pattern where the agency often stood alone against industry and much of the federal government's executive branch, all the while being harried by federal courts at the behest of environmental groups. (The EPA's early challenges are a focus of the next two chapters.)[12]

Before 1965, the lead industry itself controlled most of the scientific research on the toxicology of lead. In fact, one scientist, Robert Kehoe of the University of Cincinnati (who was also the Ethyl Corporation's chief scientific adviser), dominated lead research between the 1920s and the 1960s. Industry-funded researchers like Kehoe argued that the level of lead found in the blood of modern humans was "normal" and fell below the critical threshold of 0.8 parts per million (0.8 ppm or 80 µg/dl) associated with clear symptoms of lead poisoning. Clair C. Patterson, a geochemist at Caltech famous for work that revised estimates of the Earth's age, cracked this industry research monopoly with a controversial article published in 1965. Drawing on his own research, which required pioneering measurement techniques that differentiated between naturally occurring and human-caused rates of lead release to the environment, Patterson showed that levels of lead encountered by humans in the natural world had once been far lower. He argued that the typical amount of lead in the blood of modern humans (0.25 ppm) was one hundred times greater than normal (0.0025 ppm). He also called into question the contention that there were no health consequences from elevated blood-lead levels below 0.8 ppm. "The average resident of the Untied States," he wrote, taking inspiration from Rachel Carson, "is being subjected to severe chronic lead insult." His Caltech colleague Arie Haagen-Smit agreed, putting Patterson's work in the same league as *Silent Spring*. Studies showed that 90 percent of the lead in the air came from automobile exhaust.[13]

Patterson's article opened the door to a flood of studies conducted by scientists free from the constraint of industry funding. Although it did not happen immediately, these studies slowly undermined the industry's scientific conclusions. Eventually, researchers concluded that there was a connection between blood-lead levels in the range of 0.10–0.15 ppm and reduced intelligence in children. They also found a relation between lead exposure and high blood pressure in adults. The new research placed growing pressure on the EPA to effectively shutdown a highly profitable industry for reasons of public health. The EPA responded with its phase-down regulation, calling for a five-year reduction beginning in 1975 of the amount of tetraethyl lead in gasoline from an average two grams per gallon to a permissible maximum of half a gram per gallon. Remarkably, a survey conducted during this initial phase-down by the National Center for Health Statistics showed that levels of lead in the bloodstreams of Americans fell by 40 percent from 0.16 to 0.092 ppm. More dramatically, the blood-lead levels fell in lockstep with the reduced production and combustion of leaded gasoline. The study captured a "nationwide detoxification" that was undoing the widespread lead poisoning that Yandell Henderson had predicted would be the consequence of putting tetraethyl lead in gasoline back in the early 1920s. The EPA did not complete its series of phase-downs until 1986, and lead was not banned outright as a gasoline additive until 1 January 1996. But by 1990, the average blood-lead levels of Americans had fallen to 0.03 ppm. Ironically, the gradual replacement of older-model cars that could run on leaded gasoline with new models that could run only on unleaded gasoline to protect their catalytic converters probably had more to do with the reduced blood-lead levels than the EPA's phase-down or the public health threat. Although GM put lead in gasoline in the first place, GM's Cole played a major role in getting it out.[14]

With his SAE speech Cole in 1970 tried to deflate the pressure building in Congress to impose tough emission standards and deadlines on the automakers as part of the Clean Air Act amendments. In this effort he failed. Detroit's apparent culpability in delaying emissions technology ensured that the companies had little say in any new federal environmental regulations, even when Cole and Lee Iaccoca went to Washington to lobby Edmund Muskie on the proposed amendments. Iacocca adamantly opposed the legislation. Cole, though opposed, too, conceded that if the lawmakers really wanted emission control devices, GM and the industry could provide them. But he wanted the automakers to have until 1980 to meet the new standards. He even promised to begin installing catalytic converters on 1973 model cars and to have the device on all its cars by 1975.[15]

Cole believed that some Senate staffers like Leon Billings were out to get the industry, but the lawmakers and staffers quietly paid close attention to Cole's emissions comments and efforts. "His statements, his attitude and orientation,"

Jorling remembered, "gave members [of Congress] the confidence that they were proceeding down a direction that was reasonable and meaningful and doable." More important, Cole indicated a willingness on GM's part to defer to a conception of the national interest defined by Congress. The lawmakers also heard foreign automakers say that the proposed standards and deadlines posed no problem. In the end, members of Congress ignored Cole's appeal for a longer deadline, and the tougher Billings-Muskie standards and deadlines became law.[16]

The Clean Air Act amendments placed responsibility for reducing motor vehicle air pollution in two places: with the automakers and with states that had regions with polluted air. Under the law, the automakers had to reduce hydrocarbon and carbon monoxide emissions from new cars by 90 percent over 1970 levels by 1 January 1975 and nitrogen oxide emissions by the same amount over 1971 levels by 1 January 1976. The law's authors chose the 90 percent reduction target largely on the basis of a scientific paper written by Delbert S. Barth, the director of the Bureau of Criteria and Standards in the National Air Pollution Control Administration (NAPCA). Barth's group had been charged under the 1967 Clean Air Act amendments with determining and recommending to the states permissible levels of (and exposure times to) atmospheric photochemical oxidants, hydrocarbons, carbon monoxide, and nitrogen oxides to ensure that there was no danger to human health or welfare. NAPCA published criteria documents for the first three pollutants on 15 March 1970. Barth noted in his paper that it was now possible to determine—using simplistic calculations, he was careful to add—by how much emissions of these pollutants from automobiles had to be reduced in order to achieve the desired air quality. In 1970 Barth's calculations provided the only word on the subject from the federal government, and a member of his staff quietly passed the reduction percentages to Billings, who incorporated them into the pending legislation. Muskie later told GM's Klimisch that he had the numerical targets written into the law rather than have federal bureaucrats determine them because he didn't think the regulators could stand up to GM.[17]

The legislators and staff of the Senate subcommittee that drafted the law already knew about the catalytic converter programs at GM and Ford. They also knew that the catalytic converters worked because the catalyst companies secretly passed confidential test data from their programs with the automakers to staffers on Capitol Hill. The lawmakers chose steep reduction targets because studies suggested that reductions of this magnitude were needed to protect human health and because the lawmakers knew that the technology existed to achieve all or most of the needed reductions—at least for hydrocarbons and carbon monoxide. The automakers were unhappy that the ten-year deadline that Cole had proposed had been cut in half and with the steep 90 percent reductions, especially for nitrogen oxide. They took the law as a slap in the face, which it was. Congress meant to punish the industry. Cole's initiative to move forward on the catalytic converter came too late to stave off tough regulation.[18]

In anger the industry began speaking with an official voice that was often at variance with what it was doing behind the scenes. Public spokesmen for the Big Three protested that they lacked the technology to meet the standards by the deadlines. This claim was mostly bogus. With much research on catalytic converters behind them and with favorable results on prototype devices, all three firms had the technology to substantially meet the 1975 standards for hydrocarbons and carbon monoxide once unleaded gasoline became available. Chrysler, however, continued to believe that engine modifications held greater promise than catalytic converters. None of the automakers yet had the technology to fully meet the nitrogen oxide standard, although the Weaver and Lassen research on stoichiometry at Ford showed the direction forward. The bigger issue with the Clean Air Act was that the automakers did not want to be *forced* by federal law to meet the standards by the deadlines. In fact, there was bitter resentment at being forced. "The initial reaction [and] for many years," recalled GM's Joe Colucci, "was 'let's fight.'" "I don't think [Ford] wanted to go on record," Gagliardi remembered, as agreeing to go forward with catalytic converters.[19]

The Clean Air Act amendments gave the automakers the option of asking for a one-year extension of the reduction deadlines after the law had been in effect for a year. GM, Ford, and Chrysler all asked for an extension in meeting the standards for hydrocarbons and carbon monoxide at the earliest possible date, January 1972. Responsibility for deciding whether to grant this request fell to the EPA's first administrator, William D. Ruckelshaus, who could do so if he determined that several conditions were met: the necessary technology did not exist, the automakers were trying their best to achieve the standards, and the national interest would be served by the delay. "I had heard and been exposed to so much discussion about the need for an extension of time," Ruckelshaus remembered, "that I was convinced the automobile companies would be able to make an overwhelming case that they needed that extra year and they needed it now." Conscious that he would be making what looked to be a decision in the automakers' interest in an election year, Ruckelshaus decided that the case would have to be made before him personally in public hearings open to all. His decision would be based solely on facts established there.[20]

Ruckelshaus held three weeks of public hearings in April 1972 before making his decision. Before the hearings began, Eric Stork, deputy assistant administrator of the EPA's Office of Mobile Sources, had Karl H. Hellman and Thomas C. Austin of the EPA's Motor Vehicle Emission Laboratory in Ann Arbor, Michigan, prepare technology assessment reports based on their work with the auto companies certifying that their vehicles met the existing emission standards. The EPA also had help from the catalyst makers, Engelhard and UOP in particular, both before and during the hearings. Engelhard's Mooney had off-duty Newark police officers drive a Ford Torino equipped with their device around the Newark Airport to prove that it met the hydrocarbon and carbon monoxide standards and continued to meet them even after fifty thousand miles. In particularly compelling public testimony Engelhard repudiated Chrysler's claim that there was no device that worked. Tipped off in advance by the catalyst manufacturers that Chrysler planned dramatically to produce a melted catalytic converter during its testimony, Ruckelshaus responded when the moment came by asking Chrysler's spokesman whether it was the same device mentioned in its report that still reduced emissions by 80 percent. It was. "We had a lot of help from the catalyst industry," Stork recalled, "including John Mooney and others." "The catalyst manufacturers," Ruckelshaus remembered, "did make a very strong case that the technology was available and could be applied."[21]

There was no uncertainty. For hydrocarbons and carbon monoxide, the device worked. All that was left was for Detroit to commit to installing it. Japanese automaker Honda also indicated that its stratified-charge engine could meet the hydrocarbon and carbon monoxide standards by the deadline without using a catalytic converter. Armed with this knowledge, Ruckelshaus probed and learned that the Big Three's programs for commercializing these new technologies were still not far along and, in Chrysler's case, was nonexistent. Ruckelshaus knew that the makers still had a year before having to make final commitments on the design of their 1975 models. "After having sat there for several weeks," Ruckelshaus recalled, "I concluded they didn't make the case for an extension of time." He found the testimony of Natural Resources Defense Council (NRDC) attorney David G. Hawkins especially persuasive. A new kind of law firm whose "client" was the environment, the NRDC launched Project Clean Air to "bird-dog" the EPA's implementation of the Clean Air Act amendments, and Hawkins was the organization's air pollution point man, roaming the halls at the EPA to ensure the agency's vigorous compliance with the law. At the hearing Hawkins brilliantly summarized the industry's failure to make the case for the extension under the terms of the law. Concluding that the first two conditions for granting the extension had not been met, on 12 May 1972 Ruckelshaus denied the automakers' request for an extension, a decision that showed that the federal government was prepared to "get tough." "They really needed to be prodded by the government to take the whole thing a lot more seriously than they were," Ruckelshaus recalled.[22]

Ruckelshaus's decision produced immediate results. At GM, it reinforced Cole's view that the catalytic converter presented the only way forward. The company still worried about the cost of the converters and whether they would last the fifty thousand miles that the government wanted under real-world conditions. After finding that sulfur in gasoline compromised the effectiveness of the best base-metal catalysts, GM had to go with platinum and be able to purchase enough of it to manufacture five million catalytic converters a year for their vehicles. The American auto industry alone needed about a million troy ounces of platinum a year at a time when the total annual world output was not much more than that. The metal already sold for $150 per troy ounce. Fortunately, South African mining interests responded to GM's need, forming Impala Platinum, and offering to open a new mine to supply the 420,000 troy ounces GM needed each year. GM agreed to pay one hundred million dollars to Impala to develop the mine rapidly and purchase the first ten years' output at a fixed price. Twelve tons of rock had to be mined to produce each troy ounce, but Impala opened the mine at Merensky Reef about a hundred miles northwest of Johannesburg and hired ten thousand miners, working under apartheid-era conditions, to extract the ore from up to three thousand feet below the surface.[23]

Ford reacted even faster, albeit with greater reservations. The company chose for production the Gagliardi-Engelhard monolithic catalytic converter over a rival base-metal pelletized catalytic converter that a team headed by Ford engineer Robert Campau had developed and tested. Ford then contracted with Engelhard to supply up to 60 percent of the catalytic converters on its 1975 models. At the same time, Ford's purchasing people went to South Africa and contracted with Rustenburg Platinum Mines to provide the company with up to five hundred thousand ounces a year to be used in their catalytic converters. Calling it "Ford's High Risk 1975 Emission Lead Time Plan," Lee Iacocca bemoaned the lack of time to pursue alternatives and termed the company's four-hundred-million-dollar commitment to the catalytic converter "a lousy bet."[24]

Cole neither wavered nor bad-mouthed the catalytic converter. When contacted by White House aides in the wake of Richard Nixon's landslide 1972 re-election to enlist his support in pressuring Ruckelshaus to reverse his May denial of the automakers' request for an extension, he instead expressed confidence that GM could meet the standards, which stopped the White House effort cold. But GM chairman Richard Gerstenberg refused to let Cole commit publicly to the converter.[25]

Finding fault with the method used by the EPA in the first application for

Ed Cole explaining the exhaust conversion efficiency of catalytic converters to
the editors of *U.S. News and World Report*, 1973. *Photo by Thomas J. O'Halloran.
Library of Congress, Prints and Photographs Division, U.S. News and
World Report Magazine Collection, LC-U9-28809, frame 9.*

suspension decision, the U.S. Court of Appeals for the District of Columbia
on 10 February 1973 ordered the EPA to hold extension hearings again. At two
weeks of public hearings in March 1973 former University of California pollution control expert and newly hired GM vice president Ernest S. Starkman
stressed possible problems with the catalytic converter and asked again for the
year's extension. GM and Ford argued that production problems were possible,
while Chrysler argued that too many of the devices might fail in use. They all
wanted more time for testing. The special pleading sounded like extreme case
scenarios to observers like David Hawkins. Then there was nitrogen oxide, the
subject of a separate deadline extension request. "We didn't know how the hell
we were going to meet those nitrogen oxide standards," Klimisch remembered;
it "looked like that was not going to happen." The United Auto Workers weighed
in for the first time in support of management on the issue.[26]

On 11 April 1973 Ruckelshaus ruled again. Worried about the risk of economic
disruption should production problems be encountered in rolling out the catalytic converter to the entire national market with the 1975 models, yet confident that the technology worked and that GM and Ford were moving more ag-

gressively toward introduction on their 1975 models regardless of their public statements, he offered the industry a compromise that GM and Ford suggested: meet the standards in California and 50 percent of the standards in the rest of the country in 1975, and then bring the nation the rest of the way up to the standard in 1976. Essentially, Ruckelshaus forced the introduction of the catalytic converter on the 10 percent of the market represented by sales of 1975 models in California while permitting the makers a year to address any problems before rolling out the catalytic converter to the rest of the country. Ruckelshaus also criticized Chrysler for foot-dragging. In June the EPA ruled that one replacement of the catalytic converter at an owner's expense within five years or fifty thousand miles would be acceptable performance under the law. It also granted a year's extension of the deadline in meeting the nitrogen oxide standard. These concessions left most parties unhappy. The automakers still wanted wholesale relief, and some Californians objected to being the catalytic converter's guinea pigs. Ralph Nader suspected an effort to sow consumer unrest and drive the Californians into the automakers' camp in calling for wholesale revisions to the Clean Air Act.[27]

The still-stringent 1975 standards for California helped Cole to win his fight inside GM to commit publicly to the catalytic converter. In a letter of 10 May 1973 to the EPA, Cole announced that GM would meet the 1975 California standards with the catalytic converter. GM soon decided to install catalytic converters on all 1975 GM cars sold in the United States. The EPA's Stork, whose staff had insisted that Detroit could put catalytic converters on all its vehicles, was jubilant. "My guys are right . . . and GM just proved it," he crowed to the press, much to GM's chagrin. Cole took a very un-GM risk in doing this. Normal policy was to introduce a new technology in just one car division, not across all five at once. "It was really scary," Klimisch recalled.[28]

Cole liked the monolithic catalytic converter that Engelhard and Ford had been working on and pushed hard for it, but his research people worried that the monoliths could melt if some engine cylinders ran hot, so after a year of study GM chose a pellet-based catalytic converter, with the entire 22-pound device designed and manufactured by AC Spark Plug at a new plant at Oak Creek, Wisconsin. It purchased the catalyst pellets from W. R. Grace, the Catalyst Company, and Air Products, which in turn made the pellets with platinum and palladium that GM bought from Impala. GM reasoned that if the catalysts got fouled by the accidental use of leaded gasoline, dealers could easily replace the pellets. GM spent an estimated three hundred million dollars to get the catalytic converters on its cars.[29]

Ford went with the Gagliardi-Engelhard monolithic converter developed by Engelhard that weighed half as much as the GM device and produced less back pressure. Engelhard had moved into high gear on catalytic converter development after passage of the Clean Air Act amendments and partnered with Corning Glass Works, whose engineers Bill Armistead, Rodney Bagley, Irwin Lachman, and Ron Lewis developed a new extruded cordierite honeycomb substructure that successfully resisted thermal expansion and cracking from high exhaust temperatures. Chrysler, after an unsuccessful last-ditch effort to enlist Ronald Reagan and his California administration in opposing the catalytic converter, belatedly followed Ford in choosing a monolithic catalytic converter but contracted with UOP for its catalysts. All the catalytic converters used platinum and palladium in an approximate two-to-one weight ratio.[30]

Thanks to the industry's own criticism of the devices (Cole's voice to the contrary notwithstanding) and the human tendency to exaggerate the impact of an anticipated negative change, the public awaited the catalytic converter with dread, expecting high cost (one hundred to three hundred dollars), reduced fuel economy (up to 20 percent or worse), and a need to replace them (at fifty to seventy-five dollars) by twenty-five thousand miles (not fifty thousand). "Can you honestly imagine anyone paying for that pack of miseries," William Kease wrote to the *Los Angeles Times*, "especially in view of the fact that [a car] will get about 8 miles to the gallon? I wouldn't take such a car as a gift!"[31]

Remarkably, the EPA almost did in the catalytic converter while the industry moved forward with implementation. Gasoline naturally contained amounts of sulfur that were converted into sulfur dioxide during combustion. Catalytic converters oxidized sulfur dioxide into sulfate, which readily combined with water vapor in exhaust to produce small amounts of sulfuric acid. On 9 October 1973, the EPA's assistant administrator for research and development Stanley M. Greenfield circulated an internal memo (whose contents had already been leaked to the press) urging caution to protect the public health. "Catalytic converters," he wrote, "should not be used extensively enough to increase population exposures to acid aerosols, suspended sulfates and noble metals and their compounds. If this requires that oxidation catalysts not be utilized in 1975 motor vehicle models, so be it."[32]

Only days on the job, Russell E. Train, Ruckelshaus's successor at the EPA, had a tough call to make. In testimony before the Senate Public Works Committee on 6 November 1973 he announced his decision to go forward with the catalytic converters on the 1975 models while the EPA conducted further research. Unfortunately, this research, completed in early 1975 after the first models with catalytic converters had already been on the market for several months, suggested an even more serious sulfuric acid problem. "This was a horrible disaster,"

GM's Klimisch remembered. The more definitive studies that really needed to be done were beyond the EPA's capabilities and budget. GM Research director Fred Bowditch then approached the EPA and offered GM's Milford, Michigan, Proving Ground track, facilities, and staff to undertake a massive study of atmospheric sulfuric acid concentrations when vehicles with known emission rates were driven on the track under various meteorological conditions. These tests, which involved all the automakers and the EPA, concluded that the sulfuric acid worry was overstated. There the matter ended. "If we'd followed the advice of the public health people," Train recalled, "it would have brought the whole [catalytic converter] program to a standstill. I don't think we'd have known what direction to move at that point."[33]

In the end, Ed Cole's gamble paid off. Installed on more than 85 percent of the 1975 model automobiles sold by the Big Three at an average incremental cost of $165 per vehicle, there were no major production problems or serious failures in use. The first factory-installed oxidation, or two-way, catalytic converters were still reducing hydrocarbons and carbon monoxide about 60 percent at fifty thousand miles. Moreover, the 1975 models got 13.5 percent better fuel economy on average than the 1974 models. The EPA estimated that the Big Three spent two billion dollars from 1967 through 1973 — 75 percent after passage of the 1970 Clean Air Act — to develop and install the first catalytic converters. These first catalytic converters did not achieve the emissions goals of the act. Compared to cars with no controls, the 1975 models that met the 1975 interim standards reduced hydrocarbons by 83 percent, carbon monoxide by 83 percent, and nitrogen oxides by 11 percent. With respect to the Clean Air Act statutory requirement for 90 percent reductions in these pollutants from the levels found in 1970–1971 model cars, the 1975 models delivered only 63 percent, 56 percent, and 38 percent reductions, respectively. By then, the deadlines for meeting the standards had been pushed back, as they would be again, to give the automakers more time to improve the efficacy of the converters. In fact, as Samuel A. Leonard, a General Motors air quality expert, later observed, "Ninety-three [1993] was the first model year we ever built a model certified to seventy-five [1975] standard." Nor did the adoption of the catalytic converter cause the immediate elimination of leaded gasoline. But its adoption in 1975 marked the biggest step toward reducing smog-causing emissions and atmospheric lead.[34]

In 1975 nitrogen oxide remained a problem that lacked a solution as elegant as the oxidation catalytic converter, which reduced the pollutants without a negative impact on fuel economy. Indeed, the elimination efficiency of the oxidation catalytic converters was improved by adding more oxygen to the exhaust, a step that produced even more nitrogen oxide. But the sharp regulatory prod from the Clean Air Act forced the automakers and catalyst manufacturers to hire and fund a host of chemists and engineers whose combined efforts soon produced further results. By mid-1974, the automakers had developed proportional exhaust gas recirculation (PEGR) and dual catalyst systems (where a second catalytic converter was added solely to reduce nitrogen oxide), two approaches that showed promise.[35]

The industry achieved an even better solution by 1980, when all U.S.-made cars received a three-way catalytic converter, which reduced nitrogen oxides by converting them to nitrogen while oxidizing hydrocarbons and carbon monoxide. Working with Volvo on a 1973 Volvo station wagon along lines suggested by the Weaver and Lassen work at Ford in the late 1960s, Engelhard's Carl Keith and John Mooney pioneered the device. Volvo introduced the three-way catalytic converter on the 1977 model Volvo 244 sold to American consumers in fall 1976. As it transpired, rhodium added to platinum in a one-to-five ratio proved an even better catalyst for nitrogen oxide than platinum alone because it produced less ammonia in the process. "That's actually kind of a chemical miracle," Dick Klimisch recalled, "because one of the hardest things in chemistry is to convert nitrogen back to its elemental form."[36]

For three-way catalytic converters to work, the automakers needed to place an oxygen sensor on the exhaust system that was connected to an onboard computer that governed an electronic fuel injection system that kept the air-fuel mixture in the engine's cylinders very close to the 14.6 stoichiometric ratio. Volvo engineer Stephen Wallman, working with the German firm Bosch, developed a zirconium-based sensor, the Lambda-sond system, that monitored the level of oxygen in automobile exhaust. Running at stoichiometric eliminated the problem of some cylinders running hot and the threat that excessive heat posed to the monolithic catalytic converters. GM experienced no major problems with its pelleted converters, but this very fact obviated the major reason for choosing pellets in the first place, the ease with which they could be replaced if needed. Having already experimented with monoliths on some vehicles, GM now switched entirely to monoliths.[37]

In pursuit of hydrocarbon and carbon monoxide reduction the mechanical engineers at the Big Three took on board chemists. Now to reduce nitrogen oxide they learned about computers, and electronics came to the automobile engine for the first time as a by-product of emissions control. Ford contracted with Intel to develop the EEC-1 computer to install on its vehicles. With the three-way catalyst, the industry finally had an elegant approach that fulfilled Ed Cole's vision: steep emissions reductions without compromising engine per-

Catalytic converter, 1981. *Photo by James L. Amos.* © *James L. Amos/CORBIS.*

formance and fuel economy. Tragically, Cole did not live to see this day. After retiring from the GM presidency in 1974 at the mandatory retirement age of sixty-five, he was killed in a plane crash in 1977.[38]

When regulators, especially those with California's ARB, sought still further emissions reductions in the late 1980s, attention turned to gasoline itself. In 1989 the Big Three and fourteen oil companies started the Auto/Oil Air Quality Improvement Research Program, a six-year research project directed by GM's Joe Colucci to identify optimal fuel-vehicle systems. This program led to reformulated gasoline and the reduction of sulfur. By the 1990s, the automakers and oil companies had achieved reductions in smog-causing emissions from cars of well over 90 percent, and many researchers questioned whether pursuing further reductions from the internal combustion engine and gasoline really made sense.[39]

"We probably would never have done it on our own." The people inside the American auto industry who were most responsible for reducing automotive emissions generally concur with Jim Gagliardi's assessment of the Clean Air Act amendments of 1970. This legislation delivered the industry a sharp prod that forced it to adopt the catalytic converter for the 1975 models, the step that took the biggest bite out of smog-causing emissions. Thereafter, the industry delivered further improvements largely on its own terms and timing. Many people played important roles in pushing the American automakers to adopt the catalytic converter, especially "prodders" outside the industry, but Ed Cole deserves special recognition. His role in championing the device finally forced Detroit to get on with the technological fix that Angelenos wanted and that Eugene Houdry and the technical people at the catalyst companies and automakers had already done so much to make possible. "Ed Cole was more cooperative, more visionary, at that time than Ford or Chrysler," Eric Stork recalled, "and while, of course, everyone was adversarial, GM was less adversarial." "He was the one," Ruckelshaus recalled, "that . . . got it." Cole's efforts, like Henry Ford's waste reduction and recycling programs at the Rouge, showed just how much an industry-leading company could do when its leader decided to tackle a problem. Cole had been outraged that GM harassed Ralph Nader. Aggrieved that his company had become a whipping boy, he longed to restore its image by taking the automobile out of the pollution issue. "He was just magnificent," GM's Joe Colucci recalled. Ford's Gagliardi agreed. "He did what I was hoping we at Ford would have done, and [he] had the balls to do it."[40]

10

OUT OF MY DEAD HANDS

The struggle to get the catalytic converter on American automobiles was only half the story of the Clean Air Act and the automobile. With the 1970 amendments, Congress also made the states responsible for taking steps to change the behavior of drivers to combat air pollution from the automobile. It did so, first, by tasking the EPA with setting National Ambient Air Quality Standards (NAAQS) that ensured an adequate safety margin for public health—a pathbreaking first for federal health legislation—and, second, by requiring the states to prepare plans for achieving those standards. The law envisioned the rapid implementation of these plans in order to meet the standards and clean up the nation's air.[1]

The first part of the responsibility seemed clear. Doing the best that it could with studies done by predecessor agencies and the ninety-day deadline prescribed in the law, the EPA on 30 April 1971 announced standards for six pollutants—photochemical oxidants (later simply called ozone), hydrocarbons, nitrogen oxides, carbon monoxide, sulfur dioxide, and particulates (soot). Hydrocarbons and nitrogen oxides caused photochemical oxidants (essentially smog), a secondary pollutant. In order to protect the nation's health, the law made it illegal for the air anywhere in the United States to exceed the maximum set by the standards by 1975. The photochemical oxidant standard, for example, was set at 0.08 parts per million because research showed that at levels of 0.10 ppm or greater asthmatics began to experience a greater frequency of attacks. Automobile exhaust was a chief cause of the first four pollutants. It accounted for 70 percent of the carbon monoxide, 50 percent of the hydrocarbons, and 30 percent of the nitrogen oxides in air pollution—in all, about half of the air pollution in the United States but up to 90 percent in some cities like Los Angeles.[2]

Although critics at the time complained that there was no proven relation between the pollutant levels set by the standards and harm to human health, public health researchers over time proved that the standards were reasonable. More serious problems came with implementing them. The law required each state to write and submit to the EPA for approval a state implementation plan that detailed the steps that the state's residents were taking to ensure that they met the federal air quality standards by 1975 (or 1977 if the state's governor requested and received a deadline extension from the EPA). The 1970 amendments stated explicitly that these steps must include controls on transportation and land use if such actions were needed to meet the standards. The rationale for these actions was straightforward: additional tools beyond emission control devices and smokestack controls would be needed to achieve the air quality standards. State and local officials, in fact, originally suggested the need for these additional tools to federal lawmakers. Transportation and land-use controls meant finding ways to encourage and, if necessary, force people to drive less and possibly restricting developers from undertaking projects like building malls that encouraged rather than discouraged automobile use. The lawmakers assumed that new mass transit systems would fill the void. In other words, the lack of an adequate technological fix from Detroit by 1975 was not to be used as an excuse for failing to meet the air quality standards by the deadline and might even trigger these bold new steps that the public had not contemplated, much less explicitly endorsed.[3]

The new air standards posed an especially frustrating challenge for Southern Californians. With ten million people, six million cars, and unique geographical and climatic conditions, the region's inhabitants certainly faced the toughest time meeting the strict standards. In California responsibility for developing a plan fell to the Air Resources Board. Chairman Arie J. Haagen-Smit, the board, and its technical staff realized that they could not meet the standards in places like Los Angeles until the automobile manufacturers took far more effective emission control steps than those already taken or even in prospect with the catalytic converter. In fact, some felt that the 0.08 ppm photochemical oxidant standard, the toughest one the Southern Californians had to reach, might never be met, since it was at or near natural background levels for the region. Parts of the region violated that standard more than two hundred days every year, sometimes by as much as 600–700 percent. Los Angeles County exceeded the state's own 0.10 ppm standard for photochemical oxidants on 241 days in 1970.[4]

ARB regulators also feared that entertaining options other than a technological fix from Detroit would take the onus off the automakers for introducing more effective emission control devices for 1975—precisely the solution that they had been working so hard to attain for twenty years. Consequently, although it developed a plan that called for more substantial steps than any other state, the ARB refused to submit one that included steps sufficient to meet the photochemical oxidant standard, which would have required actions that went far beyond the board's charter and which it knew were not politically viable. Producing such a plan, in the board's judgment, was a pointless bureaucratic exercise. Instead, on 30 January 1972, the state submitted a plan that its air pollution regulators freely admitted would not meet the federal standards in the Los Angeles area. As required by law, the EPA rejected it on 31 May 1972.

The unwillingness of the Californians to do anything more forced EPA officials to exercise a response required of them under the Clean Air Act: developing a plan for California themselves. With more important issues before him in 1971 and 1972, EPA Administrator William Ruckelshaus did not initially foresee where the language of the law on the state implementation plans might lead. But soon enough the prospect of the EPA forcing land and transportation controls on cities around the country loomed before him. He and the top leadership at the EPA preferred to not exercise this authority in the hope that Congress would see the difficulties in implementing these controls and remove the requirement for them from the act. Ruckelshaus believed that the law's authors had sold it as a tough-minded experiment, but one that would be changed if necessary. Yet Congress was in no hurry to change the law. And on 6 November 1972 a federal court took the option of waiting off the table. In a lawsuit filed by lawyer Mary D. Nichols of the Center for Law in the Public Interest on behalf of the cities and residents of Riverside and San Bernardino, California (*City of Riverside et al. v. Ruckelshaus*), Judge Irving Hill of the U.S. District Court forced the EPA to produce a plan for the region by 15 January 1973.[5]

Nichols's lawsuit was part of a coordinated effort with the Natural Resources Defense Council (NRDC). In a separate action initiated by David Hawkins and the NRDC, the U.S. Court of Appeals for the District of Columbia Circuit ruled three months later in *NRDC v. EPA* that the EPA must force the states to include the land-use and transportation controls in their state implementation plans in order to meet the standards by the deadline. "David pushed hard," EPA lawyer John E. Bonine recalled, "holding EPA's feet to the fire." As difficult as this bind was for the EPA, Ruckelshaus still saw a useful side to the lawsuits if they helped to convince Congress to modify some of the problematic parts of the act. They did not, at least initially.[6]

As senior EPA officials fully realized, the court directive placed them in an impossible position — making hurried recommendations as outsiders that would be necessarily dramatic in nature with very little time to bring the Californians on board before making them public. Like their ARB counterparts, EPA officials recognized that the preferred technological solution would not be forthcoming from Detroit in the timeframe required by law to meet the air standards. It would be years before catalytic converters were on most of the cars on the road and the technology to achieve the nitrogen oxide standard did not appear to be imminent in any case. In the meantime the only option was to require sacrifice from drivers. The senior leadership at the EPA worried that this step would ignite a public backlash that harmed the environmental cause. Although the agency hurriedly funded studies to see just what lifestyle changes the public would accept, it suspected the worst.[7]

Environmentalists, in turn, worried that the EPA was not tough enough to force real cleanup. "A bunch of us," Hawkins recalled, "were very much admirers of writers like Jane Jacobs and other . . . urban visionaries who were talking about making downtown streetscapes . . . a place for celebration by human beings rather than a conduit for machines." Hawkins came by his vision from personal experience as much as idealism. He'd grown up in one of the 30 percent of American families in the mid-1950s that still did not own an automobile. After *Riverside v. Ruckelshaus* and *NRDC v. EPA*, the law and the courts seemed to be unquestionably on the side of Hawkins, the NRDC, and those inside the EPA eager to use this new power. The standards could still be met by the deadline by adopting a plan that convinced or forced Californians to drive less. It was a matter of putting the power of the federal government behind it. "It's like integration," Leon Billings observed at the time. "A lot of people were for it until it started to affect them. If the response to this is the same as to integration, I suspect the result will be the same." Billings no doubt envisioned the federal courts forcing the federal government to get on with it. With the great backlash against court-ordered school busing still to come, his prediction proved far more prescient than he knew.[8]

The top officials at the EPA were shocked when they learned just how much driver behavior would have to be changed to meet the standards. Gasoline consumption had to be reduced by 82 percent in the metropolitan Los Angeles area to achieve the photochemical oxidant standard of 0.08 ppm. People drove nearly 150 million miles per day there. If the estimates were accurate, that figure needed to be cut to thirty million. Voluntary incentives to discourage driving, the studies indicated, were likely to reduce vehicle miles traveled (VMT) only by 5–8 percent. The remainder of the reduction would have to come from mandatory gasoline rationing. "You guys are fucking nuts," Ruckelshaus reportedly told his staff when they told him what had to be done. But the EPA was under court order to produce a plan that met the air quality standards by the deadline, and there was no other step that came close to meeting the standard by the deadline. Details of the EPA plan had already been reported in the California press early

Kamikaze for clean air. William D. Ruckelshaus presenting
the EPA's Transportation Control Plan for Los Angeles,
15 January 1973. *Photo by James Ober.* HERALD EXAMINER
COLLECTION / *Los Angeles Public Library.*

that month. "Gasoline rationing . . . is about as sensible as smashing all drinking glasses in the area to conserve the Los Angeles water supply," one citizen wrote. The *Los Angeles Times* called it "unacceptably impractical." As the EPA's enforcement chief John Quarles recalled, "the court order raised blinding visions of a kamikaze mission."[9]

Ruckelshaus climbed into the cockpit for this mission. Rather than risk a contempt of court citation, on 15 January 1973 he went to Los Angeles and at a Biltmore Hotel press conference he presented the EPA's plan. The steps that the EPA required of the state, county, and city included support for public transit and disincentives for using autos, which included auto-free zones, parking fees, and preferential treatment for high-occupancy vehicles (such as buses and carpools) on the freeways, limiting the number of automobile registrations per household, and gasoline rationing to reduce VMT by up to 80–82 percent during the peak summer smog months. "Faced with a choice between my freedom and your mobility," he joked, making light of his predicament, "my freedom wins." Cautioning against emotional responses to the plan, he stood by the law and made the case for giving as many of the steps as possible a try, and failing that, at least a hearing. The debate about trade-offs necessary to achieve clean air, he thought, would be healthy. But he also reiterated several times that he presented the draconian plan solely to comply with the court order. Although he resisted the temptation to use the occasion to criticize the Clean Air Act or grandstand for Congressional action to modify it, most people involved in the issue understood that the presentation was a bit of a charade to encourage just that.[10]

Ruckelshaus and his deputy administrator Robert Fri had privately told officials in California before the presentation just that. "We were going to comply with the court order," Ruckelshaus recalled, "but in doing so we were trying to highlight the problems with the act as written, not that the objectives were wrong, but you just couldn't do it within the timeframe established by the Congress, short of getting eighty percent of the automobiles off the road, and that wasn't really a viable option." Ruckelshaus wanted to shock people in California into doing more to combat air pollution, and he wanted to send a message to Congress about amending the law. A public backlash from the shock of the plan might help accomplish the second goal. This calculation was not entirely wrong. Californians were shocked. They did complain to Congress. But Congress did not amend the law, which left the EPA stuck in the middle. In hindsight, many have questioned the wisdom of publicly making the call for 82 percent gasoline rationing. Hawkins, who opposed that step, wondered whether rationing was a "poison pill" designed to discredit the land-use and transportation controls. "He did more in that one speech to damage the Clean Air Act," Leon Billings recalled, "than anything that's been done before or since."[11]

Reflecting Ruckelshaus's efforts to set the plan in context before the public presentation, the initial reactions from California governor Ronald Reagan, ARB chairman Haagen-Smit, and the *Los Angeles Times* were circumspect. Nonetheless, the "general reaction to EPA's transportation control plan," in the words of

TRW environmental policy analyst Eugene Leong, who worked on developing the plan with the EPA's regional office, "was predictable and consistent — astonishment and disbelief." Gasoline rationing proved a lightning rod for criticism. "That's the dumbest thing I ever heard," the *Los Angeles Times* quoted a secretary, "If they don't like the smog, let 'em move out." A poll taken in Southern California by the Clean Air Constituency and funded by the EPA found that, among thirty possible antipollution steps, requiring the automakers to fix the problem was the most strongly preferred option with gasoline rationing near the bottom. With their own Clean Air Act problems, the automakers watched the drama in Southern California unfold. "As long as Californians keep on buying our cars," one cynical executive reportedly said, "we don't care whether people drive them."[12]

In public comment hearings that the EPA held in Los Angeles in March 1973 and again in August in response to a revised plan issued on 15 June 1973, representatives from nearly every government and private organization, including environmental groups, spoke in opposition to the plan, especially the gas rationing proposal. Kenneth Hahn set the unsympathetic, less than dispassionate tone of the responses. He charged that, like the smog antitrust consent decree, Ruckelshaus's plan was a sellout to the automakers that "puts the entire blame on individual motorists for smog rather than on the industry that causes it." The Greater Los Angeles Area Chamber of Commerce called the plan "unworkable and irresponsible." "The EPA's announced intent . . . ," as the Automobile Club of Southern California succinctly put it, was "to severely restrict personal independent mobility and alter life style." Nearly everyone protested that these steps would have a catastrophic socioeconomic impact on the region. Since strong evidence for a link between smog and a serious direct threat to human health was lacking, the critics argued, the proposed cure was far worse than the disease. Employing a Vietnam allusion, Los Angeles Air Pollution Control District head Robert L. Chass observed that the steps "would be a little like destroying a community to save it." One reporter at the original press conference asked, "Wouldn't it be cheaper to take the people that have asthma and send them to Arizona free?" The emotion-free debate that Ruckelshaus urged did not take place. The EPA got pilloried from coast to coast.[13]

Southern California was the extreme case, but it was by no means the only urban area in the country that would fall short of the Clean Air standards without transportation controls. The other states had until 15 February 1973 to submit their transportation control plans to the EPA if such steps would be required to meet the air standards. When they came, the EPA was not able to approve a single plan. By June 1973, nearly forty cities across the country faced the prospect of implementing transportation controls in order to meet the standards by the deadlines. On 15 June Acting EPA Administrator Robert W. Fri, who succeeded Ruckelshaus when Ruckelshaus moved to the Federal Bureau of Investigation as acting director in April, announced transportation plans for eighteen cities (including a slightly revised plan for Los Angeles), thirteen of which required gasoline rationing. In northern New Jersey, driving needed to be cut by 60 percent. "We are basically attacking the problem," Fri told a news conference, "by asking people to change their habits—their long standing and intimate relation to [the] private automobile."[14]

Thanks to the advance notice provided by the Los Angeles plan, critics of the EPA knew what was coming. "Outright hostility was the universal reaction," John Quarles recalled. Commenting that he was "not sure these are the results that the Congress intended," Fri indicated again that the EPA would welcome congressional modification of the Clean Air Act. "We are," he said, "for the first time facing up to the most profound implications of that law." Politicians, quick to sense the change in the wind, soon had congressional hearings under way. They had passed the amendments with near unanimity in 1970, but three years later many acted as if they had nothing to do with such an obviously ridiculous law. Support for the Clean Air Act had been bipartisan in 1970. So, too, now was criticism of the EPA's efforts to implement the law. California's Democratic senator John V. Tunney called the EPA's plan for Los Angeles a police-state solution, "impossible to enforce without putting a bayonet at the back of every motorist." The debate in 1970 had focused mostly on the emission control requirements forced on the auto industry, but transportation controls had been discussed in the committee report and had come up in the Senate debate. Muskie spoke to them directly: "The legislation would require new kinds of decisions with respect to transportation and land-use policies," he told his colleagues. "It would require new discipline of our desire for luxury and convenience. And it would require a new perspective on our world. . . . We have been conscious . . . that what we were considering . . . could result in drastic changes in the pattern of life we live in the urban areas of America. . . . That is going to require every State Governor and the mayor of every city in this country to impose strict controls on the use of automobiles."[15]

Staffers like Tom Jorling, who helped draft the legislation, were well aware of the implications of the transportation and land-use controls. "People knew that maybe where the cliff is on . . . environmental regulation is how to affect people's behavior." Backlash had been contemplated, too. "I can remember almost joking conversations," he recalled, "that I don't want to be a senator when this gets implemented." But the drafters of the land-use and transportation control provisions in the legislation had not acted hastily or without forethought. The controls constituted an important part of a comprehensive and expeditious solution to the nation's air pollution problem where each provision in the act contributed a necessary piece to achieving the desired outcomes within the specified timeframes.[16]

The Los Angeles plan that TRW prepared for the EPA suggested that "long range solutions to the air pollution problem must rely on better land-use planning." State and local officials, environmental groups, and the EPA shared this view. Land-use planning offered potentially the greatest lever for changing the relationship between Americans and the natural world that existed, and some younger lawyers inside EPA urged that this tool be brought to bear on the problem, although such decisions had traditionally been a matter of local and not federal authority. In the plans that it issued on 15 June and then again in revised final form on 15 October 1973, the EPA played down gasoline rationing and focused on parking with steps required from public and commercial property owners as well as developers.[17]

Parking was the Achilles' heel of driving. If you can't park when you get there, why drive in the first place? Make it harder to park, and people would drive less and demand mass transit alternatives more. Prodded by David Hawkins, the EPA proposed regulations that required parking space providers to collect a surcharge from parkers to pay for better mass transit. Hawkins also pushed EPA to seek approval authority over development projects like malls that encouraged driving. "Many of our environmental problems," new EPA administrator Russell Train noted at the time, "trace back in one way or another to man's shortsighted abuse of the land." He argued that the proposed regulations were not antigrowth. "The average American is just beginning to recognize that there are difficult choices to be made," Train argued in remarks at the National Press Club in September 1973, "that some cherished patterns of behavior may have to be modified. He is just beginning to understand that there are fundamental interrelationships between environment and the economy, energy, transportation, land use and so on . . . our society is only now beginning to recognize the truth of the basic ecological teaching, namely, that everything is related to everything else and that every choice involves a trade-off."[18]

Train arrived at the EPA in the middle of the transportation control controversy. He did not share all of Hawkins's larger vision, but he supported the parking surcharge and hoped that the regulations would force state and local officials to do more land-use planning with the environment in mind. At the same time, he carefully disassociated himself from steps like gasoline rationing that he termed "impractical of achievement." "It [was] like going over Niagara Falls in a barrel," Train remembered. "You're suddenly at the mercy of all sorts of currents pushing and pulling in all directions." He was not exaggerating. When businesses and parking lot operators learned of the surcharge proposal and when real estate developers learned that they might be forced to build office complexes and shopping malls that discouraged people from driving to them, they went bananas. And so did labor unions, especially in the construction trades. In a drive spearheaded by the International Council of Shopping Centers, developers filed two hundred lawsuits against the plans.[19]

In less than a year the 1970 Clean Air Act had forced the EPA to prod two of the most sacred cows in American life: the right to drive as much one wanted and the right to do what one wanted with one's property. Within weeks, the EPA's transportation and land-use regulations were hooted down. California communities lodged more than sixty legal challenges to the parking surcharge. The intensity of the backlash took Hawkins and the EPA by surprise. On 6 December 1973 Train indicated to Mayor Tom Bradley of Los Angeles that the EPA would reconsider the parking surcharge, and the next day the agency announced a one-year postponement and an increase from 50 to 250 in the threshold at which prior EPA approval would be required to build new parking spaces. Joking that "Los Angeles is probably the only major city in the country where I could talk about transportation control plans without the slightest danger of being ridden out of town on a rail," Train sought to mollify the Californians in a December 1973 speech in Los Angeles. Instead of prohibiting specific controls, he urged them to join him in pushing for extensions in the deadlines where necessary. But from this point forward there was no stemming the retreat.[20]

The measures proved so unpopular that not even the Arab oil embargo in the fall and winter of 1973–1974 and the urgent need to reduce oil and gasoline consumption revived them. In June 1974 Congress explicitly forbade the EPA from imposing parking surcharges to discourage driving and placed a year's moratorium on other parking restrictions. A National Academy of Sciences report on the adequacy of the Clean Air Act standards in August 1974 estimated that air pollution killed four thousand people a year in the United States, a figure the press widely cited at the time, but this claim did nothing to generate stronger support for transportation controls from the public. Pointing out that mass transit was two and a half times more energy efficient per passenger and twenty-five times cleaner than the automobile, Train pushed hard for greater support for mass transit, which clearly had to be available and attractive if people were going to accept restrictions on car use. The EPA made one last effort in Boston, where the prospects for implementing the portion of the city's transportation control plan that called for employers to eliminate 25 percent of employee parking spaces initially appeared promising, but in February 1975 the agency backed down even there.[21]

The states also challenged the EPA's transportation controls in the federal courts, and in three separate rulings in 1975 (*Brown v. EPA, Maryland v. EPA,* and *District of Columbia v. Train*) the courts essentially said that the EPA did not have the power to force the states to use state resources to carry out or enforce federal regulations. The EPA appealed the *Brown* decision to the Supreme Court. In the meantime, the agency, increasingly worried about the technical problems in forecasting the air quality impact of any of the proposed controls, formally rescinded regulations that required the implementation of many of the specific transportation and land-use controls that it had earlier proposed, including the gasoline rationing regulations, which it withdrew in October 1976. Claiming that this backpedaling meant that the EPA no longer asserted that it had the power at issue, the Supreme Court refused to hear an appeal of the *Brown* decision. John Bonine, who crafted EPA's legal strategy to defend the constitutionality of the transportation controls, was disappointed. The same federal judiciary that had forced the EPA down the transportation control road in the first place now refused to back the EPA in forcing the states to comply.[22]

In August 1977 Congress finally amended the Clean Air Act. These changes gave the states more time and flexibility to implement plans to bring their air into compliance and the EPA more discretion in determining whether the goals were achievable. Congress specifically prohibited the EPA from requiring states to review in their plans indirect sources of automobile pollution like new highways, malls, and parking lots. However, the requirement that states use transportation controls if necessary to meet the air quality standards remained in the law. By this point, however, the EPA had thrown in the towel, opting instead for "voluntary cooperation." After "discovering that when it issued commands, no one obeyed," public policy scholar R. Shep Melnick observed, a chastened EPA adopted a "sell, not tell" approach to the state implementation plans.[23]

Hawkins concluded belatedly that simply convincing insiders at the EPA that specific land-use and transportation control options had merit was not sufficient. They also needed to be sold to local officials if there was ever to be any hope of selling them to the public. "We moved too dramatically, too rapidly," Train observed, "in an area that the public never even thought about." "The most searing experience I lived through," is how Train's deputy John Quarles put it when discussing the transportation control controversy several years later. Quarles blamed Congress for forcing flawed legislation on the EPA in the first place, for pretending it had nothing to do with the law when the EPA came under criticism, and for not moving to quickly amend the law when the problems became clear. Ruckelshaus agreed: "What I didn't fully appreciate . . . was how hard it was going to be to get amendments to the law once you put the law in place." In the short run, transportation controls not only failed to contribute to reducing air pollution, they produced a backlash against the EPA that made future progress much harder to accomplish.[24]

The EPA's 1973 land-use and transportation control plan controversy was the high-water mark of proposed government regulation to address air pollution in Southern California—and the rest of the United States. In fact, much like Pickett's Charge at Gettysburg, the transportation control plans marked the farthest reach of federal government-led efforts at environmental reform in the twentieth century. Ruckelshaus had called the Los Angeles plan "a national test case in the willingness of citizens to alter lifestyles to meet air quality standards." The Los Angeles transportation control plan's failure—the plan was essentially dead on arrival—proved that federal "forcing regulations" were impotent unless the people who lived in regions that were not in compliance with the NAAQS were willing to change their lifestyles. Southern Californians refused. "I . . . am tired of feeling like a criminal every time I drive the car, whether it be for pleasure or to take the baby to the doctor," Marsha Fullmer wrote to the *Los Angeles Times* in December 1973, giving personal voice to the limit that the public placed on the extent of permissible regulation. She was not alone. "You know how I feel about the environmental situation?" Kenneth Johnson told a *New York Times* reporter that same month. "If we're all going to hell, we might as well drive there."[25]

Reaction to the transportation and land-use controls in 1973, as well as to other steps proposed during the 1970s, showed that Southern Californians and other Americans were willing to do little beyond buying catalytic converters that came factory-installed on new vehicles from Detroit. Consumers resisted any step that impinged on their accustomed automobile behavior or required them to spend money. The long list of control steps that the early EPA plans had contained (rationing excepted) remained the pallet of available (if admittedly inadequate) options to reduce air pollution from automobiles. Over the next thirty years, states, including California, took many of these steps. But in California in the years just after 1973 each time that local or state agencies proposed one of these steps, the public opposed it with a howling contrarian enthusiasm, often led by populist politicians like Kenneth Hahn.[26]

In 1975 Californians forced the repeal of a 1971 state law that required Southern Californians to retrofit their 1966 to 1970 model vehicles with emission control devices to reduce nitrogen oxide and to submit to annual vehicle inspections designed to ensure that their emission control devices were functioning properly. From March to August 1976, the state experimented with high-occupancy vehicle (HOV) lanes on the Santa Monica Freeway that were reserved for cars with more than two occupants in order to encourage carpooling and reduce VMT and emissions. But Angelenos sabotaged the program. Hahn actually drove by himself in the HOV lanes and dared the California Highway Patrol to arrest him. In 1974 and 1976, Los Angeles residents also voted down proposals to raise money for mass transit systems that most had no intention of using. In fact, one study done for the EPA showed that even a *free* rapid transit system that got all Angelenos where they wanted to go as fast as the automobile would reduce automobile VMT by only 10 percent. Apparently, even a mass transit system conceived as a highly attractive alternative to the automobile as a mode of transportation could not substitute for the other satisfactions offered by automobile ownership and use. Californians implemented some of the proposed control steps later but only after much difficulty. Hahn, in fact, reflected the views of most of his constituents faithfully: automotive air pollution was caused by the automakers and should be solved by them alone.[27]

Air quality improved in Southern California a good bit in the thirty years after 1970, but almost entirely because of technological improvements in emission control devices provided by Detroit on its automobiles and changes made to gasoline by the oil companies. Southern Californians continued to resist impositions on their accustomed freedoms with the automobile. Reacting to a 1997 state proposal to regulate emissions from older vehicles more aggressively, one irate citizen proclaimed that "if they start seizing our old cars, it amounts to tyranny. I, for one, am not going to hand over the keys. They are going to have to pry the keys out of my dead hands." Despite the improvement in emission control technology, Southern Californians at the century's end still exceeded the federal eight-hour average ozone air quality standard of 0.08 ppm eighty to one hundred days each year and were not expected to eliminate these occurrences until sometime well into the twenty-first century—if ever. And by this measure they continued to have the worst air in the United States.[28]

The health consequences of smog and of the decision by Southern Californians and other Americans to accept its continued presence also became more definitive over time. Automobile exhaust stunted lung development in children who grew up within a third of a mile of freeways. Breathing ozone from smog caused acute inflammation in the lungs. The vast majority handled this stress with no significant larger impact to their health. But some did not, and for some of these people, the exposure to elevated ozone levels pushed them over the edge and killed them. Well-crafted epidemiological studies demonstrated that death rates increased even when relatively low ozone levels rose a small amount for short periods. When most people chose to continue to drive, especially on those days when temperature and weather conditions made the formation of ozone from automobile emissions more likely, they also chose that some people in their community would die sooner than otherwise. That was an easy decision for the 99.99 percent who needed to get on with their automobile-based lives and did not need to confront their mortality for respiratory reasons as a consequence. People brought to the brink of death who wished to live another day or year might feel otherwise. But they were a tiny minority, not inclined to public protest.[29]

The transportation and land-use control controversy of 1973 raised deeper issues involving Americans, automobiles, and the environment that did not disappear simply because Americans did not wish to confront them. In 1974 Paul R. Ehrlich and John P. Holdren suggested that human environmental impact could be explained by three large variables: the size of a human population, their desired behavior, and whether the technology they used to facilitate the behavior magnified or minimized environmental impact. Southern California's population, which had been three million in 1940, grew from ten million to fifteen million between 1970 and 2000, which meant that a powerful countervailing trend worked against even the best emission control technology. But population growth was not regulated. Changes in driver behavior, absent a major health or mobility crisis, could not be easily imposed through the American political process. Therefore, the entire burden of addressing the continuing smog problem caused by population growth and unrestrained driving had to be carried by the third variable, technology, with fundamental decisions about the nature and implementation of the technology left largely to the automakers (after the initial sharp prod from the 1970 Clean Air Act amendments). By century's end, more people were driving more cars more miles in Southern California and the United States as a whole, which together worked against realizing the full benefit of the steep reduction of emissions per vehicle that the catalytic converter made possible.[30]

11

SMALL WAS BEAUTIFUL

On 9 October 1973 Richard Nixon decided to rush emergency military aid to Israel during the Yom Kippur War. Unfortunately for Americans, the war took place at a moment of important change in the world oil market. For two decades before the early 1970s Western oil companies had dictated to the oil-producing nations of North Africa and the Middle East the price that the companies would pay for their crude oil. By 1973, however, the efforts of the Organization of Petroleum Exporting Countries (OPEC), the eleven nations (Saudi Arabia, Iraq, Kuwait, Libya, Algeria, Abu Dhabi, Qatar, Iran, Venezuela, Indonesia, and Nigeria) that controlled half the world's oil production and 90 percent of the world's oil exports had reversed the terms of the relationship. The exporting nations now told the oil companies the price the companies had to pay for the oil. Although the new power of these nations had been evident for months, the sudden success of Arab arms forced Nixon to make his assistance decision in a context where he may not have fully appreciated the potential for wider economic ramifications.[1]

Nixon later claimed that he understood the stakes. When he told his reluctant defense secretary James Schlesinger to resupply Israel, he said he "would accept complete personal responsibility if, as a result, we alienated the Arabs and had our oil supplies cut off." Angered by the open American military support for Israel, the Persian Gulf oil producers on 16 October first announced a unilateral 70 percent increase in the price of their oil from $3.01 to $5.12 a barrel. When Nixon on 19 October added insult to injury by asking Congress for a $2.2 billion military assistance grant to Israel to pay for the supplies, Saudi Arabia, Libya, Algeria, Kuwait, and Abu Dhabi imposed a boycott on the sale of oil to the United States the next day. The October 1973 Arab oil embargo triggered the energy crisis of the 1970s and changed the course of America's relationship with the automobile.[2]

Oil supply worries were not new. They had accompanied the introduction of the automobile at the beginning of the twentieth century and had recurred at intervals ever since, including the 1920s, the late 1930s, and the immediate post–World War II years. Analysts, commentators, and policymakers in the United States were already in the midst of one of these periodic frets when the Arabs imposed the embargo. This time, however, the concern was especially well-founded. Before 1970 American oil fueled American automobiles. But domestic oil production peaked at 9.5 million barrels a day in 1970 and then leveled off. Over the next three years, the number of licensed drivers in America grew by 9 percent, and automobile registrations grew by 16 percent, while vehicle miles traveled increased by 18 percent and gasoline consumption by 20 percent. Output from domestic wells could not keep pace with Americans' increasing demand for gasoline. After 1970 a growing proportion of America's oil had to come from elsewhere, which meant that developments in oil-rich parts of the world now had the potential to disrupt the U.S. economy and, indeed, strike at the heart of America's automobile-dependent way of life. In previous Middle East crises, the Americans had flooded the world oil market with oil from its own reserves to counter the possibility of a price spike or an embargo by Arab nations. By 1973, Americans no longer had the reserves to exercise this power. The U.S. government could help Israel militarily, but the American people could now be punished economically by Arab oil-producers for doing so. In October 1973 Americans finally found themselves over the oil barrel.[3]

The concerns about petroleum resources in the early 1970s stemmed both from developments in the world oil market and from a growing recognition that long-term upward trends in population and economic growth had implications for the rate of consumption of critical natural resources. Books like Paul Ehrlich's *Population Bomb* (1968) and Dennis and Donella Meadow's *Limits to Growth* (1972) encouraged a much larger audience to do the basic multiplication involving growing populations and increasing standards of living predicated on energy from fossil fuels — and it was sobering. A year before the Yom Kippur War, former secretary of the interior Stewart Udall explored the implications of this new perspective for automobiles in a piece that appeared in *Atlantic Monthly* called "The Last Traffic Jam." "Unless we exercise foresight and devise growth-limits policies for the auto industry," he argued, "events will thrust us into a crisis that will lead to a substantial erosion of our domestic oil supply as well as the independence it provides us with." For Udall the prescription was clear. "The quality of our lives will be enriched if we make fewer demands on our resources.

'Less is more' is a paramount tenet of environmental reform, and it is time for us to recognize its specific benefits. Less horsepower, smaller cars, and fewer autos mean more safety, healthier urban environments, more constraints on suburban sprawl, more efficient use of fuel. Less oil consumption for fuel means more oil to share with our children and theirs."[4]

These general concerns about the adequacy of natural resources were given greater credence in the United States by otherwise unrelated shortages caused by price controls implemented by the Nixon administration in 1971 to control inflation. The environmental movement and the advent of tougher federal environmental regulations also introduced new pressures and uncertainties to the energy picture. The desire to reduce smokestack pollution pushed many manufacturers and utilities away from coal and toward cleaner-burning fuels such as natural gas and oil, while at the same time environmental concerns slowed or blocked projects such as the Alaskan pipeline and offshore drilling to increase domestic production. In fact, not a single new refinery was under construction in the United States in 1973. In the world at large there was no immediate prospect of running out of oil. But in January 1973 the United States began to experience spot fuel shortages, first with heating oil and then with gasoline, as some service stations ran out of fuel. The term *energy crisis* came into use.[5]

Fully prepared or not, when the U.S. government and American consumers found themselves suddenly confronted by the Arab oil embargo during the demoralizing winter of 1973–1974, the issue was not new. Plenty of Cassandras had warned of the need to conserve gasoline. On 7 November, Nixon announced Project Independence, which committed the United States to achieving complete freedom from foreign sources of energy by 1980. The prospect provided no solace and probably added to the quickening sense of crisis. In late December OPEC suddenly decided to more than double the price of oil from $5.12 to $11.65 a barrel, which meant that the price had quadrupled in just two months from three dollars to nearly twelve. Henry Kissinger, then U.S. secretary of state, later called it one of the pivotal events in the twentieth century.[6]

Price-controlled domestic supplies cushioned the immediate impact on American consumers, but gasoline prices rose from thirty-seven cents a gallon in January 1973 to fifty cents a gallon in March 1974 (and eventually to nearly sixty cents in December 1974). The embargo and cutbacks in production not only led to increases in the price of gasoline, it also brought outright fuel shortages due to hoarding by countries, oil companies, and, finally, by consumers in the form of topping-off gas tanks. Long lines formed at many service stations. In December 1973 Nixon's newly appointed "energy czar," Federal Energy Office administrator William Simon, asked gas stations to limit sales to ten gallons per

55 MPH. *Photo by Warren K. Leffler. Library of Congress,*
Prints and Photographs Division, U.S. News and World Report
Magazine Collection, LC-U9-34591-A, B, frame 17.

customer per stop and to close on Sundays. He also asked consumers to limit
their gasoline use to ten gallons per week, in effect calling for voluntary rationing. A year of controversy over the EPA transportation control plans had made
federal officials aware of how deeply opposed Americans were to mandatory gas
rationing. With 56 percent of the cars on the road carrying one passenger, the
government and many businesses exhorted Americans to share a ride with other
people, or "carpool." A nationwide spot-check of gasoline stations in March revealed that 20 percent were out of gasoline. On 2 January 1974 Nixon signed
legislation that forced the states to impose a national fifty-five-miles-per-hour
speed limit to reduce gasoline consumption—the only direct step that the government took to change drivers' behavior during the crisis.[7]

It is difficult to overestimate the impact that these events had on ordinary
Americans. The price increases could be grudgingly stomached, but not being
able to get gasoline at all raised a more terrifying fear—not being able to get
to work or to go shopping for food. The prospect raised by the transportation
control plans had suddenly arrived, not as a tool to reduce environmental harm
in the hands of federal bureaucrats, but as a weapon wielded by foreigners to
harm Americans. Because Americans had largely rebuilt their manmade envi-

Americans' worst nightmare. No gas, May 1974. *Photo by Marion S. Trikosko. Library of Congress, Prints and Photographs Division, U.S. News and World Report Magazine Collection, LC-U9-29538, frame 16.*

ronment around the automobile and because alternative fuels like alcohol had never been commercially developed, American life was now unworkable without gasoline. Americans had no choice but to sit in gas lines, submitting to an unpleasant experience that made many frustrated and angry. Soon, the chafing and rebellion began. Carpooling was a great idea—for everyone else. Led by truckers with citizens band radios, many also decided that they were not born to drive fifty-five. The only thing that mattered was that the gas was there when they wanted to buy it.[8]

By the time that seven of the nine Arab states decided to end the embargo on 18 March 1974, the oil exporters had inflicted a great deal of economic pain on American businesses and consumers. But the worst was still to come. The price hikes were not rescinded, and the effects of the higher price of oil and gasoline rippled through the economy. The inflation rate, already at unprecedented peacetime levels thanks to permissive monetary policy at the Federal Reserve, nearly doubled from 6.2 percent in 1973 to 11 percent in 1974. Americans reacted to the higher prices by sharply curtailing discretionary spending, opting to wait for the economic turmoil to end. This collective decision triggered the sharpest economic downturn in the United States since the Great Depression. Unem-

ployment climbed to 9.1 percent in February 1975. The combination of high inflation and high unemployment, dubbed "stagflation" by mystified economists,
was unprecedented in American history and defied obvious solution.[9]

American consumers in 1974 and 1975 experienced two of the worst years
in the twentieth century. The University of Michigan's widely publicized Index
of Consumer Sentiment plunged to depths never before registered—bottoming out at 57.6 during the first quarter of 1975. *Business Week* summarized the
mood in March 1975: "Battered by inflation and the country's worst recession
in 40 years, shaken by political scandals unparalleled in American history, and
troubled by massive strains in the world economy, the middle class is growing
angrier and more frustrated day by expensive day." "A severe lack of confidence
in the government's economic policies continues to depress consumer sentiment," the University of Michigan's Survey Research Center reported with the
release of the May 1975 Index of Consumer Sentiment. Not surprisingly, social
scientists charted significant changes in consumer attitudes, tastes, and behavior. A conservation ethic swept the American middle class.[10]

The auto industry found itself at the epicenter of this crisis. GM, whose profitability depended far more than its rivals on the sales of larger cars, saw sales
of these models cut in half. "The world's largest manufacturer," *Business Week*
reported in a March article on the company, "is rife with confusion, indecision,
even panic." GM, it was later reported, considered ending the manufacture of
large cars altogether. The indecision was nearly as bad at Ford, where a power
struggle between Henry Ford II and Lee Iacocca was coming to a head just as
the crisis hit.[11]

Americans later looked back on the energy crisis and remembered that it encouraged a shift to smaller cars (with wheelbases under 110 inches). But that shift
was already well under way. Small car sales peaked in 1961 at nearly 40 percent
of the U.S. market but then fell off sharply, bottoming out around 15 percent in
1967. Over the next six years, however, sales rebounded to 40 percent of the market. Imports accounted for nearly half of these cars. Detroit hoped that its new
compacts and subcompacts (with wheelbases under 100 inches)—Ford's Maverick (April 1969) and Pinto (September 1970), American Motors's Hornet (August 1969) and Gremlin (February 1970), and GM's Vega (September 1970)—
would repel the foreign invasion as the 1960 compacts had a decade earlier.
Although priced at less than two thousand dollars, these models expanded the
small car market by stealing sales from the domestic automakers' larger models.
By spring 1971, the small car boom was the talk of the American automobile
market. "One thing is sure," Thomas B. Adams, longtime head of Chevrolet's ad
agency Campbell-Ewald, told *Business Week*. "We aren't 'creating' the market
today."[12]

This growing interest in small cars can be explained by the intersection of demography and economics. In the late 1960s the leading edge of the Baby Boomers began to need cars, and families stretched their automotive budgets to put their older children behind the wheel. During the 1960s, consumer spending on new cars grew by half again as much as the increase in real disposable income, while the number of multicar families increased from 13 percent to 29 percent. But the American interest in small cars was not all demography and economics. Some observers discerned a deeper change in consumer tastes. Consumers no longer purchased cars for status, they argued, they bought them simply for transportation. "Discriminating consumers," *Consumer Reports* noted in its April 1970 car issue, "are beginning to make clear to Detroit that they're looking for practical, reasonably priced transportation, rather than the overpowered and overpriced extensions of childish imagination that Detroit would prefer to sell." "The days of pure snob appeal, one-upmanship on the Joneses and the gaudy chrome running rampant," *Advertising Age* observed in 1972, "seem to be passing." "Growing numbers of people," *Newsweek* reported, "had come to regard automobiles much the way they viewed dishwashers or refrigerators—as appliances that should quietly and efficiently do their jobs, not as a projection of the owner's inner self and personality." Even industry leaders agreed that Americans were less inclined to view their cars as status symbols. The moment long hoped for by critics like Keats, when American consumer tastes finally "matured," and people stopped buying cars for superficial ego-enhancing reasons, seemed to be at hand.[13]

All these factors and then some were at work in California, where by spring 1971 a considerable consumer fad for small cars was under way. Small cars constituted nearly two-thirds of the cars sold there that spring. In the Los Angeles area imports comprised 41 percent of the cars sold in the first quarter of 1971, up from less than 15 percent in 1966. Nearly all the imports were small Toyotas, Datsuns, and Volkswagens. "We're heading for a European-type market," one Southern California auto supply executive predicted, "myriads of small cars and a few Rolls-Royces." For Southern Californians in 1971 small, foreign cars were cool. Buyers on the East Coast, where sales of imports also ran ahead of the national average, shared in this fad. The "cachet of foreign origin," *Advertising Age* told its readers in a 1972 review of the history of foreign cars in America, "persists to this day for all imports." In the early 1970s, small, inexpensive imports became anti-status-symbol status symbols once again.[14]

Foreign automakers worked hard to create a market in the United States for their small cars. Volkswagen had been the leading import in America since the mid-1950s. The German automaker, which had assiduously built U.S. Beetle sales to five hundred thousand a year, appeared poised in the late 1960s to reap the full fruits of two decades of effort. It was not to be. The Japanese automakers Toyota and Datsun (Nissan) arrived just as Volkswagen sat down for the feast. Beetle sales faltered in 1970 and 1971, while those of Toyota and Datsun soared. The Japanese were not so much opportunists as careful learners. They studied American car buyers closely, and concluded that many liked a number of functional characteristics of American-made automobiles, but they also found that consumers wanted smaller cars that were economical to drive and repair. "While VW's philosophy had been 'think small,'" observed Clinton E. Frank, head of Toyota's American ad agency, "Toyota's became: 'Think big—then miniaturize very carefully." It was a shrewd take on the American market.[15]

The Japanese deliberately chose Southern California as the best place to penetrate the American market. The region's proximity to Japan minimized the freight costs involved in shipping cars to America. Southern California was also the car capital of America, the bellwether market for trends in automobiles and other new products. The Automobile Club of Southern California's member magazine attributed this interest to the high number of migrants in the region. "The westerner is less constrained by tradition . . . more interested in the present and future," the magazine observed. "He is more willing to take a chance, more curious about new experiences, less inhibited and an unusually good prospect for new products." With a beachhead in California, Toyota carefully moved into other regional markets one at a time. "Southern California today. The U.S. tomorrow?" asked *Forbes* in 1971, before concluding that "only a miracle can stop the trend from spreading."[16]

For anyone worried about the environmental consequences of the relationship between Americans and their automobiles, this moment in the early 1970s was, in one sense, about as good as it ever got. It demonstrated that consumer tastes and the cars that automakers offered in response to consumer tastes could move in more environmentally friendly directions without government regulation or painful economic pressures. Whether distinguishing oneself by driving a Toyota or Datsun or just stretching the automotive budget, the consumer motives involved remained self-serving, but the broader consequences were not. Smaller cars burned less gasoline, which meant less smog-causing emissions. They also required smaller quantities of raw materials, entailed less manufacturing pollution, and left less material to dispose of at the end of their lives. Smaller cars took up less road and parking space, which meant less congestion and less pressure for automobile-centered development.

The environmental benefits from this shift in tastes were modest in comparison with those in the works from the Clean Air Act amendments of 1970 and the Clean Water Act of 1972, but they were real. And this moment and its potential

Beachhead. Datsuns on the pier, 1972. *Photo by Gene Daniels.*
Courtesy National Archives.

should be underscored. Whether it was the right outcome for the wrong reasons or the right outcome for only partly the right reasons, it was positive nonetheless. An astonishingly high number of Californians and other Americans proved to be largely immune to Detroit's blandishments and traditional fare. By summer 1973, small cars accounted for more than 40 percent of all passenger cars sold in America. These consumers showed other Americans that you didn't need the biggest, fastest, or flashiest car on the road to be cool. As E. F. Schumacher suggested, small was beautiful. A breakthrough was at hand. Then came the energy crisis.[17]

Although auto sales rebounded slightly after the embargo ended, the spiraling cost of raw materials and parts also coincided with the arrival of catalytic converters and safety enhancements on the 1975 models. The increased cost per vehicle exceeded the profit margin on most vehicles, which forced Detroit to increase the prices of its 1975 models substantially, an average of $450 per vehicle, the largest single-year average price increase in history. In fact, in the fifteen months between October 1973 and January 1975, the average price of an American automobile rose by a thousand dollars, or about 20–35 percent depending on the model. As part of its public relations war against federal regulation, Ford released data showing why the suggested retail price of its Pinto had increased

$1,010 (or 42 percent) from $2,382 in 1971 to $3,392 in 1975. Ford president Lee Iacocca attributed 28 percent of the increase to government regulations, 24 percent to product improvements, and 48 percent to inflation.[18]

Whatever the reasons, consumers were stunned. Raising prices sharply during a sales slump had the predictable effect: sales plummeted further. Even demand for small cars evaporated. By January 1975, 314,000 or 40 percent of the auto industry's hourly workers had been laid off. In 1974 Americans bought 25 percent fewer cars than the year before, the industry's worst downturn since 1958. At the market's year-end 1974 low point, cars sold at an annual rate of 6 million, down nearly 50 percent from the industry's 1973 peak. The first eight months of 1975 were worse. While sales of small foreign imports revived, expanding their share of the market to 22 percent, Detroit's cars languished. Although Detroit priced its small cars comparably, the imports offered better gas mileage and, Americans now discovered, provided better quality, which translated into lower operating costs for hard-pressed consumers.[19]

The arrival of the energy crisis with the Arab oil embargo and the recession of 1974–1975 made automobile fuel economy a matter of national security, national prosperity, and personal economic concern. With 40 percent of its oil coming from imports, the United States groped to find ways to reduce the vulnerability that came with dependence on foreign suppliers. Before the embargo, fuel economy had been one of several appeals of small cars but not the most important one. Consumers simply saw fuel economy as a part of keeping overall ownership costs manageable during budget-stretching and inflationary times. Fuel economy was not linked in the average American's mind with the need to conserve oil. After the embargo and the experience of soaring gas prices and lines at service stations, fuel economy, conservation, and small cars became fused in the popular mind. Thereafter, Americans bought small cars when the price of gasoline increased. Naturally, they no longer bought small cars for the same mix of reasons as before 1973. "It was the extraordinary scare of the fuel shortage," Gertrude McWilliams argued in 1975, "that drove people in hordes to little cars, not the free exercise of their own preferences." Her comment, which ignored the 40 percent of new car buyers who bought small cars in some sales periods before the energy crisis, showed how quickly the upheaval began to change Americans' views of small cars.[20]

Before the energy crisis, American cars had been getting smaller, but they had not been getting more fuel efficient, largely because Detroit tried to sell small cars with profitable but gasoline-consuming options that consumers loved such as power steering and air conditioning. Steps to increase the safety of American automobiles also added to the weight and therefore reduced the power of vehicles. So, too, did the pre–catalytic converter engine modifications that the companies made to meet the early federal emission standards and that reduced fuel economy by about 10 percent, according to the EPA. Detroit compensated for the weight by increasing engine power. These steps worsened rather than improved the fuel economy of cars in the years just before the energy crisis. In 1973 the average American car got 11.8 miles per gallon. Confusion reigned initially about whether the catalytic converter, more or less forced on Ford and Chrysler by the EPA and GM, would hurt or help fuel economy. The catalytic converter improved fuel economy, just as the EPA and GM promised, but the improvement was a modest one in light of the nation's now much greater need.[21]

With the energy crisis, fuel economy became the paramount automobile policy goal for the U.S. government, superseding emissions reductions and safety. In fall 1974 Detroit had promised the new president Gerald Ford that it would voluntarily improve the fuel economy of its cars by 40 percent by 1980 in exchange for a five-year moratorium on the implementation of the Clean Air Act's automobile emissions standards. The president, in turn, called on Congress to pass a comprehensive energy bill that would put the nation on the path to energy independence by removing price controls from domestic oil production in the hope that higher prices would encourage greater output while reducing demand. But Congress, not the president, made the laws. Between 1973 and 1975, members of Congress introduced more than a hundred bills that encompassed motor vehicle fuel economy. Many of these bills actually called for the use of taxes to cut demand by making gasoline and "gas guzzling" vehicles more expensive.[22]

With enactment of the Energy Policy and Conservation Act (EPCA) on 22 December 1975, Congress decided to take Detroit's promise and better than double it, while ignoring its request to extend the emissions deadlines for the moment. Although much of the law dealt with the domestic oil supply, creating what the *New York Times* called a "hideously complex oil pricing structure" that continued price controls, EPCA imposed mandatory corporate average fuel economy (CAFE) standards on the auto industry, starting with 1978 models. By the 1985 model year all the passenger cars that a company made during a model year would have to get a "corporate average" of 27.5 miles per gallon. A company could sell cars that got worse mileage, so long as it sold a compensating number that got better mileage to bring the company's average up to the requirement. Like the Clean Air Act amendments of 1970, EPCA required the automakers to achieve a specified numeric standard by a deadline. While Congress allowed the executive branch to negotiate with the automakers over the specific CAFE standard that would apply in the 1981–1984 model years, it permitted no discretion with respect to the final 1985 goal. It did allow the secretary of transportation to set different fuel economy standards for light trucks. Congress chose 27.5 miles per gallon as the ultimate passenger car goal, not so much because it wished to punish the automakers by calling for a figure that represented a doubling of the average fuel economy of American cars, but because the 1974 joint Department of Transportation–EPA *Report to Congress on the Potential for Motor Vehicle Fuel Economy Improvement* suggested the feasibility of that standard. Unlike the 1970 Clean Air Act, EPCA asked nothing of consumers, mainly because the widespread public anger at the perceived profiteering of the oil companies during the energy crisis made any step that raised the price of gasoline further politically unwise. Over the next decade and a half the automobile industry slowly improved the average fuel economy of all passenger vehicles (cars and light trucks combined) to 21.2 miles per gallon in 1991.[23]

The urgency of improving automobile fuel economy during spiraling prices, slumping car sales, and high unemployment triggered a bitter public debate about trade-offs and compromises among three highly desirable policy ends. Many viewed the matter as a "no-brainer": emissions and safety regulations had to give way before fuel economy, which meant intense pressure to reopen the Clean Air Act and relax the 1970 standards and deadlines. Although foreign automakers appeared to take U.S. safety, emissions, and fuel economy regulations in stride, the American automakers complained that they could not make more fuel-efficient cars while trying to meet the emissions and safety standards at the same time. The automakers, who had argued all along that they could not meet the emissions standards mandated by the Clean Air Act by the specified deadlines—especially the standard for nitrogen oxide that the 1975 two-way catalytic converter did not address, now argued that they should be excused from meeting the safety and emissions demands. The 1974–1975 recession won the automakers an important new ally in this fight, the United Auto Workers (UAW). Big business and big labor made a persuasive team. For the first time in a decade federal legislators, especially the Democrats who controlled Congress, began to listen sympathetically to the automakers (now often accompanied by the autoworkers). Led by Dearborn's powerful Democratic congressman, John Dingell, ironically an early and important supporter of strengthening the nation's water pollution laws, many Democrats now argued that emissions laws had to yield before the economic needs of the auto industry and its workers.[24]

The auto industry also gained new leverage in the emissions debate due to the inaccurate but common public perception that the EPA was a clumsy and unrealistic implementer of the Clean Air Act's emissions standards. The transportation control controversy, with the prospect of sharply reducing the amount of driving that Americans did, and the possibility of an about-face on catalytic converters due to sulfuric acid undermined the agency's reputation. Moreover, the EPA had quickly grown into the largest federal regulatory agency, with nine thousand employees and a budget of more than seven hundred million dollars a year. Implementing and enforcing regulations that touched nearly every business and individual in the United States won the agency no friends. Russell Train wasn't joking when he told the National Press Club in September 1973 that his first week on the job had been unusual because the EPA wasn't sued. In the early years someone sued the agency nearly every day. Many of the controversies made front-page news. No surprise then when market researcher J. David Power reported in 1975 that "a growing feeling among customers that government control of the auto industry, once oriented strictly to customers, was messing up the auto market."[25]

It fell to Russell Train and his staff to prevent the wholesale gutting of the Clean Air Act as a result of the energy crisis, and the fight they waged between fall 1973 and early 1977 may have been the EPA's finest hour. Train and the EPA had to prevail against an opposition that included many parts of the federal government—the White House, the Federal Energy Office, and the Office of Management and Budget, much of Congress—and nearly all of big business. Critics worked hard to set up the EPA and the Clean Air Act as targets of the nation's energy frustrations. The oil industry waged an advertising campaign to blame the energy crisis on environmental controls. The coal, electric utility, and steel industries joined the automakers and oilmen in pushing to get the law modified. And the EPA had little support from the media and the public. But instead of getting steamrolled, Train and the EPA fought and won inside the Washington bureaucracy.[26]

A lifetime Washington insider and natural public servant, Train took the initiative to get the facts about energy-economy-environment trade-offs before the public, underscoring the multiple programs that the EPA had to reduce energy consumption and increase energy efficiency. "We will never come to grips with either our energy or our environment ills," he told a California audience in December 1973, "until we understand that they stem, essentially, from the same source: from patterns of growth and development that waste our energy resources just as surely and shamefully as they lay waste our natural environment." More important, he understood the importance of good information, and he ensured that the EPA had the best environmental, public health, scientific, and economic data in town. He carefully ensured that economic data could not be used against environmental regulations.[27]

When the push came in early 1974, the EPA was ready to push back, and in the end the proposals to amend the Clean Air Act that went to Congress from the administration were largely Train's proposals to give the EPA greater flexibility to extend deadlines for meeting the standards in special circumstances. The 1974 Energy Supply and Coordination Act pushed back the deadlines for meeting the automobile emissions standards by one year and gave the EPA administrator the option of pushing back the compliance date an additional year, which Train subsequently exercised in March 1975 during the sulfuric acid controversy. By the end of 1975 independent analyses commissioned by the EPA and other federal agencies had unanimously concluded that environmental expenditures to meet the new regulations added less than 1 percent to increases in wholesale prices. "Our approach was we fought every inch of the way . . . ," Train recalled, "and we [only] gave in here and there in small ways." Alvin L. Alm, the man Train tapped to lead the numbers fight, believed that the EPA "had the best economic analysis shop in government." Train put it more bluntly: "Our data was just a damn sight better." The EPA, with crucial help from the Clean Air Act's supporters in Congress and the catalyst manufacturers, held the line against relaxing environmental standards. It was a victory, but one tempered by the fact that the federal emission reduction effort never regained the regulatory leverage or public support that it had before 1973.[28]

The EPA's responsibility for cutting automobile emissions had gotten the agency deeply involved in automobile fuel economy well before the oil embargo. The engine modifications that the automakers used to meet the pre-1975 emissions standards had reduced fuel economy. When car owners complained, dealers blamed government regulations. Consumers then complained to their congressional representatives, and Congress turned to the EPA for an explanation. The EPA conscientiously admitted that engine modifications reduced fuel economy by 10 percent, but it also pointed out that air conditioning, automatic transmissions, and, by far, vehicle weight had much greater effects on vehicle fuel economy. Tired of being pestered, Eric Stork asked Tom Austin and Karl Hellman in Ann Arbor what data they had that might be useful. Austin and Hellman realized that while testing vehicles for emissions certification they had collected sufficient information to calculate each car's fuel economy. When the White House learned from the EPA that it had this data, Nixon ordered the fuel economy ratings published in the *Federal Register*. Journalists quickly recognized that the ratings could be arranged in rank order and pleaded with Stork to publish the rankings. Stork used the threat of doing this to convince the automakers to add the fuel economy information to the price sticker on each new 1974 model car. When Train announced this program at a September 1973 press conference, he called it "voluntary." "We blackjacked them," Stork recalled. "They didn't volunteer for that." But dealers of big cars just took the labels off. Consequently, Congress added mandatory fuel economy labeling to EPCA.[29]

Although the EPA successfully defended the Clean Air Act against substantial modifications and played an important role in improving automobile fuel efficiency, Detroit kept up the pressure for relief from regulation, even after the benefits became indisputable. Instead of forthrightly meeting the challenges posed by regulation and embracing safety, emissions reduction, and fuel economy as worthy ends that served the public and the national good, the industry complied only grudgingly. In author Jack Doyle's withering assessment, Detroit spent years "picking away at rules and regulations at every turn, mounting public disinformation campaigns, filing petitions and appeals, and going to the courts and Congress to overturn or change the law in any way it could." Russell Train remembered Lee Iacocca as an especially outspoken opponent of regulation. "He was very antagonistic and I thought unnecessarily so on the whole . . . and incorrectly." The automakers did not just fight harder. They became far more effective at doing so, especially after they lost the protection from even stricter regulation provided by a veto-wielding Republican president in the White House when Jimmy Carter defeated Gerald Ford in 1976. General Motors's Washington office staff grew from three to twenty-eight between 1968 and 1977. But effectiveness required more than just money and manpower; it required finding a conception of the public interest that trumped cleaning up the environment. Fuel economy failed in this role, thanks to Train, the EPA, and congressional supporters of the 1970 Clean Air Act amendments.[30]

In 1977, with the Clean Air Act up for renewal and the automakers threatening to stop production (and lay off workers) rather than violate the mandated exhaust standards, the industry backed Congress into a corner and forced Senator Edmund Muskie and his House counterpart Paul G. Rogers to "legalize" its 1978 models by agreeing to extend the deadlines for meeting the standards, already pushed back to the 1978 model year, a further two to three years. The nitrogen oxide standard was also relaxed from 0.4 to 1.0 grams per mile. In this fight the UAW stood squarely with management, as the automakers played the "jobs card" against Democrats in Congress, a fight led again by Dearborn's Democratic congressman John Dingell joined by James T. Broyhill, a Republican congressman from North Carolina. "The auto companies never got to first base in persuading Congress to relax auto emission standards," Leon Billings observed, "until they got the support of the UAW on the issue of jobs." But when they did, they found a public interest that trumped the environment, and they got the upper hand in the fight with Congress over regulation, gaining at least greater control over

"This here country ain't big enough for both of us,"
from *Herblock on All Fronts* (New American Library, 1980).
Library of Congress, Prints and Photographs Division,
LC-USZ62-126936.

the timing of meeting the standards. The auto industry did not get more in the 1977 Clean Air legislation in part because Engelhard used the introduction of its three-way catalyst to lobby wavering members of Congress against adopting any relaxed standards.[31]

"We say, 'we can't do it, we can't do it, it's gonna cost too much, it's gonna cost too much,'" Joe Colucci of GM Research recalled years later, but then "we do do it, and it doesn't cost as much as we said it's gonna cost. It's the history of the whole emission control thing." What he meant was that his company said it could not be done and fought against having to do it, while he and his colleagues actually went out and did it. The industry delivered under regulation, but it fought so hard against regulation that by the end of the 1970s it was able to deliver increasingly on its own terms. When it discovered that by linking regulation to job loss it could trump emissions control as a policy priority, it found a strategy that rendered unlikely the prospect of further substantial regulation. The link was tenuous at best, but it had plausibility (thanks, in part, to the automakers' own threats to close plants) in a context of growing concern about the ability of the federal government to solve problems. More important, no other public interest group had the resources to match the auto industry (and big business in general) and argue otherwise. The industry turned this corner before the market disaster of 1979–1982. [32]

Events at the end of the 1970s and beginning of the 1980s confirmed the conclusions about the inevitability of smaller cars that many people drew from the turmoil of 1973–1975, and they did so in an especially disheartening way. Auto sales picked up in fall 1975 and then boomed for three successive years—1976, 1977, and 1978. The most notable aspect of this resurgence was that Americans drifted back to larger American cars, even though both automobile and gasoline prices remained much higher than they had been before 1973. This interest was explained in part by deferred demand and by the general rebound in the economy. Banks and finance companies helped, too, by lengthening the typical car loan from thirty-six to forty-two or forty-eight months, which cushioned the price increases of recent years by holding down the increases in monthly payments on car loans. Although Congress took an important step when it mandated fuel economy improvements under EPCA, it could not resist pandering to consumers and punishing the oil industry. Congress not only extended price controls on domestically produced oil until 1979, it actually forced a rollback in the price of domestic oil for the year 1976 as a kind of consumer "rebate" to compensate for the oil industry's alleged price gouging and windfall profits during the 1973–1975 energy crisis. Congress thus intervened in the market to keep the price of gasoline lower than it would have been, dampening an incentive for Americans to switch to smaller, more fuel-efficient cars.[33]

Though not unwelcome in Detroit, consumers' renewed interest in larger vehicles caught the industry by surprise. GM had just completed a $1.1 billion crash program to "downsize" and lighten its 1977 vehicles. Automakers and regulators alike concluded that substantial additional improvements in fuel economy to meet the EPCA 1985 CAFE standard could be met only by further reductions in vehicle size and weight. Still, there was more money to be made per car in larger cars, and the automakers happily sold them, although they now had to be careful not to sell too many gas guzzlers or the imbalance would wreck the maker's CAFE figure and subject the company to large federal fines. Who was going to force the public to buy the smaller, fuel-efficient cars when the public wasn't buying enough of them? The answer was Detroit, and the automakers actually began rationing sales of their larger models, despite their own economic interest in selling as many of these models as possible.[34]

In 1976 market observers also noted a resistance to small cars on the part of consumers that had not been evident before. Now that both Detroit and Washington had concluded that the small car was the car of the future, many Americans stubbornly resisted the idea of buying one. The cachet that sold small cars in California and elsewhere in the early 1970s—in part as a contrarian act of free will—vanished. Small cars began to carry the stigma of Detroit, Washington, and the energy crisis. If most American consumers understood the need for smaller cars, many nonetheless balked at buying them when they felt forced to do so. And with both Detroit and Washington actually helping them to buy larger cars, many consumers concluded that they could buck the trend. For a time large and small car buyers alike hoped with some reason that the events of 1973–1975 had been a painful aberration.[35]

When gasoline prices again soared in the wake of the Iranian Revolution in 1979, the terror of spot shortages and lines at gas stations instantly ended the renewed consumer interest in larger cars. Although Detroit now had plenty of subcompact and compact models and had made most of their larger models smaller and more fuel efficient, the second oil crisis caught the industry, especially Henry Ford II and the Ford Motor Company, leaning back toward intermediate and large models. By April 1979, sales of these models had collapsed, and once again there was a run on small cars that Detroit could not meet. By year-end 1979 nearly 60 percent of the cars sold in the United States were subcompacts and compacts. With the ten most fuel-efficient cars sold in America all foreign made, sales of small Japanese imports soared, and foreign sales—70 percent of them by Japanese makers—approached the 25 percent market-share barrier for the first time. In June 1979 a *New York Times–CBS News* poll found that 60 percent of Americans would accept gasoline rationing if it ended shortages and curtailed skyrocketing prices. Sales of American cars rallied during the summer but collapsed again in the fall, driving the U.S. economy into a recession and the automobile industry into a depression worse than the 1974–1975 downturn. Seventy percent of Americans reported driving less, according to J. David Power, a figure borne out by government statistics that showed that VMT fell slightly and gasoline consumption dropped 8 percent between 1978 and 1980, the most dramatic evidence of changed driver behavior since World War II.[36]

Gas line, 1979. *Photo by Warren K. Leffler. Library of Congress,
Prints and Photographs Division, U.S. News and World Report
Magazine Collection, LC-U9-37733, frame 8A.*

American automobile owners and the American auto industry faced an unprecedented set of problems in 1979. Inflation reached 11.3 percent, which
eroded the purchasing power of millions. Car loan interest rates eventually
soared to 17 percent to keep ahead of inflation, resulting in monthly payments
the likes of which Americans had never seen before. Since the typical auto loan
had recently been lengthened, lenders lacked that option to cushion the new
shock. Banks and loan companies in states where usury laws capped the interest
rate simply stopped making auto loans when the rates rose above the cap. As
the recession rippled outward from the auto industry, unemployment started to
climb, dealing another blow to consumer purchasing power. Decreased sales,
especially of larger vehicles that contributed most to profitability, meant that
the automakers had less money to build the fuel-efficient cars required by 1985.
As one industry analyst put it, "Detroit's dilemma is that it must sell cars no one
wants to get the cash to make the cars that people will buy."[37]

If 1979 was bad, 1980 was worse. Inflation for the year reached 13.5 percent.
By May automobiles sold at an annual rate of five million cars, the lowest level
since the Buyers' Strike of 1958. Consumer confidence, as measured by the
monthly University of Michigan Index of Consumer Sentiment, reached its all-

time low of 54.4 in spring 1980. Layoffs idled more than three hundred thousand of the industry's 750,000 workers. Perhaps another five hundred thousand to six hundred thousand American workers had been laid off because of the auto industry's slump, accounting for one-third of the 2.3 million increase in unemployment during the 1979–1980 recession. Some dealers reported losing nearly half their potential sales because would-be buyers could not find financing. In the second quarter of 1980 the four American automakers lost a combined $1.5 billion, the largest quarterly loss ever experienced by a U.S. industry. Ford, it appeared, might join Chrysler in sliding toward bankruptcy and requiring a federal bailout. The third quarter was even worse. The four companies lost $1.8 billion. The smaller, more fuel-efficient, front-wheel drive 1981 models like Chrysler's K-cars and the Ford Escort came with substantial price increases, a desperate gamble to generate revenue that did not pay off. Consumers simply balked at the combination of high price tags and high interest rates on car loans. Chrysler lost $1.8 billion for the year. Ford lost $1.7 billion (and another billion in 1981). GM reported its first annual loss—$762 million—since 1920–1921. "The fast, hard fall of the U.S. auto industry," *Time* glumly reported in September 1980, "will doubtless rank as one of the most remarkable collapses in the annals of American business."[38]

The Japanese automakers, in contrast, cleaned up. In 1980 they sold a record two million cars in America, 25 percent of the market. Imports even briefly broke the 30 percent barrier for the first time. In spring 1981, under threat of congressional action, the Japanese automakers "voluntarily" agreed to limit their 1982 automobile exports to the United States to 1.68 million vehicles a year. (This agreement lasted until 1985.) They also accelerated plans to assemble their cars in the United States so that they could sell "American" cars in America. Honda, moving aggressively into the American automobile market, opened the first assembly plant the next year. The American economy experienced another and deeper recession in 1982, as Federal Reserve Board chairman Paul Volcker deliberately tightened the nation's money supply to wring inflation out of the economy. Detroit had another horrendous year. The American makers did not recover until 1983. About the only silver lining for the industry was that its woes, together with the 1980 election that placed Republicans in the White House and in control of the Senate, ended the prospect of further regulation and even opened the door to the possibility of regulatory rollback. But by then the industry had more serious problems than regulation.[39]

Given the 1974–1975 heads-up that the American industry had been given, the told-you-so recriminations from 1979 to 1982 were large. American automakers bore the brunt of this criticism. Much of it boiled down to the entirely too simplistic charge that they made the wrong cars. "The American built car," *Time* wrote, "has become an object of derision and jokes. All too many American drivers now consider the cars they once fawned over to be simply too big, too heavy and too expensive." Pessimism—indeed demoralization—pervaded this discussion. American automakers—managers and workers alike—simply could not compete against Japanese automakers that were now setting the world standards for automotive excellence. "Many Detroit executives," *Business Week* wrote at the nadir, "fear that the industry will simply not be able to make small cars whose quality, price, and performance will prevent further erosion of U.S. market share." A survey of American auto engineers by *Ward's Auto World* found that nearly half felt that Japanese cars were better made than comparable American models.[40]

Nearly all the criticism downplayed or ignored the role that the fickle tastes of American consumers had played in Detroit's problems. Well before the debacle of 1979–1982 American automakers had accepted the need to build smaller cars—indeed, they were committed—at least to a point—to building fleets of smaller, more fuel-efficient vehicles. If not jumping with the crisis-inspired alacrity that suddenly seemed necessary in 1979 and 1980, the automakers were steadily moving in this direction. They believed that there would continue to be a decent-sized market for larger automobiles, but they knew that most of the cars they had to sell in the 1980s would be small, fuel-efficient ones. Like 1958, the automakers found themselves out of step with consumers and unfolding events only to the extent that it was impossible to produce either highly fuel efficient cars or the right mix of smaller and larger cars overnight in response to sudden shifts in consumer demand.

But Detroit's heart was not really into building small cars. Henry Ford II called them "little shitboxes." Perceptive critics such as Ford and Chrysler executive Hal Sperlich later admitted a deeper truth to author David Halberstam: "The small cars that Detroit had produced, [Sperlich] believed, had been bad ones, reluctant efforts at best. If anything the companies had brought out inferior, unappealing small cars to prove their own thesis that small cars were in fact inferior and unpopular. The cars were not very good, were weak and underpowered, and the companies had not pushed them, and that had proved, the industry argued, that Americans did not like small cars."[41]

With 65 percent of the cars sold in United States compacts and subcompacts, the consensus about the future of the automobile in America in 1980 was nearly absolute. "There is now no longer any doubt that big cars are dead," *Time* wrote. "Cheap energy is . . . gone for good, and with it the big car," *Business Week* concluded. "Unlike the panic of 1973, which produced only a short-lived obsession with economy," *Fortune* wrote in June 1980, "the upheaval of 1979 seems to have changed forever the American consumer's tolerance for fuel-inefficient cars." Only very small, light, fuel-efficient or alternative-fuel cars would be sold in the future. "By the mid-'80s," GM chairman Thomas A. Murphy proclaimed, "there won't be any big cars left at all." Lee Iacocca, who had been fired by Henry Ford II only to be hired by Chrysler, said of his new company's cars, "We're going to be building three types—small cars, smaller cars and smallest cars." *Forbes* magazine, self-proclaimed "capitalist tool," lauded the federal government and Congress for the CAFE standards. The magazine told its business-class readership that auto executives admitted that their cars would not be nearly so far down the path to fuel efficiency if not for the regulations. But regulation had been rendered unnecessary by the market—at least for the time being.[42]

The impact of these events on the design of the typical American car was large. The first oil shock forced Detroit to lop off a foot or so of overhang and seven hundred pounds of weight from their vehicles, but did not result in any major steps to redesign the basic American car. Even before the second crisis, the automakers knew that more fundamental changes would be required to meet the 1985 EPCA CAFE standard. The shock administered to the industry in 1979–1980 hurried these efforts. The first casualty was the V-8 engine, which powered 76 percent of the American cars sold in the United States in 1977. The V-8 was replaced by four- and six-cylinder engines, mostly the former. The next major step that the industry took was front-wheel drive with the engine mounted transversely, or cross-wise, which permitted manufacturers to reduce the space under the hood and maintain interior space by removing the drivetrain hump that ran down the car's center to the rear wheels. Finally, and most important, pushed by the need to run at the stoichiometric point so that the three-way catalytic converter could reduce nitrogen oxide, the automakers replaced carburetors with electronic fuel-injection systems, a step that also facilitated, as a now very welcome side-effect, substantial gains in fuel economy.[43]

As these changes were made, American cars began to look much more like one another. Observers began calling them "commodity" products. When consumers also got the impression that Detroit was not in the business of making the best cars at the lowest prices and the Japanese companies were, they chose cars accordingly. After all, few people wanted to be rolling advertisements for their own stupidity. The arrival of front-wheel drive and the voluntary import quotas on Japanese cars gave Detroit a reprieve to arrest its deteriorating market position, but the quality of the new American cars of the early 1980s remained highly suspect. As the 1980s began, the American industry was on the ropes but still standing, and American automobile consumers were dazed, weary, and confused. If a new era had dawned in their relationship with automobiles, there were many who were not sure they liked it.

From the late 1960s to early 1980s, Americans bought small cars in substantial numbers, the only time that they did so in their twentieth-century relationship with the automobile. In retrospect most Americans associated this development with the energy crisis of the 1970s. Yet the preference for small cars that developed between 1967 and 1973, especially that part of consumer demand encouraged by fashion rather than by the "push" of demographic or economic need, suggested that the story was not so simple. As a largely unintended side-effect, the growing appeal of small cars to Americans between 1958 and 1973 proved that it was possible at times for consumer tastes and manufacturer offerings to evolve in a direction that reduced the impact of automobiles on the environment without government intervention to regulate the nature of the car. The voluntary shift in consumer tastes toward small cars produced modest environmental benefits. When the fad ended, the environmental benefits would have gone, too, unless environmental advocates had successfully fused in the popular mind a link between small cars and environmental improvement and made environmental concern an important continuing positive motivator in consumer behavior. The development of markets for "green" products in other industries in the last quarter of the twentieth century suggested that this possibility should not be entirely dismissed, but for automobiles it did not happen. The environmental benefits from government regulation of the automobile far exceeded those from the voluntary shift to smaller cars. By the 1980s, cars in America were much safer, cleaner, and fuel efficient than they had been in 1965, but mostly because government had forced the industry to move in these directions and not because consumers had collectively asserted their market power and encouraged the industry to make these changes.

Rather than permanently fixing this environmentally positive alignment of consumer tastes and manufacturer offerings, as everyone assumed in 1980 when they proclaimed that the American love affair with the big, powerful, gas-guzzling automobile was finished, the twin energy shocks of the 1970s actually destroyed the positive connotation that small cars had for consumers before 1973 and transformed the cars into symbols of painful sacrifice, diminished expectations, and national humiliation. Americans found themselves driving smaller, less powerful cars at slower speeds on their nearly completed Interstate Highway System. Breaking the law to drive sixty-five miles per hour on roads designed for seventy-five, while anxiously trying to avoid state troopers, was not fun. Perhaps the best way to turn off a sizeable number of American consumers to small cars was to stamp them with the approval of both big government and big business. The inflation, energy shocks, and recessions that came with—not because of—the regulations nonetheless compounded the problem. The association of small cars with extraordinarily negative external events and pressures—indeed with a profound sense of national impotence—left a sour taste in the mouths of many Americans even as they bought smaller cars.

Although no one posed it at the time because everyone believed the shift to smaller cars to be permanent, the real question was what kind of cars would American consumers buy should conditions change? Would they still buy small cars when gasoline prices stabilized or fell? Would they accept the sameness in size, performance, and styling that resulted from downsizing sedans? Would color alone be sufficient to express their sense of individuality now that 50 percent of American households contained more than one vehicle, each ever more likely to be a personal car? What would happen as Americans became wealthier and more of them could afford multiple higher-priced vehicles? How far would concern for the environment and concern for one another influence their choices? Far from being closed, these questions remained open as the 1980s began.

12

THE RIDDLE OF THE SPORT UTILITY VEHICLE

The relationship between Americans and automobiles over the last two decades of the twentieth century proved quite different than predicted in 1980. The future belonged, not to the small car, but to big cars—very big cars—or, more precisely, to light trucks. Largely because of the Baby Boomers' enthusiasm for pickups, minivans, and sport utility vehicles (SUVs), half of all the vehicles sold in America by the end of the 1990s were light trucks—a milestone that marked the culmination of a styling revolution in the American automobile market.

But this revolution in consumer tastes had its consequences. When millions replaced their cars with light trucks, progress toward improving the safety, cleanliness, and fuel efficiency of American motor vehicles began to slow and, by some measures, was reversed. Americans, who had grown up listening to Ralph Nader crusading for safer automobiles, who knew that automobiles caused smog, and who lived through the energy crisis in the 1970s, certainly understood that larger, heavier vehicles burned more gasoline and posed a threat to smaller, lighter vehicles in collisions. Yet it was as if these Americans had not lived through the fifteen years before 1980 or suddenly repudiated these goals. Exploring why these people did not care enough to choose differently, even when they knew better in light of their experiences in the 1970s—the riddle of the sport utility vehicle—provides a telling end to both the century and this story.

Pickup trucks had been around almost as long as cars. And as the 1980s began, SUVs had been around for twenty years, a tiny niche cultivated by Willys Jeep and International Harvester and, from 1965, by Ford and Chevrolet. Together, pickups and SUVs constituted about 20 percent of passenger vehicle sales in the United States—although nearly all of these were pickups. Traditionally, consumers bought pickups for work and commercial purposes. A far smaller group of outdoor enthusiasts bought SUVs as specialized recreational vehicles and, indeed, often took them off-road. Naturally, some people had always used pickups and SUVs chiefly as a substitute for cars, and the automakers' marketing efforts actively encouraged this use in addition to their traditional uses. Evidence in the 1970s showed a substantial shift toward use as car substitutes. But pickups and SUVs were large, heavy vehicles that got poor gasoline mileage, so sales plummeted after 1978. When everyone concluded that high gasoline prices were permanent and the future belonged to the small car, it seemed that using a pickup or an SUV as a car would be a luxury of the past.[1]

Responding to the energy crisis, the automakers introduced compact pickups and compact SUVs that got better gas mileage. And they aimed these new models even more squarely at the suburban market by improving the passenger comforts offered. American Motors, General Motors, and Ford all introduced lighter 1983 and 1984 model four-wheel-drive SUVs (Jeep Cherokee, Chevrolet S-10 Blazer, and Bronco II). Chrysler also introduced the minivan in 1983, a concept that Lee Iacocca had unsuccessfully pushed for when he was at Ford. With the compact pickups and SUVs, the automakers simply made product adjustments to existing niche products within the larger context of an American automobile market forever changed by the energy crisis.[2]

When consumers began to buy these downsized vehicles in greater than expected numbers in 1983, followed a year or so later by an upturn in sales of the full-sized pickups and SUVs, these developments took the automakers by surprise. The sudden popularity of the SUVs was especially anomalous. "Safari-style transportation picks up appeal in the U.S. of all places," headlined a March 1984 *Wall Street Journal* blurb on the new trend. Light trucks not only survived the energy crisis, they now seemed to be thriving, with sales growing from 20 percent of the market in 1981 (two million) to nearly a third in 1987 (4.9 million). Minivans, compact pickups, and SUVs paced these sales. Industry research showed that two-thirds of light truck buyers bought them primarily for personal use. SUV sales jumped six-fold from 132,000 to eight hundred thousand from 1982 to 1985. "Barring a return to the gasoline lines of the 1970s and sharp rises in the price of fuel," auto reviewer Charles E. Dole predicted in April 1984, "the full size sports-utility vehicles should be around for some time."[3]

Analysts called them "crossover" buyers, mostly Baby Boomers—then younger consumers—who substituted light trucks for cars as their basic vehicles. The median age of SUV buyers in 1987, for example, was thirty-five when the median age of buyers for many domestic car lines was over fifty. Nearly 80 percent bought the vehicles for personal use. The SUV phenomenon began with male Baby Boomers, especially among that wealthier subset of Boomers who epitomized Reagan-era conspicuous consumption—the young, urban professionals, or "yuppies." At the same time, a companion boom in large pickups occurred among working-class Boomer males. In the 1980s, men drove more than 80 percent of the light trucks sold in America most of the time, and SUVs and pickups became important accessories to Boomer conceptions of masculinity. Commenting on the demographics of the early SUV buyers, then Chrysler executive Robert A. Lutz expressed surprise. "They were young, with good education and high income, like import buyers." After a decade of being preached to about limits and lowered expectations, these Boomers turned their backs on the concerns and cars of the 1970s.[4]

The suggestion made by observers later that falling gasoline prices triggered the light truck boom was too simple. Oil prices peaked in the first half of 1980 at nearly forty dollars a barrel for West Texas intermediate crude and remained at high levels (roughly twenty-five to forty dollars) through 1985. But with the price of oil under strong downward pressure from the growing production of non-OPEC nations, OPEC's cartel collapsed. In eight months between November 1985 and July 1986 the price of oil fell by nearly two-thirds from about thirty dollars to eleven dollars per barrel. Yet the groundswell of interest in SUVs began in 1983–1985. The real bane of large vehicle sales was uncertainty—how high would the price of gasoline go, how long would it remain there, would gasoline be readily available at all. When these were open questions consumers did not buy large vehicles. Waning uncertainty, and not a sharp fall in the price of gasoline, initially made the context congenial for large vehicles again, a reminder that people who buy large, expensive automobiles have usually not been too concerned with the price of gasoline.

The end of high inflation and falling interest rates also opened the door to the light truck boom. By 1983, Federal Reserve Board chairman Paul Volcker had substantially reduced the rate of inflation and car loan interest rates had started down again. As the economy revived from the sharp recession of 1982, consumer confidence surged. Beginning in spring 1983 the University of Michigan's Index of Consumer Sentiment registered eighteen consecutive quarters above 90—the longest and highest sustained level of consumer optimism since the 1960s. The first ten of these quarters occurred before the sharp fall in oil prices. Yet the affordability and availability of gasoline, falling interest rates, and consumer optimism did not explain why Americans bought SUVs, only that they could buy larger, more expensive cars if they desired.[5]

There was no special marketing magic behind the light truck boom. The automakers mostly benefited from the improved economic climate and the unanticipated change in consumer tastes. The companies had always marketed pickups and SUVs in a manner that appealed to a variety of needs, including off-road recreation and suburban living. After 1980, the automakers changed their generational focus to the Baby Boomers, who were beginning the twenty-year process of becoming the segment of the market with both the largest numbers and the most money. The idea of rugged, off-road recreational activities resonated much more strongly for Baby Boomers in the 1980s than it had for their parents. To underscore this connection, the automakers designed SUVs to go off-road, a fact repeatedly underscored by vehicle reviewers in the press.

However, the automakers and these new SUV buyers had no illusions that the vehicles would ever leave the road. Market research in the 1980s suggested that no more than 10 percent of four-wheel-drive vehicles did so. Yet the companies and their ad people certainly worked with Boomer males to create a potent Walter Mitty–style fantasy around SUV ownership and the wild outdoors. To many, this advertising seemed an egregious effort to "greenwash" the SUV's environmental problems, but this angle should be pushed carefully. Automobile consumers, in contrast to the consumers of many other products, got only 20 percent of the information about the cars they bought from paid advertising. Indeed, some marketing consultants scoffed at the effectiveness of advertising at all. "Television," sales training consultant D. Forbes Ley told *Automotive News* in 1986, "is like background music."[6]

On the other hand, psychologist Timothy D. Wilson argued that advertising often has its most powerful motivational impact when it is indirectly experienced, that is, when it is background music. If so, SUV marketing and sales undoubtedly benefited from the emergence of the modern environmental movement and the fact that nearly all late-twentieth-century Americans considered themselves to be — in the broadest sense — environmentalists or outdoor people. The outdoors had long been one of the most important associations with American masculinity, and outdoor imagery always presented a great place for marketers to rendezvous with anxious American males. The names that the automakers gave to their SUVs tapped into this older outdoor masculine imagery: Scout, Bronco, Blazer, Range Rover, Trooper, Yukon, Pathfinder, Wagoneer, Cherokee, Durango, Explorer, Expedition, Navigator. On direct reflection, the "cowboys and Indians" associations evoked by these names can make the connection between automobiles, masculinity, and the outdoors sound silly. But the Baby Boomers' experience of the modern environmental movement may well have given this appeal greater unconscious force. "Even though they are used mostly on the road," market researcher Thomas F. O'Grady noted, "sports utilities fit people's images of themselves. They project a rugged, tough image that some people like."[7]

Whatever the deeper motives, Americans bought SUVs largely for reasons of fashion. As Alfred P. Sloan and so many others before him observed, cars were subject to the same "laws" of fashion as women's dresses. Many who followed the American automobile market in the 1970s concluded that the childish days of automobile fashion were over. Developments in the 1980s suggested that far too many people embraced this conclusion. The automobile remained the most expensive, publicly used, personally associated consumer product. So it retained the capacity for doing important work beyond providing transportation. This obvious and familiar potential made it hard for Americans to accept the one-size-fits-all notion of the car.

As it transpired, the American automobile market in 1980 was ready for a major shift in tastes. Not just large, but also tall and boxy, the SUV offered the most significant departure in popular motor vehicle styling since the 1920s. For decades the automakers followed Harley Earl's aesthetic, building long, low sedans whose length slowly increased or decreased and whose shape oscillated back and forth between curves and angles. Such incremental changes reflected the cautious, defensive nature of automobile styling, a consequence of the automakers' great fear that a radical departure from an industry styling norm might result in lost market share and punishing losses. With time, consumers viewed these changes as infinite minor variation on the same style—and thus with boredom. When commentators in the late 1970s and early 1980s argued that cars had "turned into commodity products," they noted, albeit implicitly, the potential for a significant portion of American consumers to decamp to a different style should a major alternative present itself. The SUV emerged as the great new alternative.[8]

"For Drivers, the Fun Is Back," the *New York Times* proclaimed in 1986. Fun was certainly the leitmotif of the early years of the SUV boom. Baby Boomers frolicking in the great outdoors . . . a cool new vehicle style . . . cheap gas . . . no more doom and gloom from the bad old 1970s. It was, well, "morning again in America," as a campaign ad for Ronald Reagan proclaimed in 1984. Consumers, the automakers, and the media all joined in striking this upbeat chord. As Jack Kadden noted in a *New York Times* review, Suzuki marketed its small, low-budget four-wheel-drive Samurai to nineteen to thirty-five-year-olds "as an antidote to the boredom of the subcompact econobox" that promised, in the words of Suzuki's own advertising, "more smiles per gallon." The fun was also back for Detroit. "Domestic trucks never developed a poor reputation for power and reliability," auto analyst Maryann N. Keller observed. In other words, they had not been stigmatized by the 1970s. "This shift in consumer taste," wrote the *New York Times*'s John Holusha, "appears to be presenting General Motors, Ford Motor and Chrysler with a second chance to reclaim some of the buyers they have lost to Japanese companies."[9]

For five years the light truck boom unfolded, invoking little more than wonder. But in June 1988 two clouds appeared on the horizon. On 2 June Consumers Union, publisher of *Consumer Reports,* held an unusual press conference to demand that the federal government recall the 150,000 Suzuki Samarais on the road because its testers had found that the car had a tendency to roll over if maneuvered quickly to avoid an obstacle. Then on 23 June James E. Hansen, director of NASA's Institute for Space Studies in Manhattan, told the Senate Energy and Natural Resources Committee that "it is time to stop waffling so much and say that the evidence is pretty strong that the greenhouse effect is here," an event that more than any other triggered the emergence of global warming as a significant public policy issue in the United States. Periodically since the nineteenth century, scientists and engineers had raised the prospect that humans burning carbon-based fossil fuels like coal, oil, and gasoline might release enough carbon dioxide into the atmosphere to raise the Earth's temperature. By the end of the 1980s, this possibility had gained considerable credence among scientists. And with every gallon of gasoline consumed producing twenty pounds of carbon dioxide, the automobile was an ever-growing part of the problem.[10]

Rollovers and global warming made little impact on consumers. Instead, their infatuation with SUVs grew. In the 1980s, the SUV appealed primarily to men and childless couples interested in tapping into and projecting an outdoor image, but as the popular minivan came to be seen as a stereotypical badge of responsible suburban family life, families looked to the SUV as a vehicle that downplayed the family image and permitted the projection of images less one-dimensionally domestic. The sporty, outdoorsy, semiadventurous imagery associated with the SUV, combined with its practicality, made this possible. The automakers responded with changes to their SUVs that catered to the family market, especially by challenging Chrysler's four-door Jeep Cherokee with four-door models of their own that offered additional passenger car features.[11]

The Ford Explorer, introduced in April 1990 to replace the Bronco II, epitomized the adaptation of the SUV to this broader market. Ford's designers gave the Explorer the more luxurious, feature-laden interior of an upscale passenger car. At the same time, they provided the vehicle with a number of safety enhancements. The Explorer was an instant hit, eclipsing the Cherokee to become

Style, status, and symbol for a new century. SUV, 2006. *Photo by Lisa Dumont.*

the best-selling four-door SUV in America in its first year. Within a year and a half, it outsold Chevy's new four-door Blazer and the Cherokee combined. Many analysts considered it the breakout vehicle that sparked the great SUV boom of the 1990s. At mid-decade, Ford sold nearly four hundred thousand Explorers a year, and the vehicle remained the best-selling SUV into the new century. In 1992 light trucks accounted for 37.4 percent of all passenger vehicles sold in America, up from 19 percent in 1975. The Big Three dominated this market. In fact, by 1993, Chevrolet and Ford both sold more light trucks than cars, and the high profit margins on light trucks made them the American industry's great money maker.[12]

Contemporary analysis of the SUV boom argued that American men bought the vehicles for the macho outdoor image and women—a rapidly growing percentage of buyers in the 1990s—bought them for safety, but there was also evidence that 1990s' "soccer Moms" found other satisfactions in SUV ownership. "For years men drove around in big cars and trucks and looked down at women, at their legs," SUV-owner Jill Hedstream noted. "Now I think a lot of women are enjoying riding around and looking down at the little men." As Headstream suggested, American women appropriated a symbol of American masculinity and used it to strengthen their own sense of self-worth. The ready appropriation of the SUV's symbolism by women for their own purposes marked, probably as well as any indicator, that the SUV had achieved the coveted role of status symbol.[13]

The success of the Ford Explorer, and the much larger community of SUV owners that resulted, motivated some owners to differentiate themselves from one another. Detroit met and encouraged this desire by offering larger SUVs and SUVs from car divisions that had not traditionally sold light trucks, including the luxury car divisions. Chevrolet's Suburban and Ford's successively larger SUVs, the Expedition and the Excursion (which critics dubbed "the Excessive"), illustrated the former trend and the Lincoln Navigator the latter. Even the king of the car era surrendered, and Cadillac offered its own SUV, the Escalade. In this consumer community distinctive meant bigger, pricier, and more luxurious. For Detroit it meant stupendous profit margins. Harley Earl, no doubt, spun in his grave, but his patron, Alfred P. Sloan, could only be smiling.

There was more to the appeal of the SUV's styling than its different shape and large size. The SUV offered owners a form of conspicuous display with some egalitarian cover. As automotive journalist Trip Gabriel put it, "They do not automatically shout out an owner's income bracket." The vehicle was practical. It was outdoorsy. It said that its owner was successful but down to earth. Despite growing evidence to the contrary, late-twentieth-century America still expected the wealthy to display the common touch and show a respectful deference to egalitarian norms. "Buying one," longtime industry observer Jerry Flint noted, "is a way of being trendy without seeming to want to be." Thus, the Chevrolet Suburban made the perfect vehicle for 1990s taste-maven Martha Stewart to be chauffeured around Manhattan and Connecticut. It also appealed to a growing number of CEOs, worried perhaps that as their compensation soared through the stratosphere, they would lose touch with the rest of their fellow Americans.[14]

When asked why they purchased their vehicles, SUV owners produced the same rational-sounding, practical reasons that car buyers had always articulated to justify their decisions. They needed four-wheel drive for safety in bad weather. They needed a big vehicle for towing. They needed the space for their families. Most probably believed they did. But when surveys measured behavior, the facts contradicted the statements. Only 6 percent of SUV owners ever used four-wheel drive. The industry's own studies found that 15 percent of owners used their SUVs for towing, including less than half of the owners of the largest SUVs. Similarly, 75 percent of the owners of the largest SUVs lived in households with three or fewer members. Often unaware of the deeper reasons involved, Americans had always used similar rationales to defend their vehicle choices. No American has ever admitted that he or she purchased a status symbol to advance or shore up his or her self-esteem, yet few vehicles have ever made the point as directly as the SUV. Jerry Flint had no illusions on this score: "As they always have, people chase status. Only today they call it 'reliability' and 'ruggedness.'" Automobile market researcher Christopher W. Cedergren was not fooled either. "Americans make up all sorts of excuses—they like sitting up high, they want all-wheel-drive for bad weather. But they really want S.U.V.'s for one reason. They're cool." By the mid-1990s, commentators, struggling to explain the SUV's appeal to a growing number of market segments, called them "unbeatable fantasy machines." For a consumer product, it did not get any better than that.[15]

In the 1980s and 1990s Americans created community—if that was still the right word—in SUV ownership in a time of anxious prosperity, much as they had in the early 1920s around the simple fact of automobile ownership or in early 1950s around the Big Three's be-chromed and tailfinned Cadillac and Cadillac wannabes. But the prosperity of these two decades was also a new kind of prosperity. Success and failure in this new era had a sharper edge. Junk-bond king Michael Milken earned $550 million in 1987—as an employee! Yet white-collar managers found themselves vulnerable to dismissal as never before. A master of the universe today. Downsized tomorrow. The traditional middle-class redoubts of greater education, competence, and hard work no longer guaranteed basic economic security.

In this anxious context many Americans reached for a consumer product that allowed each person to reassert in symbolic terms an unmistakable claim to status and respect. Sue Zesiger probably said it best. Writing about the message conveyed by the largest SUVs, she observed, "They are not about downsizing, fitting in, making do with a little less." Caught up in a whirlwind of change now beyond the ability of even the most successful to control, anxious Americans used SUV ownership to assert mastery, individuality, and defiance. As Matthew L. Wald observed, "For the commute between high stress job and hectic home, it may be more mastery than people get at either destination." In SUVs American automakers and consumers again found common ground, and their mutual interests in pursuing this common ground outweighed any qualms that some may have had about imposing greater social and environmental costs on the larger community. The "to hell with you, we're having fun" contrarianism in this market turn could be smiled upon indulgently if not for the problems.[16]

There never was any mystery among automakers, government regulators, or consumer and environmental groups about the safety, emissions, and fuel economy problems with light trucks. After the 1970s anyone who knew anything about automobiles and how they were regulated knew that there were problems with this massive change in American consumer tastes. Yet public concern about light trucks, and SUVs especially, grew slowly, even reluctantly, during the 1990s, a reminder that, while there generally are always people who perceive and criticize the problems, it is quite another thing to rouse the larger public to demand that they be addressed. By century's end, however, the specific problems of the SUV had made it into the mainstream media and become better known, thanks in part to reporting by *New York Times* journalist Keith Bradsher, who covered the industry between 1996 and 2001 and then wrote about SUVs in *High and Mighty*. The problems that Bradsher detailed involved all three major areas of motor vehicle regulation—safety, fuel economy, and emissions. They stemmed from the fact that the automakers wrung regulatory concessions for light trucks from Congress and federal regulators and because the automakers chose not to do as much in these areas as they might have.

The automakers designed and marketed SUVs to convey the impression of greater safety and security for their drivers and passengers. And many consumers joined the SUV boom truly believing that they had chosen safety as a benefit that outweighed the SUV's problems. The boom itself frightened many Americans into buying SUVs because as the number of light trucks on the road increased, driving a car became more and more perilous. As larger vehicles, SUVs posed a lethal threat to the occupants of smaller vehicles in multivehicle collisions. "SUV buyers who do not need to go off-road," wrote Bradsher, "are literally putting their neighbors at mortal risk for the sake of fashion." But the impression of SUV safety was part illusion. Normally, a larger, heavier vehicle like an SUV is safer, but the height of the SUV (which made it easier to see the road) raised its center of gravity. Height combined with the poor handling of the vehicles to make them much more prone to rollover. This problem, which Consumers Union had raised in 1988, came back to bite apathetic consumers and the Ford Motor Company in the firestorm that followed the August 2000 announcement by the National Highway Traffic Safety Administration that a fatal mismatch between the Ford Explorer and its factory-installed Firestone tires was killing Explorer drivers and passengers in rollovers after high-speed tire blowouts.[17]

Light trucks presented two emissions problems. First, regulatory exceptions permitted them to emit more hydrocarbons and nitrogen oxides. As a consequence, some SUVs emitted as much as five and a half times more smog-causing pollutants per mile than cars. By 1997, when half the new vehicles sold in America were less stringently regulated light trucks, the emissions from the sixty-five million light trucks already on the road equaled those from 125 million cars. On a per vehicle basis, both cars and light trucks produced 90 percent (or better) less smog-causing emissions in 1999 than in 1970. But since more people drove more miles in more vehicles—an increasing percentage of them light trucks—this improvement resulted in only a 30 percent reduction in American ozone levels since 1978.[18]

The second and bigger emissions problem with light trucks was also a fuel economy problem. Under EPCA, Congress permitted light trucks to meet less stringent fuel economy standards than cars. As larger, heavier vehicles, they consumed more fuel, produced more carbon dioxide, and thus made a greater contribution to global warming per mile traveled than cars. The relationship between SUVs, carbon dioxide, and global warming became by far the greatest environmental concern with the automobile during the 1990s. The basic connection was unassailable, but the automobile's role in global warming requires some context. Passenger vehicles accounted for only about 12 percent of all manmade greenhouse gases, and American cars accounted for only a third of these emissions, or about 4 percent of the global total. Although SUVs approached 20 percent of the American passenger car total, their contribution to the world's global warming problem, if growing, remained minor, despite the impression conveyed by some critics.[19]

When negotiators from more than a hundred countries met in Kyoto, Japan, in December 1997 to initial the Kyoto Protocol, which called for thirty-eight industrial nations to take steps to reduce their emissions of greenhouse gases to an average of 5 percent below 1990 levels by 2012, carbon dioxide emissions from light trucks were the fastest growing source of new greenhouse emissions in the United States. Vice President Al Gore initialed the agreement on behalf of the United States, but President Bill Clinton refused to submit it to the Senate for approval because he knew it lacked the votes for ratification. When George W. Bush became president in 2001 he publicly repudiated Kyoto altogether, saying that it was not in the interest of the United States. Whether Bush and his political adviser Karl Rove knew, as market research has shown, that SUV owners tended to be red-state Republicans, is not known. As the rest of the world began to grapple in a modest way with the prospect of global warming, Americans seemed to move in just the opposite direction. The rejection of Kyoto and the simultaneous embrace of SUVs and pickups that worsened global warming became primary examples of the "to hell with the world" arrogance of Americans.[20]

After the terrorist attacks of 11 September 2001 on the World Trade Center and the Pentagon, some Americans saw a connection between Islamic terrorists funded by Arabs enriched by gas-guzzling American SUV owners. Comedian Bill Maher made this point in his book *When You Ride Alone You Ride with bin Laden*. So, too, did the anti-SUV television ads run by activist-columnist

Arianna Huffington's Detroit Project in January 2003. These spots parodied federal antidrug ads that linked American drug use with support for overseas terrorist groups. "This is George," began the voiceover in one. "This is the gas that George bought for his SUV. . . . And these are the terrorists who get money . . . every time George fills up his SUV." Although the intentionally over-the-top campaign incensed SUV owners, it also bothered many American technocrats, who preferred a more dispassionate, fact-based style of problem-solving. Indeed, over the past generation a small army in academia and the think tanks has deplored the emotion-based exaggerations made by activists in defense of human health and the environment. These researchers have called for more accurate assessments of risk, clear evidence for chains of causation, and comprehensive cost-benefit analyses before regulation. Their general point is well taken. Societies can be ill served by emotion-driven policy. Unfortunately, some analysts have had probusiness, antiregulatory, or libertarian agendas. They often overlooked similar emotion-laden exaggerations by businesses that, in fact, suggest a larger systemic shortcoming.[21]

While the Huffington approach could be criticized, it was also defensible. Those who worried about America's developing taste for light trucks were frustrated that there had been no significant public discussion about the fuel economy problems presented by light trucks and the unlikelihood that either the automakers or the federal government were prepared to take major steps to address them any time soon. In 2002 Americans bought nearly four million SUVs, 25 percent of all passenger vehicles. But even after 9/11, Congress refused to support a bill introduced by Senators John McCain and John Kerry that would have required raising the CAFE standards. The SUV, Stanford cultural anthropologist Sarah Jain told the *New York Times*, "represents the inability of Americans to make a connection between consumption decisions and their social impact. . . . [T]he Huffington ads—are giving voice to that frustration."[22]

The fuel economy problems with SUVs and light trucks that finally became subject for debate in the late 1990s and early 2000s could be traced to the regulatory war between the auto industry and the federal government in the 1970s. The automakers, casting about for any and every argument to use against any and every regulation, simply claimed that they would have to pass the high costs of meeting the emissions and fuel economy regulations on to the beleaguered American farmer and worker. Pickup-driving Americans were not without friends among politicians, who were only too happy to jump into a fray that pitted hard-working real Americans against pointy-headed Washington bureaucrats. As the automakers well knew (and the automobile press reported at the time), the industry already marketed pickup trucks and SUVs as suburban passenger vehicles, and, as its own research at the time showed, Americans used more than two-thirds of these vehicles as such.[23]

But the bureaucrats were also beleaguered. Reeling from the credibility-sapping problems with implementing the Clean Air Act, the EPA in the early 1970s did not want to be blamed for putting an American automaker out of business. Therefore, it allowed American Motors, then heavily dependent on Jeep sales, to classify Jeeps as light trucks, since they were built on a pickup truck chassis, the precedent-setting ruling that allowed all the automakers to lump their SUVs with their pickups. When Congress amended the Clean Air Act in 1977 it decided to regulate light truck emissions under separate, less stringent standards. The same thing happened in 1975 when Congress directed the Department of Transportation to come up with separate CAFE standards for light trucks under EPCA. When former Ralph Nader protégé and National Highway and Traffic Safety Administration (NHTSA) administrator Joan Claybrook proposed the first fuel economy standards for light trucks in 1977, her Democratic superiors in the Carter administration forced her to promulgate less stringent standards after Chrysler played the jobs card and threatened to shut down its inner-city Detroit Jefferson Avenue plant, provoking an outcry from African-American leaders. With the Big Three fighting every regulation tooth and nail, no matter how small or how tangential to its core business, the EPA and NHTSA compromised because they no longer had the strong backing of the press, the public, and Congress. The resulting regulatory differences persisted to the end of the twentieth century.[24]

Efforts to mandate improved fuel economy in the 1980s and 1990s underscored the central weakness in federal fuel economy regulation and the basic problem with regulating the American automobile industry. EPCA placed decision-making authority for setting fuel economy standards for light trucks and for cars after 1985 in the hands of a political appointee in the executive branch, the NHTSA administrator. When Republicans who opposed government regulation on ideological grounds controlled the presidency and the executive branch of government (as they did after 1980), the administrator set standards that Detroit could live with comfortably and that did not push the industry to improve fuel economy. Facing fines running into hundreds of millions of dollars for failure to reach the mandated requirements, GM and Ford in 1985 and 1986 again raised the specter of layoffs and pushed NHTSA hard for permissive fuel economy standards for future models. So instead of raising the 1985 fuel economy standard for passenger cars above the congressionally-mandated 27.5 miles per gallon for future models, Transportation Secretary Elizabeth Hanford Dole and NHTSA administrator Diane K. Steed exercised an option permitted under EPCA and rolled back the passenger car standard to 26.0 for the 1986–1988 model years and to 26.5 for 1989.[25]

In 1990 Republican president George H. W. Bush and Democratic representative John D. Dingell agreed to support amendments to the Clean Air Act that tightened emission standards for the first time since the 1970s, so long as the legislation did not include regulation of carbon dioxide or require improved fuel economy for motor vehicles. The growing dependence of the United States on foreign oil, dramatically underscored by the Gulf War in 1990–1991, encouraged many in Congress to believe that the prospects for passing legislation to raise the CAFE standards looked good. But a slump in auto sales in 1991 allowed opponents to raise the jobs issue again. The industry also claimed that selling more small cars to meet higher fuel economy standards presented a safety threat to Americans (which was increasingly so now that there were many more light trucks on the road), and this legislative effort failed. The Democratic administration of Bill Clinton that took office in 1993 also disappointed the hopes of proponents of better fuel economy. Eager to demonstrate to the American electorate that the Democratic Party could deliver economic prosperity as effectively as the Republicans and therefore disinclined to create an adversarial relationship with major corporations or the business community, the Clinton administration chose not to use NHTSA to pursue higher standards. Clinton, in fact, agreed to a moratorium on tighter standards in return for Detroit's participation in a ten-year joint industry-government project to develop cars that got eighty miles per gallon, the Partnership for a New Generation of Vehicles. When the Republicans took control of Congress in 1995 they explicitly banned NHTSA from tightening fuel economy standards. When they regained the White House in 2001 they ended the Partnership for a New Generation of Vehicles. With the help of one and a half billion dollars in federal support, the automakers had developed prototype vehicles that exceeded seventy miles per gallon. Yet rather than work with Detroit to commercialize these developments, the Bush administration replaced the partnership with a much more long-term research program directed toward hydrogen fuel cells. The only bright spot for the environment on the regulatory front grew out of the California waiver. With the regulatory power of the federal government blunted, Californians exercised their right to impose stricter standards, including steps to limit carbon dioxide emissions. In the 1990 Clean Air Act amendments Congress gave the states the right to choose the stricter California emission control standards or the federal government's standards. By 2007, eleven other states—Washington and Oregon, New York, Pennsylvania, New Jersey, and Maryland, and all of the New England states except New Hampshire—had followed California's lead.[26]

In the 1970s, when members of Congress and their aides had written emission and fuel economy standards, they had been careful not to overreach while still pushing the automakers hard. Nonetheless, mandating technical standards that the automakers had to achieve by specific deadlines had always been subject to the criticism that standards and deadlines simply might not be realistic in terms of what could be accomplished in the specified timeframes, hence the escape clauses in the Clean Air acts and EPCA that allowed administrators at EPA and NHTSA to modify standards and extend deadlines. As Edmund Muskie had recognized, the automakers' greater command of the relevant technical knowledge gave them great leverage in these negotiations. Critics of the constant negotiation and bickering that this approach engendered argued instead for taking more market-oriented approaches to regulation, usually by placing or raising federal taxes on the price of gasoline and/or gas-guzzling automobiles. Proponents of such taxes maintained that these steps, if taken, would encourage consumers to buy and automakers to make more fuel-efficient vehicles with far less conflict between industry and government. This approach also eliminated the byzantine accounting required to calculate CAFE under EPCA. But proponents of these ideas were either naive or disingenuous. Absent a national emergency, few democratically elected representatives (especially after the Republicans made middle-class tax aversion a commanding issue in American politics) would vote for such taxes. So long as consumers hated taxes and voted accordingly, this approach had no chance of being enacted. Only when Congress mandated the achievement of specific numerical fuel economy standards by specific dates did the automakers make substantial improvements in fuel economy. The approach was crude, contentious, and bureaucratic, but it worked. When this mandate lapsed after 1985 and the automakers and their Republican (with some Democratic) supporters combined to block tightening the standards and the tax alternatives, improvements in the overall fuel economy of American vehicles stopped. As American consumers bought more light trucks, the average fuel economy of passenger vehicles, which peaked at 21.2 miles per gallon in 1991, began to drift downward again.[27]

The lack of support for more demanding fuel economy regulation continued into the new century. With exceptions such as Bradsher and the *New York Times*, the media largely ignored what was happening. Few newspapers, magazines, or television networks were financially secure enough to play the classic watchdog role assigned to the press in a democracy for any length of time. The *New York Times* could and did at times. The rest had to compromise and peddle content that held an audience large enough to bring in advertising revenue. Why drive your audience away with the details of public policy? Why launch a crusade against a large source of advertising revenue when Americans could clearly give a damn? People dropping dead from global warming would be newsworthy. Protesters on the Beltway would be interesting. But hand-wringing environmentalists bored people, even when they were right. "People don't really care," a DaimlerChrysler senior vice president told Bradsher. "They just presume we meet the standards and the standards are stringent."[28]

The approach taken in the Huffington ads reflected a deep frustration with this larger context, but it showed a realistic appreciation for the way that American politics worked, too. Emotion-charged, simplistic overstatement made the media and the public pay attention. In democratic societies like the United States this tactic often provided the most direct path to reform. It was not pretty, and the outcome, if any, usually fell short of what experts held to be preferable, but it was effective politics that rested on a sound appreciation of human psychology. Reformers, if they wanted reform, had no choice, especially now that they faced a business community far more effective at defending its interests than existed in the 1960s and early 1970s. And that was the problem. Business interests understood all too well the connection between exaggeration, psychology, and democratic politics, and they had more money to play this game than would-be reformers. "It is difficult to be optimistic about the prospects for confronting such an entrenched system of power," Stan Luger concluded in *Corporate Power, American Democracy, and the Automobile Industry.* "Business interests so thoroughly dominate campaign giving, lobbying, the media, and think tanks that few dissenting voices challenging the corporation can be heard." Indeed, the two decades of failed efforts to tighten fuel economy standards strongly suggested that, absent some calamity that directly touched large numbers of Americans, reformers would not win many battles over regulation in direct conflict with the industry—so long as the U.S. automakers remained economically important.[29]

Even after the spasm of SUV controversy in 2002 and 2003, most Americans still seemed impervious to the larger social and environmental problems, or at least unwilling to act in ways that contributed to their solution. What were they thinking? Their behavior said that they were satisfying desires more important to them than the safety of others and planetary health. The auto industry's own market research supported this conclusion. Bradsher summarized it in what may well be the harshest criticism of the motives of Americans who bought SUVs. "They are apt to be self-centered and self-absorbed," he wrote, "with little interest in their neighbors or communities." In Japan, the public snuffed out an SUV boomlet in the early 1990s by heaping social opprobrium on SUV owners for their transgressive behavior. Yet something in American culture encouraged this behavior, and there was also a certain tolerance for it. Americans with higher levels of self-involvement and aggression took advantage of this freedom to pursue an idealized image of themselves through SUV ownership, largely oblivious to the wider ramifications. These people were not evil. They just reflected a deeper problem.[30]

"Not only is the mind a black box for scientists," wrote psychologist Timothy D. Wilson in *Strangers to Ourselves* (2002), "it is often a black box to the person who owns that mind." Unconscious motives not only powerfully influenced human behavior, Wilson found, but people had a hard time discerning their own motives directly and, therefore, controlling their own behavior. Ironically, according to Wilson, outside observers appeared to be as good, and often better, at spotting the motives behind behavior as people were themselves. The psychological research in behavioral economics discussed by Wilson had clear implications for the relationship between Americans, automobiles, and the environment. "Free" consumer choices, meaning those that are highly influenced by unconscious motives—and the stimuli of peer behavior and corporate marketing and public relations—would continue to produce SUV booms with their attendant social and environmental problems, even when people had knowledge of connections between their choices and the problems, unless producer and/ or consumer behavior was regulated to address the problems. Regulating automakers against their will became exceedingly difficult after 1975. Regulating a group of consumers—often a large minority and sometimes the majority in a democratic society—against its will, especially when many truly believed that they bought those SUVs for safety and not status, was even more difficult, especially when the automakers recognized that they had the power to plant fears in the minds of consumers and showed no compunction about using it. American automakers and consumers together largely foreclosed the possibility of further regulation.[31]

"For as far as anyone can see into the future," Keith Bradsher concluded in *High and Mighty*, "America's roads and the world's are likely to become ever more crowded with SUVs." His pessimism was understandable in 2002. But the future may not belong to the SUV any more than it belonged to the small car in 1980. And for much the same reasons: fashion and the continuing worries about gasoline supply. Light truck sales plateaued at about 50 percent of the U.S. market between 1997 and 2004. Yet even before gasoline prices climbed sharply again in 2005 and 2006, light truck sales actually fell, as growing numbers of American consumers began to embrace small cars again. The new interest in small cars had been evident in the reception for Volkswagen's New Beetle, introduced in March 1998. It was even more evident in the reception that met

BMW's Mini Cooper in 2002. These small cars were not economy cars. They bore no trace or stigma of the 1970s. The appeal was in the styling—in fashion—and they turned heads as no small cars had during the twenty-year rise of light trucks. In fact, BMW put Minis on top of SUVs and drove the SUVs around U.S. cities to introduce the new car.[32]

New styling pervaded far more than just the market for small cars. The introduction by DaimlerChrysler of the PT Cruiser in 1999 and Dodge Magnum in 2004 and Chevrolet's SSR in 2003 suggested that the American market had entered a period in which, arguably, the American consumer had never had so many different options in automotive styling to choose from nor embraced so many at the same time. Indeed, if the wide range of gorgeously styled convertibles was a valid indication, the industry had entered a new golden age of styling. In this new market the styling of SUVs changed considerably, too. In 2000 Ford introduced the Escape, a small or "crossover" SUV built on a car platform. The shapes of SUVs, station wagons, and minivans also began to blur together. By the time that Bradsher published *High and Mighty* the intensity of the bias in American automobile tastes toward larger and more powerful vehicles had started to ebb once more. Bradsher underestimated the importance of fashion in the automobile world.

Then came higher gasoline prices in 2005 and 2006, and suddenly it was the 1970s all over again—and perhaps then some. As in 1973, the sharp increase in the price of oil and gasoline occurred while a trend toward smaller cars was already under way. Between 2004 and 2006 the price of a barrel of oil more than doubled from thirty to seventy-eight dollars per barrel, and gasoline prices, as low as a dollar a gallon as recently as 1999, rose to above three dollars a gallon— the highest real price for gasoline in American automotive history. Transportation took 67 percent of the oil consumed in the United States in 2004, as opposed to 52 percent in 1974, most of it consumed as gasoline in the nation's more than two hundred million motor vehicles—11 percent of the world's daily oil output. At the time of the Arab oil embargo in 1973 the United States imported 30 percent of its oil, up from 18 percent at the beginning of that decade. In 2005 Americans imported 65 percent, half of this from the Persian Gulf. So much for energy independence. Oil industry analysts worried that global oil production might not be able to keep up with the demand from economic growth around the world.[33]

James R. Schlesinger, who served as the first secretary of energy in 1977, summed up U.S. energy policy over those three decades: "We have only two modes—complacency and panic." The 1970s had seen near panic—at times. But this time there was a good deal more complacency. Nonetheless, buyer behavior discernibly changed. Sales of SUVs plummeted. "We are seeing people who are driving $40,000 Suburbans trading them in on $15,000 Corollas," a Texas Toyota dealer told the *Washington Post* in September 2005. Sales of big SUVs fell by more than 50 percent. For the first time since 1992 car sales increased their share of the U.S. market in 2005. After a quarter century of focusing on light trucks, American automakers were not prepared for a market in which light trucks gave ground. Their share of total U.S. sales slipped in 2005 to a lowest-ever 57 percent. General Motors and Ford had their credit ratings reduced to junk. Both companies announced major restructuring plans that included plant shutdowns and considerable workforce downsizing. Bankruptcy was in the air. Toyota, Honda, Hyundai, and Nissan, careful to prepare for all market contingencies—especially those congruent with long-term trends—took the market turn in stride.[34]

Rising gas prices and worries about the adequacy of petroleum supplies meant that it was time for the media to orchestrate another public discussion of alternative vehicles. There have always been people working on alternative vehicles, and since the 1970s, there have been quite a number of them. Periodically, reporters surveyed the pros, cons, and future prospects of each basic technology. Naturally, more attention got paid to the alternatives during periods of anxiety about gasoline prices and supplies. In part these reviews reassured the public that help was just around the corner. But they were also a reminder that the basic automobile had not changed much in the century since William K. Vanderbilt, Jr., and Henry Ford. Almost all vehicles on the road in the United States still had gasoline-powered internal combustion engines.

As the twenty-first century's first large-scale discussion of alternatives got under way the perennial favorite, electric vehicles, received little attention because their central shortcoming—limited driving range and lengthy battery recharging times—had recently been reaffirmed. In 1990 the California Air Resources Board mandated that 2 percent of the vehicles sold in California in 1998 be zero-emission—which meant electric—vehicles. GM opposed the mandate (and California relaxed the deadline), but the company decided to spend a billion dollars to "go out and find the market" for electric vehicles nonetheless. In December 1996 in Southern California and Arizona it introduced its EV1 electric-powered vehicle, the first commercially available car designed as an electric from the Big Three. Despite substantial efforts to overcome the drawbacks, GM leased only one thousand, even after replacing the lead-acid battery with nickel-metal-hydride batteries that doubled the car's range to about 150 miles. The company stopped production in 2000, only to discover an unfixable safety problem in the early vehicles it leased. It then repossessed the vehicles and destroyed them to prevent liability issues. By 2002, the automakers had "pulled the plug" on this latest attempt at making a go of electric vehicles, despite protests from many lessees who had grown attached to their vehicles, the subject of Chris Paine's documentary film *Who Killed the Electric Car?* (2006). The electrics were zero-emission vehicles in use, but the technology was not emission free. Producing the electricity to charge their batteries and run them in the United States still required burning a lot of oil and coal and the production of carbon dioxide. In fact, comparison studies suggested electricity production resulted in 127 to 198 grams of carbon dioxide per megajoule of work energy expended compared to 15 to 26 grams of carbon dioxide for gasoline.[35]

Two alternatives that promised to reduce but not replace gasoline received the most attention: gasoline-electric hybrids and flexible-fuel vehicles. Hybrids, which accounted for 1.2 percent of U.S. vehicle sales in 2005, used both an electric motor and an internal combustion engine. The vehicle relied on the electric motor at low speeds in stop-and-go traffic, permitting a large reduction in the amount of gasoline consumed per total miles traveled. Braking the vehicle recharged the batteries for the electric motor, so there was no need for recharging while parked as with electrics. To many scientists and engineers, hybrids seemed a halfway alternative. Yet Toyota's Prius (Latin for first) found a niche and had steadily expanding sales after being introduced to the U.S. market in 1997, broaching the hundred-thousand-sales mark in 2005. Hybrids sold at a premium of three thousand to five thousand dollars, which Toyota promised to halve with improved technology and economies of scale from larger production runs. But the higher purchase price meant that owners had to drive them a lot of miles before the gasoline savings covered the premium and made the purchase a sound economic one. Nonetheless, many Prius owners paid this premium for the opportunity to drive a car that produced far less carbon dioxide.[36]

In 2006, American politicians, automakers, and third-party ethanol producers spent considerable effort hyping ethanol or biofuels production, especially the manufacture of E85, an 85 percent ethanol–15 percent gasoline mix for use in flexible-fuel vehicles. This boosterism suggested that the longtime impediments to alcohol use—high production cost and nonexistent distribution—might finally be overcome. GM and Ford, in particular, rushed to embrace this alternative, touting their flexible-fuel vehicles as both green and fuel efficient. But these efforts met a skeptical reaction. Biofuels (chiefly from corn) provided only 3 percent of the total of gasoline production, and only 850 of America's 180,000 gas stations, nearly all in the Midwest, offered E85. Nearly 40 percent of the gasoline sold in the United States already contained 10 percent ethanol (E10). Nor was there anything new in the technology to run E85, and that was really the point. Flexible-fuel vehicles were the closest thing to environment friendly vehicles that the American automakers had to offer, having fallen far behind the Japanese in hybrid technology. And many questioned even this friendliness. First, the flexible-fuel vehicles offered by Detroit tended to be large gas-guzzlers designed to run on either E85 or gasoline. In fact, sixteen of the eighteen models that GM offered were pickups or SUVs. Detroit pushed these vehicles, not only because sales were down, but because a loophole written into the Alternative Motor Fuels Act of 1988 allowed the makers to assume that the vehicles would be run on ethanol half-time and therefore have a much higher miles-per-gallon of gasoline rating for CAFE calculation purposes, even if a drop of ethanol never came near the vehicle's gas tank. Flexible-fuel vehicles did not represent Detroit "going green"; they meant that the American automakers still played the same old CAFE games to sell big vehicles. Second, independent tests showed that fuel economy dropped when the vehicles used E85 because a gallon of ethanol had only 75 percent the energy content of a gallon of gasoline. Thus, consumers had to pay more and tank up more often to do the same amount of driving. Finally, although biofuels were renewable, scientists and analysts split on the question of whether ethanol required more energy to make it than it contained. Cornell's David Pimental argued that it took 30 percent more energy to grow corn and make ethanol. More recent studies suggested a positive energy balance for ethanol of 23–40 percent.[37]

Most automakers, scientists, engineers, and analysts viewed hybrids and flexible-fuel cars as bridge technologies to everyone's ideal alternative, hydrogen fuel cells, which convert hydrogen, the most abundant element on earth, and oxygen into electricity, producing only water as its waste. Fuel-cell-powered vehicles already existed. In 2005 one California family, the Spallinos, even leased a prototype from Honda, the FCX. The problem for fuel-cell vehicles was cost and fuel availability. The FCX cost a million dollars. And hydrogen, like alcohol, lacked a nationwide distribution system and a corporate-sponsor to create one. Moreover, producing the pure hydrogen necessary for fuel-cell cars required using a lot of energy, which for the present meant making electricity from non-renewable oil and coal.[38]

This most recent look at alternatives — indeed the lesson from the thirty years since 1975 — suggested that there was no obvious near-term replacement for gasoline that was cheap, safe, renewable, and clean. The alternatives all had positive characteristics that addressed one or more of the problems with gasoline. But no alternative solved them all. Several simply shifted the problems elsewhere. One tool that energy economists use to compare the alternatives is energy returned on energy invested, or EROEI. Oil's EROEI used to be around 100, meaning that it took the energy in one barrel of oil to produce one hundred barrels of oil. The ratio for Saudi Arabian oil today is 10:1. Oil from many other places has a much lower ratio. The problem with oil is not so much that it is a finite resource that will be exhausted someday but that its EROEI is falling toward 1:1, after which it will cost more energy to get the oil than the energy obtained from it. For oil the windfall age is ending. The problem with many alternatives was that, while they might be renewable and cleaner, they did not have great EROEIs. The EROEI for ethanol, for example, was about 1.34:1. Unless people were prepared to spend a lot more money for clean, renewable energy, gasoline would be around for a while. The best near-term strategies were those like hybrids that reduced gasoline consumption without requiring an additional energy source, for producers to make vehicles that got better gasoline mileage, and for consumers to use less gasoline by buying such vehicles and driving less. But these could only be stopgap measures.[39]

When the environmental impacts of the automobile across the product life cycle were considered, the zero-impact automobile seemed even more a pipe-dream. Even if Americans found a cheap, safe, clean, renewable fuel, they still bought automobiles made from materials taken from the natural world, manufactured with processes that invariably produced gas, liquid, and solid wastes, and that were disposed of in ways that also had environmental effects. The zero-impact alternative to the automobile was walking. During the twentieth century, Americans showed a decided preference for driving over walking. And having remade their built environment around the existence of the automobile, Americans no longer had that option either. The distances between home, work, and shopping were too great. The light truck boom at century end showed that most Americans were still not much concerned with the environmental impact of the automobile in any case. History suggested that it would take something far larger than knowledge of the problems to change any of this any time soon.[40]

EPILOGUE

So what did the twentieth-century American experience with the automobile suggest about the larger relationship between humanity, consumer goods, and the environment? The American answer—the SUV and light truck boom of the 1980s and 1990s following hard on the heels of the safety, environmental, and fuel economy debates of the 1960s and 1970s—seems to be that consumer decisions consistently served the personal desires of individuals that for most rarely, if ever, gave precedence to matters involving the wider social and environmental consequences of automobile use. When we recall our own shortcomings as consumers, this behavior was not altogether surprising. But it meant that showing people the negative consequences of their behavior was not a sufficient condition for changing their behavior when those consequences did not obviously affect them. Those concerned about the impact of human behavior on the health of the planet must accept this troublesome fact as they work to find solutions, especially in societies that prize freedom of choice for consumers and businesses.

In the United States many people share the powerful cultural preference—a stubborn hope—that problems will solve themselves as individuals make free choices in the service of personal desires. The importance of the automobile to individual Americans for practical, psychological, and symbolic reasons was not necessarily incompatible with them seeking more environmentally friendly forms of it through the exercise of consumer choice in free markets. Consumers could force automakers to move in this direction by buying more cars with more environmentally friendly features. The automakers would respond, as they did with small cars in the late 1950s, early 1970s, and late 1990s, by changing the cars that they offered so that they could keep making sales. If consumers placed environmental concern at the top of their agendas and kept it there, producers would do likewise, and do so profitably.

Unfortunately, the American experience with consumer choice offered little evidence that this approach worked. Most of the time, consumers and producers found common ground in automobiles that made the environmental impacts worse simply because consumers did not know about the impacts, or the effects did not concern them enough to influence their choices. That consumers and producers ever found common ground that coincided with lightening the automobile's environmental impact was mainly a matter of luck: an unintended by-product of occasional shifts in taste to smaller, plainer cars. How realistic was it to hope that consumers would make environmental concern a primary motive when buying a car, rather than cost, safety, functionality, style, or symbolic import? If automobile exhaust gave a lot of people cancer, then Americans would have been concerned. Since it did not, they pursued more important ends than reducing environmental impact. So the most desirable solution for Americans, while possible in theory, proved to be the least effective in practice.

The second approach, producer choice, potentially offered the most direct solution to the problems. Producers could offer only automobiles that represented their best efforts to reduce the environmental impact of their product. If producers freely chose to make this commitment (the aim of industrial ecologists in the 1990s), found ways to do so profitably, and did not succumb to the temptation to win market share and greater profits at the expense of their rivals by moving in unfriendly directions, consumers would have no choice but to buy environmentally friendly vehicles. Over time they would not experience this range of offerings as a loss of choice and would find satisfactions among the products offered them.

The American experience with the automobile demonstrated that, when executives like Henry Ford, George Romney, and Ed Cole showed leadership by taking a risk and committing their companies to a goal, improvements could be made. The degree of improvement depended on whether the other companies followed their lead, which boiled down to whether the step contemplated promised clear bottom-line benefits in greater sales or reduced costs. Henry Ford's commitment to waste reduction and recycling at the Rouge and elsewhere produced real environmental benefits, but his idiosyncratic motives, insulation from board and shareholder oversight, and aloofness from rival firms discouraged peers from following his lead because the activities did not always result in clear cost reductions. In the other two cases, consumer demand or government regulation played powerful prodding roles. In the late 1950s the Big Three grudgingly followed George Romney's lead and introduced compact cars, but this step required both a discernible change in consumer tastes and Romney's success with the Ramblers (on top of Volkswagen's impact) before the Big Three offered cars with smaller profit margins. GM's embrace of the catalytic converter under Cole's leadership forced the rest of the industry to follow. By putting the industry on a path that dramatically reduced smog-causing emissions, while removing lead from gasoline and increasing fuel economy, Cole achieved by far the greatest improvement in the automobile's environmental impact. Yet Cole and GM ultimately acted, not of their own volition, but under severe regulatory pressure from the Clean Air Act and the EPA and with knowledge that their competitors had to meet the same requirements.

Thus, at century's end producer choice by itself had not accomplished a great deal, despite the efforts of industrial ecologists who urged their companies and industries to take upfront responsibility for reducing the environmental impact of their products and some small positive steps in this direction by automakers in the 1990s. Whether Toyota's announced commitment to hybrid vehicles in the twenty-first century provides the first great success for this approach remains to be seen. Jim Press, the Midwesterner who runs Toyota Motor North America, suggested recently that his firm tries to anticipate what society will need as well as what consumers will want. Since the company has decided that the societies in which it sells cars will need cars that use far less gasoline, it is preparing to offer them, a reason for optimism. The American automakers could follow suit. They also could join the other ten major corporations (including General Electric, BP, Du Pont, and Alcoa) in the recently formed United States Climate Action Partnership and get on with reducing the carbon dioxide emissions of their vehicles and operations.

The American experience with the automobile certainly underscored the challenges of protecting the environment in a society with both free market capitalism and democratic government. Unfortunately, free market capitalism, operating through consumer and producer choices alone, did not produce substantial environmental improvements. Automobile recycling, where "consumers" were other businesspeople, was an exception. And hunters and anglers, homeowners, and other businesspeople did complain when environmental impacts from the automobile harmed their interests, providing an outer limit to environmental harm. But the big story here is that government regulation, far more than any other approach, helped the environment. Americans made tremendous environmental progress between 1945 and 2000, but most of it was mandated by legislation and regulation adopted by democratic governments. The people at the automakers most responsible for reducing environmental impact believed almost unanimously that they would not have done as much as quickly if not for regulation. Yet ironically, Americans soured on regulation just as it proved its worth.

Polls from the late 1960s forward consistently showed that Americans overwhelmingly considered themselves friends of the environment. They said they favored strong environmental laws and seeing them vigorously enforced. The problem was that nearly all environmental problems were the result of human economic activity, including consumer choice. To make an impact, most environmental laws came down to telling some people that they could not pursue an activity that they perceived to be in their self-interest or found to be just plain fun. The automobile stood at the very center of American economic activity, indeed, at the center of American beliefs, behavior, and fun. Legislating significant changes in who Americans were and what they wanted was no small matter. More often than not, it could not be done.

Nowhere did this prove to be truer than efforts to regulate the behavior of automobile owners and drivers. Americans may have said that they wanted an improved environment, but they also consistently said that they didn't want to change their behavior to achieve it. Nor did they want to pay for it or be bored with the details. This attitude pervaded the public's reaction to the EPA's 1973 efforts to implement the transportation controls required under the 1970 Clean Air Act amendments, but it was evident, too, in opposition at the state level to steps such as mandatory vehicle and emissions inspection programs. From the public's standpoint, solving the automobile's problems without involving them always constituted the best approach. Consumers had more votes than automakers. So elected officials from Kenneth Hahn to Edmund Muskie, and regulators all leaned toward forcing the automakers rather than consumers to take the steps necessary to help the environment—to provide, that is, the manufacturer-provided technological fix.

But the automakers had more money than the public interest groups that spoke for consumers collectively or for the environment. That counted for a great deal when it came to promoting and defending the industry's economic interests in an American political context that gave corporations extraordinary leeway to use their money to influence public policy. Auto executives shared the primal fear of most private sector producers—losing some of the decision-making freedom that they deemed essential for ensuring company profitability and survival. They instinctively and tenaciously fought every encroachment on this prerogative by legislators and regulators. This behavior, though explainable, proved problematic, since the public policy goals at issue—improved safety, reduced emissions, better fuel economy—were worthy ones. More important, the goals fell well within the realm of the possible. The regulations conferred no significant competitive advantage on any one automaker (and regulators made allowances when necessary for the weakest competitor, American Motors). The Japanese automakers, although not in favor of government regulation per se, took the regulations in stride, tried to anticipate future requirements, and strove to meet them. The American automakers, by contrast, screamed bloody murder. The American tendency to view business and government as natural adversaries and to treat the question of government regulation as part of a life-or-death struggle over sacred principles rather than as a matter-of-fact attempt to reduce unwelcome side-effects ill served the American people.

Once organized and funded, the opposition of the American automobile industry to regulation proved potent. The automakers, autoworkers, and auto dealers represented nearly a fifth of American GNP. When the largest and wealthiest industry in America put up a fight—and sustained it year after year and decade after decade, the approach proved highly effective. The automakers may have lost major fights between 1966 and 1975, but they learned from those fights. By 1980—before the Republican Party won control of the presidency and the Senate that year—the industry had fought the federal government to a draw and convinced many that it was preferable to leave the industry alone rather than continue to be adversaries.

Sometimes, too, nothing succeeds like failure. In the early 1980s, when Japanese automakers appeared poised to administer the coup de grace to the Big Three, politicians in Washington became even more solicitous toward Detroit. And there matters have stood for a generation: no major rollback of the regulations but no major new regulations and little or no tightening of the existing federal regulations. Add to this the rightward turn in American politics that accelerated in 1980, and the federal government problem-solving of the 1960s and 1970s that included automobile regulation seems ever more an exception in U.S. history. It remains to be seen whether the growing number of states embracing the stricter California emissions standards, including reductions in carbon dioxide from automobiles and the dramatic 2 April 2007 Supreme Court decision forcing a reluctant Bush EPA to regulate carbon dioxide emissions, mark a breakthrough and a renewed period of regulation.

Even with Democratic majorities in Congress bent on using the federal government to solve problems, there would not have been environmental regulation had not the press taken up the cause. The press was the best publicist that the automobile and the automakers ever had. But beyond just reporting and explaining the latest developments in the automobile industry, the press also thought of itself as an essential institution of democracy, a watchdog finding and shedding light on problems so that informed citizens and legislators could take the actions necessary to fix them. In this role journalists saw themselves as friends of common citizens in their battles with powerful interests. The press also viewed itself as a good friend of the environment, and, indeed, the fourth estate played midwife to the major environmental legislation that the United States adopted during the second half of the twentieth century. Conservation and environmental interest groups simply lacked the resources—the power—to go toe-to-toe with the largest corporations in the world in the realm of special interest lobbying unless the press seized an issue and created an aroused body of voters who also brought pressure to bear on legislators. This dynamic among activists, the press, citizens, and legislators—with an essential role played by the press—has accounted for much of all the reform that has been accomplished in America. Unfortunately, the automakers learned during the 1970s how to become much better at countering this dynamic before it resulted in regulation.

The press was not a reliable friend of the environment either. As a profit-making industry, the press survived by building and holding the attention of an audience and then selling access to this audience to commercial advertisers in order to earn a profit. Readers and viewers liked to discover new issues or, to put this more plainly, new viewing and reading thrills. After the drama of discovery, outrage, confrontation, and (possibly) reform had played out, people lost interest. Therefore, the press could not hold the searchlight on any one subject for long without loss of audience and doing itself financial harm. The press and the public found issues that involved human victims, preferably bloodied or in obvious emotional anguish, much more riveting than policy questions. The environmental ramifications of the automobile, as varied and large as they were, rarely produced such obvious victims. On the other hand, crime, terrorism, and war provided subjects that the press—and politicians—knew commanded the attention of the public. These issues tended to crowd out other problems from the press and the agenda of legislators.

The twentieth-century American experience with the automobile thus showed how hard it was to address the related environmental impacts, even after various groups made those impacts widely known. This difficulty also suggested that implementing further regulation at the national level will be a real challenge—unless the U.S. automakers weaken further. The unsuccessful efforts to improve the fuel economy of American vehicles to help with global warming certainly showed how daunting it was. Unlike smog, smokestack pollution, and abandoned cars, global warming did not directly affect people's senses. Earlier problems could be seen and, with smog, even directly felt. The contribution of carbon dioxide to global warming, in contrast, could not. Moreover, for Americans, the predicted consequences still seemed remote in time and space. Even if a global warming–caused problem struck the United States and the press made a big deal about it, an aroused public still had to overcome the vastly more powerful lobbying efforts of the automakers, the strong antiregulatory bias of Republicans, and the jobs protection bias of many Democrats. The prospects for reducing automobile reliance through the development of attractive mass transit systems and land-use reform seemed even more remote. By the end of the century, Americans lived in an utterly automobile-based society. They had made this choice and, notwithstanding the safety, emissions, and fuel economy reforms of the 1960s and 1970s, had intensified their commitment to this course. Whatever the problems, there was no going back—at least anytime soon.

'I'hese difficulties forced one to ask how well Americans made use of experts, careful analysis, and foresight in their free-market capitalist, democratic society. How did such societies satisfactorily address problems when the solutions required short-term sacrifice and pain in return for the promise of future benefits? As we have seen, experts who understood the problems and who offered wise long-term solutions have not been in short supply. A certain amount of healthy skepticism toward experts is a good thing. So, too, is debate among experts. And certainly experts must be held accountable to the public they serve. But the U.S. experience with the automobile suggested that Americans could have done more to heed experts who offered advice in the long-term public interest. That experts consistently had so little influence on the course of events, that their advice was often overwhelmed by emotion-laden arguments from both activists and business spokespeople obligingly amplified by the press, appeared to be a serious systemic shortcoming. Indeed, a great paradox of the American system is that the use of experts by corporations in the interest of ensuring future profitability is viewed as necessary and legitimate, whereas their use by government to chart wise paths for society is viewed with profound suspicion. For all the growth in federal power that took place during the twentieth century, the American experience with the automobile suggested that public policy expertise was still too little heeded in the halls of government. Emotional appeals cannot be eliminated from policy debates, but surely the public interest would be better served if Americans took steps to change the policy-making process to reduce or counteract the influence of these appeals.

Consequently, the future that Americans made across the twentieth century largely reflected the cumulative outcomes of near-term private decisions by individuals and businesses. But a society that persistently belittled public experts, one that worked systematically to reduce their power by subordinating them to political appointees or coopting them to serve narrow private interests, might also prove to be a foolish society. Choosing to accept a "whatever" future in return for the freedom to try to carve out a better future for ourselves as individuals and as private businesses may be fine. But it may not be a wise choice for most people, for a society, or for the natural world. For those concerned about the impact of consumer and business behavior on the health of the planet—and who also wanted the freedom of capitalism and democracy—this message from the twentieth-century American experience with the automobile should be worrisome to say the least.

The American experience with the automobile ultimately brings us back to the Great Experiment and some thoughts about individuals within this larger context. By the end of the twentieth century, Americans—as well as many other people around the world—had made some important choices about the parameters of this experiment. There would be little compulsion forced on either consumers or producers through government regulation because in a democracy both groups could muster great power when they perceived vital interests to be threatened. In effect, absent a compelling crisis or catastrophe, Americans had chosen to accept what many scientists warned was likely to be a slow, cumulative deterioration of the global environment over time in return for mostly continuing to do what they most wanted to do today.

Would this choice be a poor one for humanity and the planet? The honest answer was that Americans did not know. The few people who took the question seriously divided into two sharply opposed groups that held opinions in both cases largely on faith. One side believed that present environmental trends portended dire consequences. Humanity must make some significant changes in behavior today to avoid these consequences. Better to make them today before the problems have grown much worse and the solutions far more expensive and disruptive. The other side said that it was folly to make costly changes today to address problems that might not develop. If the problems materialized, humans would rise to the challenge and successfully solve them. Public debate, such as it was, over this unresolved issue had effectively ended in the United States by century's end. The issue no longer occupied a significant part of public discourse because Americans had sheepishly chosen what they perceived as the lesser of two possible evils: the approach that offered less disruption to their agendas today, a decision that dovetailed neatly with the hope that tomorrow would not be as bad as some feared. Nowhere was this decision more evident than the connection between the automobile and global warming. Those concerned about the issue still struggle to arouse their fellow citizens to action, despite substantial consensus in the scientific community about the causes, consequences, and ne-

Role model for the world. Rio summit parking, 1992. *By permission of Mike Luckovich and Creators Syndicate, Inc.*

cessity for action, even as polls show that 80 percent of Americans acknowledge the problem.

The American experience with the automobile and the stakes involved in the Great Experiment raised deeper questions about the role that consumer goods played in the lives of twentieth-century Americans. John Muir told them that if they looked closely enough, they would see that everything was hitched to everything else. There was (and is) a connection between what went on in the minds of consumers and the environmental problems associated with the consumer products that they purchased, used, and discarded. Consumer motivations, often unknown or hazy to consumers themselves and usually having nothing to do with the environment, were hitched to environmental problems that occurred across the product life cycle. Ignoring these connections or missing them altogether actually made dealing with the problems harder. That was why Americans did not readily see connections between their motivations for purchasing automobiles and clear-cutting Michigan forests, air and water pollution at the Rouge, smog in Southern California, or abandoned automobiles in rural and urban areas. But by the end of the 1970s, far more Americans understood these connections.

The perspective on consumer behavior that is emerging from psychological research offers little solace for those worried about the environment. It does offer an explanation to the riddle of the SUV or why knowledge of the problems had so little impact on consumer behavior. The motives in play when we buy an important consumer good like an automobile—for achievement, for affiliation, for power—are important to us as individuals. Cues from the outside world, including the behavior of others and advertising, easily arouse emotions that trigger these motives. Thus, the automakers do exert a powerful influence on our behavior—even more so, ironically, when we are not paying attention directly. But the actual emotions and motives behind our behavior are very difficult for us to discern. If we can't recognize that an external cue has aroused our need for affiliation with a community, how can we control our behavior before acting on that motive to find a different and less problematic way to satisfy it? We know that buying gives us pleasure. We even know that the pleasure we derive from a product will wane over time. Yet we have a tough time recognizing that no matter how many times we buy or how much money we spend, we will eventually end up no happier than we were before we made that first purchase. Where happiness is concerned, most consumer spending is a waste rather than an investment. Concerns about the larger implications of our behavior as consumers rarely play a determining role in our choices. As consumers, most of us are half-blind bulls in a china shop, charging this way and that, with scant prospect of learning how to either see or stand still.

Then there is the inescapable fact that the automobile actually played useful roles for people as transportation and beyond. A symbolic dimension to automobiles was unavoidable, given the high cost of the product, the public nature of its use, and its personal association with a family or an individual. Someone once said that people were never as naked as when they dressed for a party. The same was true when Americans made substantial symbolic, emotional, and financial investments in automobiles. Whether the car was a Model T in the 1910s, a Chevrolet in the late 1920s, a Cadillac or a Volkswagen in the 1950s, or an SUV in the 1980s and 1990s, the odds were that its owner at the time was anxious about his or her place in the world and using the car in question—in part—to ask a larger community for some indication of affirmation and respect to allay that anxiety. People may have used particular models to ask for recognition of a higher status, to disassociate themselves from this practice altogether, or simply to say that they were regular folks successfully keeping up with their community. But they were always saying something—or some set of things—about themselves and how they wished to relate to the larger world.

To use consumer goods in this fashion may seem superficial, even silly. But they actually performed this function quite well. In his famous speech at the Lincoln Memorial in 1963, Martin Luther King, Jr., spoke of a world in which his children would be judged, not by the color of their skin, but by the content of their character. Few Americans argued with the worthiness of that goal. But consider the practical difficulties involved in getting to know the content of a person's character in twentieth-century America, even when people were able to see past skin color or gender. To know someone's character meant knowing that person well. Normally, knowing someone well required knowing that person for an extended time and seeing him or her in a variety of roles. How did Americans ask for and receive respect from their communities if many of them now lived in circumstances where it was difficult, perhaps impossible, to earn it the old-fashioned way? How many Americans unreservedly extended their highest level of respect to strangers without any clues about who they were?

Long before the advent of the automobile and modern America, human beings learned that bearing, speech, grooming, clothing, jewelry, and possessions—in short, the myriad ways that they managed their physical presentation to other people—offered effective tools to elicit specific kinds and degrees of respect. The geographic and economic mobility of Americans greatly increased the number of times that they encountered others as strangers. As these encounters grew, Americans turned to those consumer goods whose use they shared with strangers to communicate with them, to establish common ground with them, to ask one another for respect. Indeed, Americans actively built their society around this practice. Some people worried that if the content of their character was known, it would be judged unworthy, but many more recognized in consumer goods a short-cut to higher levels of respect—and multiple forms of respect. In America undeserving and average folks alike could successfully lay claim to respect that might not have been readily forthcoming in other human societies. Indeed, the chance to do so defined the essence of the American Dream.

Respect comes from communities. Daniel J. Boorstin has argued that Americans created communities of consumption to replace the older forms of community. Membership in these ersatz, often ephemeral, communities required, not just the ownership and use of particular kinds of consumer goods, but a demonstrated mastery of the symbolism associated with them. Cultivating this mastery helped individuals to find, relate to, and command respect from a larger community. Similarly, newcomers and marginalized members of American society long recognized that the acquisition of the right consumer goods and the successful mastery of the symbolic language associated with them offered a shorter path to greater inclusion and respect. Although serving multiple ends, symbolically potent consumer goods such as the automobile functioned rather effectively as shorthand forms of visual symbolic communication by which people in mass societies who were anxious about their places in the world asked one another for affirmation and respect. Most important, the use of consumer goods allowed Americans to make the request without the public loss of face involved in directly asking other people for respect. In short, Americans' need for dignity and respect—to be recognized as valued members of a community—and their use of consumer goods to ask for that respect proved an important cause of their environmental problems as well as an important barrier to solving them.

These connections do not bode well for the health of the planet or humanity. But they do offer choices for the lives we lead as individuals. Automobile ownership is an almost inescapable requirement for life in the United States. Nonetheless, we can choose to limit the amount of driving that we do. We can choose cars that get the best mileage and those that place a smaller total burden on the natural world. Generally, a small car means a smaller burden, a choice that also entails accepting a greater degree of safety risk for ourselves and those that we care about. These things can be done without completely sacrificing the freedom, fun, and other satisfactions that we derive from automobile ownership and use. More important, we need to use pressure as consumers and citizens to force automakers to offer vehicles that are safer for ourselves and the environment, which includes turning to government regulation, if necessary, to do so.

Americans concerned about the impact of the automobile on the environment could chastise the owners of Cadillacs or SUVs for selfishly privileging their needs for affirmation and respect over ensuring that their choice of vehicle did not contribute to environmental or social problems. But we could also do more to understand and address the need of those automobile owners for affirmation and respect. Above all, we need to recognize that these owners probably had only the haziest recognition of what was going on and a limited ability to do anything about it. Thus, a better response to the environmental problems posed by consumer products like SUVs was to find other ways to confer the highest level of respect on their owners so that they would be less likely to make recourse to troublesome consumer products to find it. We must do this together as a group for each other because individuals cannot do it well themselves, our cherished notions of individual responsibility for behavior notwithstanding.

Yet achieving this goal presented an extraordinary challenge, given the way in which we organized our society. Affirmation and respect became more important to twentieth-century Americans because we created a society characterized by a rapid rate of change. Continuous change constantly challenged and often undermined our sense of mastery. It made us anxious, and it accentuated our need for affirmation and respect. Americans were an optimistic people, precisely what we had to be in these circumstances. But a world characterized by a high rate of change, even when resolutely fronted with an attitude of optimism, required us to shoulder more anxiety—or, at least, more status anxiety. Only the very few who could let go of everything without despair and accept whatever might come could transcend it. For the rest, anxiety was the necessary price to be paid in the hope that change would improve our lives. Even those who successfully adapted to change and pursued lives as producers, consumers, and human beings that enhanced their feelings of self-worth could not entirely escape the anxiety because they could not fully control change and eliminate the threat that it posed to their lives. In short, an economic system based on change, or "creative destruction," though capable of delivering many marvelous things, was not conducive to human contentment. Of course, evolution probably encouraged us to be anxious about our fit with the group and to be proactive in manipulating the natural world to reduce our anxiety (and meet our needs for food, clothing, and shelter). The question is whether these innate proclivities (if such they be) have become more problematic—dysfunctional even—with the growth in human numbers and our improved ability to bend the natural world to our will.

This relationship between change and contentment offered a final and more fundamental opportunity to make a positive contribution to the natural world—voluntary humility. We can demand less respect for ourselves while unconditionally offering more of it to others in turn. We also can consciously try not to exalt ourselves over others through buying consumer goods. If we can do these things, we can simultaneously move toward contentment and away from doing further harm to other people, other forms of life, and the health of the planet in general. This behavior is too demanding to be legislated, but more of us can choose it, one person at a time, and persevere in it, despite the great difficulty. We can also do more to encourage each other to make this choice, helping one another become true "masters of the universe" by becoming masters of ourselves.

In twentieth-century America we learned through the automobile that we needed to do a better job looking out for the health of the planet. But we only just started to appreciate that doing so required finding better ways to look out for each other. Ultimately, these others have to include the rest of the people with whom we share the planet, whose legitimate aspirations for dignity and respect will necessarily involve problematic activities like buying and driving cars until together we find other paths to satisfactory levels of dignity and respect for all. So, as the new millennium began, there were still many rivers to cross. In the meantime, six going on nine billion anxious, restless human beings reached for the car keys. The earth trembled at the prospect.

NOTES

1. THE ARROGANCE OF WEALTH

1. On the adoption of the automobile in America, for starters see James J. Flink, *America Adopts the Automobile, 1895–1910* (Cambridge, MA: MIT Press, 1970); Clay McShane, *Down the Asphalt Path: The Automobile and the American City* (New York: Columbia University Press, 1994); and Gijs Mom, *Electric Vehicle: Technology and Expectations in the Automobile Age* (Baltimore: Johns Hopkins University Press, 2004). William Kissam Vanderbilt, Jr. (1878–1944), leaps from the pages of newspapers and early automobile magazines, but he has been strangely neglected by automobile historians. For one exception, see Beverly Rae Kimes, "Willie K.: The Saga of a Racing Vanderbilt," *Automobile Quarterly* 15:3 (1977): 312–327. The background of these men can be found in the usual places. For their exploits, see esp. the *New York Times*. In addition, see Edwin P. Hoyt, *The Vanderbilts and Their Fortunes* (Garden City, NY: Doubleday, 1962), 317–325, 344–348, 381–389, 397–398; and Alden Hatch and Foxhall Keene, *Full Tilt: The Sporting Memoirs of Foxhall Keene* (New York: Derrydale, 1938). Thorstein Veblen, *The Theory of the Leisure Class* (New York: Penguin, 1899, 1979), esp. chaps. 3 and 4 on "Conspicuous Leisure" and "Conspicuous Consumption." For examples of conspicuous displays by the speeding sportsmen in nonmotoring contexts, see Hoyt, *Vanderbilts*, 317–318; "American Outshines Nobles," *Washington Post*, 9 April 1911, 15; and "Bostwick, the Magnificent," editorial, *Washington Post*, 14 April 1911, 6.

2. M. C. Krarup, "The Prospects for Economical Automobiling," *Iron Age*, 25 April 1901, 11, as reprinted in *Scientific American Supplement*, 11 May 1901, 21201; M. H. Daley, "Speed and Safety," *Horseless Age*, April 1896, 8; Mom, *Electric Vehicle*, 43; C. E. Duryea, "Argues for Higher Power," *Horseless Age*, 24 May 1899, 14 (quotation); and "Retaliatory Measures," *Horseless Age*, 10 July 1901, 326. The six major American automobile magazines in the early 1900s were *Horseless Age*, *Automobile*, *Motor*, *Motor Age*, *Motor World*, and *Cycle and Automobile Trade Journal*, usually classified as trade journals today, but in the first decade of the twentieth century they were actually enthusiasts' magazines aimed at the automobile owner as well as producer.

3. See "High Speeds," *Automobile*, April 1900, 58; P. I. E., "The Speed Craze," *Horseless Age*, 24 Oct. 1900, 10; "Sowing the Wind, Reaping the Whirlwind," *Horseless Age*, 14 May 1902, 572; "The

Old Year and the New," *Horseless Age*, 7 Jan. 1903, 1; "Life Insurance Companies Investigating Speed Mania," *Horseless Age*, 12 Sept. 1906,320; "Responsibility for the Speed Craze," *Horseless Age*, 13 May 1908, 550; "Automania in a New Light," *Horseless Age*, 29 Apr. 1903, 519–520.

4. Mom, *Electric Vehicle*, 34, argues that the fin-de-siècle fetish for record-setting was a way to quantify progress for a mass audience; "Mr. Vanderbilt's New Locomobile," *New York Times*, 2 June 1900, 1; "Adventures of W. K. Vanderbilt and His Automobile," *Chicago Daily Tribune*, 12 Aug. 1900, 47.

5. "The Latest Sporting Fad," *New York Times*, 10 June 1900, 19; "Among the Automobilists," *New York Times*, 5 Aug. 1900, 21; "New Automobile Records," *New York Times*, 19 Sept. 1900, 9; "Polo Player Buys Racing Auto," *New York Times*, 5 June 1901, 7; "Auto Races at Newport," *Washington Post*, 31 Aug. 1901, 2; "Newport's Automobile Races," *New York Times*, 7 Sept. 1900, 7; "Bostwick Will Not Race," *New York Times*, 8 Sept. 1900, 5; "New Automobile Records," *New York Times*, 11 Oct. 1901, 6; "Bostwick After Automobile Records," *New York Times*, 19 May 1903, 7.

6. "Motor Show Is Opened," *New York Times*, 4 Nov. 1900, 10, as quoted in Kimes, "Willie K.," 318; T. H. Moore, "The Auto or the Man," letter to the editor, *Horseless Age*, 8 Oct. 1902, 388.

7. On hill climbs, see "Eagle Rock Hill-Climbing Contest," *Automobile*, 5 Dec. 1903, 579; "Beats Record for Hill Climbing," *New York Times*, 27 Nov. 1903, 2; "Auto Crashes o'er Precipice," *Atlanta Constitution*, 27 Nov. 1903, 11; "Fourth Annual Eagle Rock Hill Climb," *Automobile*, 3 Dec. 1904, 611–612, 614; "Mountain Auto Record," *New York Times*, 13 July 1904, 8; and "Lofty Mount Washington Conquered by the Climbing Automobile," *Los Angeles Times*, 31 July 1904, B3. "Romp right up" is from "Responsibility for the Speed Craze," *Horseless Age*, 13 May 1908, 550. "Test hill" is from "Features of the 1906 Models," *Horseless Age*, 13 Sept. 1905, 309–310.

8. "Lessons of Eagle Rock Hill Climb," *Automobile*, 3 Dec. 1904, 630; "The Trend toward Economy," *Horseless Age*, 5 Oct. 1910, 454.

9. Hatch and Keene, *Full Tilt*, vii.

10. For a discussion of early automobile touring that nicely captures the sense of curiosity and adventure that motivated these drivers, see Marguerite S. Shaffer, *See America First: Tourism and National Identity, 1880–1940* (Washington, DC: Smithsonian Institution Press, 2001), 130–168; "Automobile Touring," *Horseless Age*, 2 Sept. 1903, 233.

11. "The Touring Season," *Horseless Age*, 11 Sept. 1901, 490. For a similar discussion, see Mom, *Electric Vehicle*, 42–43. Shaffer draws a sharper distinction between the "speeders" and the "tourists" than I would. Speeders tended to be younger and male, but there was some overlap between the two groups.

12. Mom, *Electric Vehicle*, 32–35.

13. As quoted in ibid., 27, who notes the ubiquity of this quotation in the automobile history literature. He points out, 53, that touring was also viewed as a sport.

14. "For and against the Auto," *Washington Post*, 17 Aug. 1902, 17.

15. Ibid. Among automotive historians, Mom, *Electric Vehicle*, 45–47, stands apart in appreciating that opposition to the automobile was nowhere near this simplistic.

16. "Road and Track Racing," *Automobile*, November 1900, 200.

17. "Clearing the Air," *Automobile*, January 1902, 12; "Irrational Use of the Horn," *Horseless Age*, 27 Aug. 1902, 210. The *Washington Post* also made this connection in "For and against the Auto," 17.

18. See "Automobile Accidents," *Horseless Age*, 10 Sept. 1902, 265, where the magazine estimated

that 90 percent of accidents involved reckless speeding. For a summary of infamous accidents, see "For and against the Auto," 17; "The Persecution of Conservatism," *Automobile*, July 1900, 117; and McShane, *Down the Asphalt Path*, 176.

19. On Vanderbilt, see "For and against the Auto," 17, and Hoyt, *Vanderbilts*, 344–348. For the travails of Edward R. Thomas, see his coverage in the *New York Times* between 1902 and 1908.

20. "The Safety of the Automobile," *Horseless Age*, 28 Oct. 1908, 580; "Traffic Accidents in a Big City," *Horseless Age*, 9 Feb. 1910, 213. See "Record Breaking 'Tours,'" *Horseless Age*, 23 July 1902, 81, for the assertion that automobiling was being "'boomed' by the daily press." "The Annual Slaughter," *Horseless Age*, 10 June 1908, 680.

21. "Retaliatory Measures," 326. For examples of similar actions, see "Wisconsin Maliciousness," *Automobile*, 17 Dec. 1904, 683; "Nails on the Highway," *Automobile*, 14 Dec. 1905, 674; and "Crusade against Glass Throwing in the Oriole City," *Horseless Age*, 23 Nov. 1910, 727.

22. On stonings, see "Automobile Party Mobbed by Boys," *New York Times*, 25 May 1902, 1; "Mrs. Thomas's Assailant Committed," *New York Times*, 26 May 1902, 12; "Hits Woman in Auto," *New York Times*, 23 May 1904, 1; "Hilton Hit by Stone Thrown at His Auto," *New York Times*, 25 May 1904, 1; "A Climax in Hoodlumism," *Automobile*, 28 May 1904, 579; "Current News from New York," *Automobile*, 11 June 1904, 644; "Youthful Diversion in Missouri," *Automobile*, 23 July 1904, 93; "Reward for Conviction of Stone Throwers," *Horseless Age*, 3 Oct. 1906, 426; J. Stuart Blackton, "Street Dangers in New York," *Horseless Age*, 11 June 1904, 634; and McShane, *Down the Asphalt Path*, 176–177. On the geographic range of the incidents, see "Provincial Prejudice," *Horseless Age*, 27 Apr. 1904, 446; and "Youthful Diversion in Missouri," 93. Also, "Reward for Conviction of Stone Throwers," *Horseless Age*, 3 Oct. 1906, 426; Cholly Knickerbocker, "A Chappie and a Horseman Try the New Horseless Carriage," reprinted from the *New York Journal* in *Horseless Age*, March 1897, 15; Horace Burns, "Anti-Stone Throwing Petition," *Horseless Age*, 12 Dec. 1903, 624.

23. "New York Street Fatalities," *Automobile*, 12 Dec. 1903, 625; "Wisconsin Maliciousness," *Automobile*, 17 Dec. 1904, 683; and "For and against the Auto," 17; J. H. T., "Assault upon Automobilists," letter to the editor, *Horseless Age*, 2 Aug. 1905, 177–178; Burns, "Anti-Stone Throwing Petition," 624.

24. The *Washington Post* made this observation in "For and against the Auto," 17. "Enterprising Tramps" reprinted from the *Philadelphia Record* in the *Washington Post*, 19 Aug. 1904, 6.

25. "New York Street Fatalities," *Automobile*, 12 Dec. 1903, 625; McShane, *Down the Asphalt Path*, 179.

26. For the Thomas incident, see "Automobile Party Mobbed by Boys," *New York Times*, 25 May 1902, 1. "The Automobile and the Public," editorial, *New York Times*, 16 June 1902, 8; "New Turn in Hoodlumism," *Automobile*, 2 July 1904, 27.

27. Wilson's remarks at the dinner were partially reported the following day as "Wilson Blames Speeders," *New York Times*, 28 Feb. 1906, 2, and reprinted in *The Papers of Woodrow Wilson*, vol. 16 (Princeton, NJ: Princeton University Press, 1973), 320. The second quotation appeared in a newspaper report published four days after the first, "Motorists Don't Make Socialists, They Say," *New York Times*, 4 Mar. 1906, 12.

28. On stonings and street games, see "To Prevent Ruffian Attacks on Automobilists in New York City," *Horseless Age*, 1 June 1904, 585; "New Turn in Hoodlumism," 27.

29. "Road Improvement a Legislative Problem," *Horseless Age*, 10 Feb. 1904, 152.

30. Burns, "Anti-Stone Throwing Petition," 624.

31. The classic treatment remains Samuel P. Hays, *Conservation and the Gospel of Efficiency* (Pittsburgh, PA: University of Pittsburgh Press, 1959; reprint ed., 1999). Hays briefly discusses oil lands in the context of public land policy, 87–90, but does not address the issue of petroleum conservation. On the importance of foresight and planning for the future among conservationists, however, see Hays, xii–xiii, 2, 123. A good discussion of emerging pollution consciousness is found in Adam W. Rome, "Coming to Terms with Pollution: The Language of Environmental Reform, 1865–1915," *Environmental History* 1:3 (1996): 6–28.

32. Albert L. Clough, "Conservation Idea in Its Application to Motoring," *Horseless Age*, 29 June 1910, 958. For an article on rubber, see "Danger to the Rubber Supply," *Current Literature*, April 1900, 51. For hardwood, see "The Waning Hardwood Supply," *Horseless Age*, 22 Jan. 1908, 91. Apart from the search for new oil fields, the standard histories of the of the oil industry say little about concern with the gasoline supply in the first decade of the twentieth century. See, e.g., Daniel Yergin, *The Prize: The Epic Quest for Oil, Money, and Power* (New York: Simon and Schuster, 1991), 90–95; and Harold F. Williamson et al., *American Petroleum Industry: The Age of Energy, 1899–1959* (Evanston, IL: Northwestern University Press, 1963), 184–195, 299–338. See the 1904 U.S. vehicle production figures in Mom, *Electric Vehicle*, 113, where gasoline cars accounted for 86 percent of production. The definitive studies of this outcome are Mom, *Electric Vehicle*, and David A. Kirsch, *The Electric Vehicle and the Burden of History* (New Brunswick, NJ: Rutgers University Press, 2000).

33. Clough, "Conservation Idea in Its Application to Motoring," 958.

34. Ibid.

35. Ibid., 959.

2. FORESIGHT AND EMOTION

1. "The Oil Famine Bugaboo," *Horseless Age*, January 1897, 1; "Price of Gasolene Advancing," *Motor Age*, 7 Nov. 1899, 162; Thomas J. Fay, "Motor Fuel," *Horseless Age*, 13 Dec. 1905, 753.

2. "The Oil Famine Bugaboo," 1; A. L. Clough, "How to Operate a Gasoline Car—Its Future Development," *Horseless Age*, 2 Mar. 1904, 266. For similar sentiments, see Elihu Thomson, "Alcohol and the Future of the Power Problem," *Cassier's Magazine*, August 1906, 310–312; Albert P. Sy, "Tax-Free Alcohol," *Journal of the Franklin Institute* 163:1 (1907): 65; and Thomas L. White, "Alcohol as a Fuel for the Automobile Motor," *Automobile*, 2 May 1907, 737.

3. "Gasoline and Its Substitutes," *Scientific American*, 22 Mar. 1913, 276. The higher end of this percentage range is given in Daniel Yergin, *The Prize: The Epic Quest for Oil, Money, and Power* (New York: Simon and Schuster, 1991), 111. Yergin, 14, makes the claim about dumping, but see, too, Alexander Johnston, "Motor Car Industry as a Consumer of American Products," *Motor*, December 1922, 28.

4. For the claim that the oil industry had a hand in developing uses for gasoline, see Fay, "Motor Fuel," 753. "Supply and Demand in Motor Fuels," editorial, *Scientific American*, 2 Nov. 1912, 362; "Gasoline and Its Substitutes," 276.

5. On price, see E. N. B., "Facts and Kinks," *Horseless Age*, 14 Mar. 1900, 16; "The Rising Price of Gasoline," *Horseless Age*, 13 Aug. 1902, 158; "Automobile Topics of Interest," *New York Times*, 30

Nov. 1902, 14; Fay, "Motor Fuel," 754; Henry Norman, "The Coming of the Automobile," *World's Work,* April 1903, 3305; "Economy in the Purchase of Fuel," *Horseless Age,* 24 June 1908, 743.

6. "Review of the Year," *Horseless Age,* 3 Jan. 1906, 8. The monopoly was also noted in V. Lougheed, "The Fuel Problem of the Near Future," *Motor,* May 1905, 47; and T. Byard Collins, "Potato Alcohol vs. Standard Oil," *New York Times,* 18 June 1905, SM3. Such appeals were regularly repeated. See, e.g., the *Scientific American* editorial "Supply and Demand in Motor Fuels," 362.

7. For competing alternatives, see "The Rising Price of Gasoline," 158. On the history of American efforts to advance the use of ethyl alcohol for fuel down to 1980, see Hal Bernton, William Kovarik, and Scott Sklar, *The Forbidden Fuel: Power Alcohol in the Twentieth Century* (New York: Boyd Griffin, 1982). On German efforts, see "The Alcohol Motor," *Scientific American Supplement,* no. 1356, 28 Dec. 1901, 21738; "Encouraging Use of Alcohol," *Washington Post,* 10 Aug. 1902, 16; "Automobile Topics of Interest," *New York Times,* 24 Aug. 1902, 12; "Progress of Invention," *Washington Post,* 23 Nov. 1902, 38; "The Free Alcohol Question," *Scientific American,* 4 Mar. 1905, 178; Collins, "Potato Alcohol vs. Standard Oil," SM3; "Tax-Freed Alcohol Assured," *Scientific American,* 30 June 1906, 538; H. Diederichs, "The Use of Alcohol as a Fuel for Gas Engines," *International Marine Engineering,* July 1906, 264–270; "The Use of Denatured Alcohol in Internal Combustion Engines," *Horseless Age,* 4 July 1906, 1; Charles Edward Lucke and S. M. Woodward, "The Use of Alcohol and Gasoline in Farm Engines," *USDA Farmers' Bulletin No. 277* (Washington, DC: GPO, 1907), 13; "Alcohol Now Urged as a Motor Fuel," *New York Times,* 31 May 1914, pt. 8, 7; and Harold B. Dixon, "Researches on Alcohol as a Fuel for Internal Combustion Engines," *Automotive Industries,* 3 Feb. 1921, 211. For an early American prediction that alcohol was the fuel of the future, see "Automobile Topics of Interest," 24 Aug. 1902, 12. See, too, "Kerosene Fuel," *Horseless Age,* 21 Feb. 1906, 311. Fay, "Motor Fuel," 753.

8. The U.S. Department of Agriculture published several studies on industrial alcohol. See H. W. Wiley, "Industrial Alcohol: Sources and Manufacture," *USDA Farmers' Bulletin No. 268* (Washington, DC: GPO, 1906); H. W. Wiley, "Industrial Alcohol: Uses and Statistics," *USDA Farmers' Bulletin No. 269* (Washington, DC: GPO, 1906); and Lucke and Woodward, "The Use of Alcohol and Gasoline in Farm Engines." See "Government Encouragement of Industrial Alcohol," *Horseless Age,* 8 Aug. 1906, 174, which refers to the 6 Aug. 1906 issue of *Daily Consular Reports;* Sy, "Tax-Free Alcohol," 62–63; John Preston, "Alcohol Engines to Replace Gasoline Engines," *Scientific American,* 11 May 1907, 396; and "The Need of Cheaper Fuels," *Horseless Age,* 11 Mar. 1908, 281. On the U.S. Geological Survey, see "Gasoline and Alcohol as Motor Fuels," *Engineering Magazine,* May 1908, 262–264; and Thomas L. White, "Alcohol as a Fuel for Internal-Combustion Engines," *Engineering Magazine,* August 1908, 739–747. On the auto club's interest, see "Alcohol for Motor Use," *New York Times,* 29 Nov. 1905, 11; "Automobile Notes of Interest," *New York Times,* 22 Dec. 1905, 7; "Alcohol for Autos May Displace Gasoline," *New York Times,* 18 Mar. 1906, 12; "Alcohol for Auto Use Meets Popular Favor," *New York Times,* 18 Apr. 1906, 12; "Auto Club Aroused over Alcohol Bill," *New York Times,* 26 Apr. 1906, 12; "Alcohol for Auto Fuel," *New York Times,* 15 May 1906, 7; "The A.C.A.'s Alcohol Contest," *Horseless Age,* 25 July 1906, 110; and "Concerning the Tax on Alcohol," *Motor,* June 1904, 36.

9. On the amount of the excise, see "The Free Alcohol Question," 178; "What 'Free Alcohol' Might Do," *Current Literature,* June 1906, 590; "The Hope for Free Alcohol," *World's Work,* June 1906, 7599; Charles Baskerville, "Free Alcohol in the Arts and as Fuel," *American Monthly Review of*

Reviews, August 1906, 211; and John Preston, "Alcohol Engines to Replace Gasoline Engines," *Scientific American*, 11 May 1907, 396.

10. On the U.S. exception, see Sy, "Tax-Free Alcohol," 59–61. On repeal efforts, see, in addition to Sy, Fay, "Motor Fuel," 754; "In the Big Industries," *Los Angeles Times*, 13 Apr. 1906, II4; and Baskerville, "Free Alcohol in the Arts and as Fuel," 212. For those who prefer to see corporate perfidy behind America's automobile dependence, Standard Oil's lobbying against the repeal of the alcohol excise tax provides more grist for the mill. Most commentators at the time named the wood alcohol industry as the chief opponent of repeal, but some hinted at Standard Oil machinations in the background.

11. On Thomson and General Electric, see "Cheap Alcohol for Motors," *Scientific American*, 7 Apr. 1906, 282; and Baskerville, "Free Alcohol in the Arts and as Fuel," 214. See also Thomson, "Alcohol and the Future of the Power Problem," 310–312. For support from automobile interests, see "Alcohol for Auto Fuel," *New York Times*, 15 May 1906, 7; "Automobile Notes," *Scientific American*, 23 June 1906, 511; and "The A. C. A.'s Alcohol Contest," 110. For claims about the cheap production of alcohol, see the untitled editorial, *New York Times*, 13 Feb. 1907, 8. This belief persisted for a number of years. See, e.g., "Alcohol Now Urged as a Motor Fuel," *New York Times*, 31 May 1914, pt. 8, 7. The belief that the price of alcohol would fall was general. For examples, see "Alcohol as Fuel for Automobiles," *New York Times*, 12 Feb. 1907, 6; untitled editorial, *New York Times*, 13 Feb. 1907, 8; and H. H. Franklin, "Alcohol Motor Now a Reality," *New York Times*, 17 Jan. 1909, pt. 6, 9. For confidence in consumer shift, see "Poor Gasoline Troubles," *New York Times*, 29 July 1906, 7. On Standard Oil, see Ida M. Tarbell, *History of the Standard Oil Company* (New York: McClure, Phillips, 1904). On Roosevelt and Standard Oil, see "What 'Free Alcohol' Might Do," *Current Literature*, June 1906, 589; "Denaturized Alcohol," *Los Angeles Times*, 1 Sept. 1906, II4; and Sy, "Tax-Free Alcohol," 64. On passage, see "Free Alcohol Bill Passed," *New York Times*, 25 May 1906, 7; "Free Alcohol Bill Signed," *New York Times*, 9 June 1906, 1. For the act itself, which focused mainly on the denaturing requirement, see 59th Cong., sess. 1, chap. 3047, in *The Statutes at Large of the United States of America from December, 1905, to March, 1907*, pt. 1 (Washington, DC: GPO, 1907), 217–218.

12. On the competitive price, see Diederichs, "The Use of Alcohol as a Fuel for Gas Engines," 270; and "Alcohol as Fuel for Automobiles," *New York Times*, 12 Feb. 1907, 6. For the actual price, see "An Alcohol Test for Motor Cars," *New York Times*, 6 Mar. 1907, 7; "Three Months of Denatured Alcohol," *Scientific American*, 6 Apr. 1907, 286; Preston, "Alcohol Engines to Replace Gasoline Engines," 396; and White, "Alcohol as a Fuel for Internal-Combustion Engines," 739. The actual price did not fall much more in the years to come. See "Praises Alcohol as a Motor Fuel," *New York Times*, 24 May 1914, pt. 7, 11, which indicates that a gallon of alcohol was still selling in the range of 30–33 cents a gallon. On the distillers, see "Launching a Great Industry: The Making of Cheap Alcohol," *New York Times*, 25 Nov. 1906, SM3; "Alcohol Company Dividends for Distillers Securities," *Wall Street Journal*, 15 July 1907, 2; "Denaturized Alcohol," II4. On the farmers' interest, see Lucke and Woodward, "The Use of Alcohol and Gasoline in Farm Engines," 2; "Tax-Freed Alcohol Assured," *Scientific American*, 30 June 1906, 538; "Free Alcohol Distilleries," *New York Times*, 13 Sept. 1906, 6; "Launching a Great Industry," SM3; "The Blue Barrel's Eclipse?" *New York Times*, 6 Mar. 1907, 8; and "Gasoline Finds Rival," *Washington Post*, 25 Mar. 1907, 9. On availability, see White, "Alcohol as a Fuel for Internal-Combustion Engines," 739.

13. There was a large technical literature on the problems with alcohol. For more sources, see Tom

McCarthy, "The Coming Wonder? Foresight and Early Concerns about the Automobile," *Environmental History* 6:1 (2001): 70, n. 41. On carburetor modification, see "Alcohol Tests with Auto Engine," *New York Times*, 20 Oct. 1907, S3. On alcohol engines, see Lucke and Woodward, "The Use of Alcohol and Gasoline in Farm Engines," 2, 14.

14. For disappointment, see "The Denatured Alcohol Situation," *Scientific American Supplement*, 21 Dec. 1907, 392; "Gasoline and Alcohol as Motor Fuels," *Engineering Magazine*, May 1908, 262–264. For the Heldt quotation, see "The Need of Cheaper Fuels," *Horseless Age*, 11 Mar. 1908, 281. See David A. Kirsch, "The Electric Car and the Burden of History: Studies in Automotive Systems Rivalry in America, 1890–1996," *Business and Economic History* 26:2 (1997): 304–310, which concluded that the electric car had ceased to be competitive with gasoline-powered vehicles by 1902. For a nuanced analysis of this competition, see Gijs Mom, *Electric Vehicle: Technology and Expectations in the Automobile Age* (Baltimore: Johns Hopkins University Press, 2004).

15. For examples in the growing literature on this period, see Joel A. Tarr, "Searching for a 'Sink' for an Industrial Waste: Iron-Making Fuels and the Environment," *Environmental History Review* 18:1 (1994): 9–34; Katherine G. Aiken, "'Not Long Ago a Smoking Chimney Was a Sign of Prosperity': Corporate and Community Response to Pollution at the Bunker Hill Smelter in Kellogg, Idaho," *Environmental History Review* 18:2 (1994): 67–86; Adam W. Rome, "Coming to Terms with Pollution: The Language of Environmental Reform, 1865–1915," *Environmental History* 1:3 (1996): 6–28; David Stradling and Peter Thorsheim, "The Smoke of Great Cities: British and American Efforts to Control Air Pollution, 1860–1914," *Environmental History* 4:1 (1999): 6–31; and Angela Gugliotta, "Class, Gender, and Coal Smoke: Gender Ideology and Environmental Injustice in Pittsburgh, 1868–1914," *Environmental History* 5:2 (2000): 165–193.

16. For a definition of nuisance, see Rome, "Coming to Terms with Pollution," 16–18; and Christine Meisner Rosen, "Noisome, Noxious, and Offensive Vapors, Fumes and Stenches in American Towns and Cities, 1840–1865," *Historical Geography* 25 (1997): 49–82. On the then still tenuous connection between public health concerns and nuisance regulation, see Rosen, 63–66.

17. For the public health context, see Nancy Tomes, *Gospel of Germs: Men, Women, and the Microbe in American Life* (Cambridge, MA: Harvard University Press, 1998), esp. 6, 95–100; and Suellen Hoy, *Chasing Dirt: The American Pursuit of Cleanliness* (New York: Oxford University Press, 1995), esp. 59–86. For favorable public health comparisons with the horse, see, e.g., "Horseless Carriages and Sanitation," *Scientific American*, 18 Jan. 1896, 36; "The Future of the Motor Car," *Scientific American*, 17 July 1897, 35; "The Automobile a Blessing," *Automobile*, May 1900, 80; *Automobile*, October 1900, 177; "The Horse vs. Health," *American Review of Reviews*, May 1908, 623–624; and "Foul Exhaust Competition," *Motor*, March 1907, 96. A nice summary of this view is found in P. G. Heinemann, "The Automobile and Public Health," *Popular Science Monthly*, March 1914, 284–289. See also Joel A. Tarr, "The Horse—Polluter of the City," in Tarr, *Search for the Ultimate Sink: Urban Pollution in Historical Perspective* (Akron, OH: University of Akron Press, 1996), chap. 12, for a good overview of the horse as an urban problem. Mom, *Electric Vehicle*, 8, suggests that automobile proponents made more of this point than others. There were a number of articles on the automobile as panacea. For the baldness and dandruff claim, see "The Healthfulness of Motoring," *Literary Digest*, 10 Feb. 1912, 284. On the disinfectant claims, see "Motors Reduce Mortality," *New York Times*, 2 Aug. 1908, pt. 3, 1; and "Do Auto Fumes Kill Germs?" *Chicago Daily Tribune*, 3 Aug. 1908, 3. This argument was also made on behalf of industrial smoke. See Gugliotta, "Class, Gender, and Coal Smoke," 170.

18. "Foul Exhaust Competition," *Motor*, March 1907, 96; "Automobile Lubrication One of Greatest Problems," *Atlanta Constitution*, 7 Nov. 1909, F10. On excessive oil, see "Smoky Exhaust," *Motor*, July 1906, 86, quoting J. W. G. Booker in a paper delivered before the Auto-cycle Club of England. On carbon monoxide, see W. E. Smith, "Poisonous Effect of Exhaust Gases," *Horseless Age*, 4 Feb. 1903, 206–207; and C. H. Lemon, "Gasoline Intoxification," *Horseless Age*, 10 Aug. 1904, 128. On imperfect combustion, see "Foul Exhaust Competition," 96; Herbert L. Towle, "The Gentle Art of Making Smoke," *Motor*, April 1907, 56; and "New York's Anti-Smoke Crusade," *Horseless Age*, 13 Apr. 1910, 535–536.

19. Sir Dugald Clerk, FRS (1854–1932), patented the Clerk two-stroke cycle engine in 1881. Clerk was the foremost British automotive engineer of his generation. He was knighted for his wartime services to the British government during World War I. See Dugald Clerk, abstract of paper presented to the Institution of Automobile Engineers, London, 11 Dec. 1907, published as "The Principles of Carbureting as Determined by Exhaust Gas Analysis," in *Horseless Age*, 1 Jan. 1908, 13–14, and 8 Jan. 1908, 38–39. On the impact of Clerk's paper, see Albert L. Clough, "Exhaust Gas Analysis," *Horseless Age*, 7 July 1909, 1. This same finding was reported in "The Exhaust Gases from a Petrol Motor," *Engineering*, 9 Aug. 1907, 197.

20. For Clerk's concern, see "The Principles of Carbureting as Determined by Exhaust Gas Analysis," 14. Clough, "Exhaust Gas Analysis," 1. "Stopping the Smoke Nuisance by Law," *Horseless Age*, 26 Aug. 1908, 249–250.

21. For mention of efforts in other cities, see "Smoking Autos Not Necessary, Says Leland," *Atlanta Constitution*, 24 Oct. 1909, E5. The District of Columbia adopted an automobile smoke ordinance. See "Rules for Auto Users," *Washington Post*, 2 June 1909, 16; "No More Auto Gases," *Washington Post*, 23 July 1909, 14; and "Thirteen Automobilists Fined," *Washington Post*, 1 Nov. 1911, 2. Philadelphia banned excessive automobile smoke in Fairmount Park. See "Speed Traps for Motorists," *New York Times*, 13 Oct. 1912, X14. On the park ordinance, see T. O'R., "Automobiles in the Park," letter to the editor, *New York Times*, 24 May 1906, 8; "Automobiles in the Park," editorial, *New York Times*, 29 July 1906, 8; "The Smoky Exhaust Nuisance," *Horseless Age*, 29 July 1908, 139; "The Ruling Park Board," *New York Times*, 2 Aug. 1908, pt. 4, 4; and "Ill-Smelling Autos Barred," *New York Times*, 3 Aug. 1908, 5; and "Upholds Auto Smoke Law," *New York Times*, 1 Sept. 1908, 8. For a claim that the ordinance was successful, see "Doctors Protest against Auto Smoke," *New York Times*, 4 Feb. 1910, 5.

22. "Automobiles Should Not Smoke," *New York Times*, 7 Mar. 1910, 8. "Automobiles Mustn't Smoke," *New York Times*, 28 Apr. 1910, 12. On efforts before the aldermen and the failure to pass an ordinance, see "To Bar Smoking Autos," *New York Times*, 12 Mar. 1910, 16; and Edward S. Cornell, "Automobile Smoke Nuisance," letter to the editor, *New York Times*, 30 Apr. 1910, 8.

23. On the failure, see "Smoking Automobiles," *New York Times*, 30 Sept. 1910, 12; "Nuisance in Smoking Autos," *New York Times*, 30 Sept. 1910, 13; and "Smoking Motor Cars," *New York Times*, 3 Oct. 1910, 8. These editorials or articles indicate that to the end of September 1910 not a single arrest had been made for transgressing the ordinance, apparently because the Board of Health declined to enforce it. Pressure from the *Times* resulted in periodic spasms of enforcement over the next five years. See "Smoking Motor Cars," *New York Times*, 2 Dec. 1910, 8; "Raid Smoking Autos," *New York Times*, 2 Jan. 1911, 4; "Arrests for the Auto Smoke Nuisance," *New York Times*, 11 May 1911, 2; "The Traffic Smoke Nuisance," editorial, *New York Times*, 12 June 1911, 10; "Thirty-Five Autoists Fined," *New York Times*, 25 June 1911, 8; untitled editorial, *New York Times*, 25 June

1911, 10; untitled editorial, *New York Times*, 30 Oct. 1912, 12; "Fined for Smoking Autos," *New York Times*, 10 Nov. 1912, pt. 2, 11; "Fine Fifty for Smoking Autos," *New York Times*, 13 Nov. 1912, 12; and "Fined for Letting Autos Smoke," *New York Times*, 28 Oct. 1915, 9. Difficulty with enforcement due to distinguishing between light and dark smoke is implied in "Police to Stop Smoking Autos," *New York Times*, 29 June 1910, 2. New York City health commissioner Dr. E. J. Lederle apparently had been skeptical about the wisdom of revising the sanitation code from the outset for precisely this reason. "Lederle Cuts Expenses," *New York Times*, 5 Feb. 1910, 2. On the matter of smoke being unavoidable, see "Raid Smoking Autos," 4; untitled editorial, *New York Times*, 21 Jan. 1911, 12; "Auto Smoke Nuisance," *New York Times*, 2 July 1911, C9; and "Causes and Remedy of Engine Smoking," *New York Times*, 17 Nov. 1912, X16.

24. "New York's Anti-Smoke Crusade," 535–536.

25. See Cleveland Moffett, "Automobiles for the Average Man," *American Monthly Review of Reviews*, June 1900, 704, for an early explicit description of envy.

26. "Enterprising Tramps," reprinted from the *Philadelphia Record* in *Washington Post*, 19 Aug. 1904, 6; "The Rambler: Pleasures of Pedestrianism," *Los Angeles Times*, 4 Dec. 1906, II4; "Topics of the Times," *New York Times*, 11 Mar. 1903, 8. A news story about the speech appeared the day before: "Prince Henry Advises Automobilists," *New York Times*, 10 Mar. 1903, 3.

27. A. B. Tucker, "The Social Status of the Chauffeur," *Horseless Age*, 2 Sept. 1903, 240. Julian Smith found this same problem with assessing the social status of the chauffeur in a number of early motion picture plots involving the automobile. See Julian Smith, "A Runaway Match: The Automobile in the American Film, 1900–1920," in David L. Lewis and Laurence Goldstein, eds., *The Automobile and American Culture* (Ann Arbor, MI: University of Michigan Press, 1980, 1983), 184–185. For a different take on the "chauffeur problem" that focuses on the power relations between owners and chauffeurs, see Kevin Borg, "The 'Chauffeur Problem' in the Early Auto Era: Structuration Theory and the Users of Technology," *Technology and Culture* 40:4 (1999): 797–832.

28. "The Drunken Driver," *Horseless Age*, 16 Dec. 1908, 859. Joyriding was a national problem that occurred everywhere people owned automobiles. For some typical newspaper accounts, see "Drunken Chauffeur out on a Wild Ride," *Atlanta Constitution*, 14 May 1911, 2; "Smash-Up on a Joy Ride," *New York Times*, 20 Oct. 1911, 15; "Rogers a Second Victim of Driver," *Chicago Daily Tribune*, 7 May 1912, 1; "Smash Mondell's Car," *Washington Post*, 10 Apr. 1913, 2; "'Jitney' Joy Ride Is Disastrous," *Los Angeles Times*, 8 Jan. 1915, II3; and "Joy Ride Ends in Death," *Washington Post*, 3 Apr. 1911, 1.

29. "Chauffeur Is Convicted," *New York Times*, 17 May 1911, 13, makes mention of the first conviction under the New York law. See "Stop Joy Riding, Car Owners' Aim," *Chicago Daily Tribune*, 25 July 1912, 20, for mention of the Illinois law. "Blames Driver for Fatality," *Los Angeles Times*, 29 Aug. 1915, II1. "The Status of the Chauffeur," *Horseless Age*, 8 July 1908, 39.

30. Columbia economist Edwin R. A. Seligman recognized this status impact. See Seligman, *The Economics of Instalment Selling*, vol. 1 (New York: Harper, 1927), 243. This dynamic surfaced again as the wealthiest Chinese adopted automobiles in the early twenty-first century. See Jim Yardley, "Chinese Go Online in Search of Justice against Elite Class," *New York Times*, 16 Jan. 2004, A1, A6. For one take on masculinity anxieties in early-twentieth-century America, see Gail Bederman, *Manliness and Civilization: A Cultural History of Gender and Race in the United States, 1880–1917* (Chicago: University of Chicago Press, 1995).

3. A MONSTROUSLY BIG THING

1. Philip Van Doren Stern, *Tin Lizzie: The Story of the Fabulous Model T Ford* (New York: Simon and Schuster, 1955). The literature on the Ford Model T is large. The best place to begin remains Allan Nevins with Frank Ernest Hill, *Ford: The Times, the Man, the Company* (New York: Scribner's, 1954) (hereinafter Nevins and Hill I). For a more recent treatment that places the Model T in the larger context of popular culture, see Douglas Brinkley, *Wheels for the World: Henry Ford, His Company, and a Century of Progress* (New York: Viking, 2003), 99–133. Although its scope is broader than the Model T and its rural owners, Reynold M. Wik's *Henry Ford and Grass-Roots America* (Ann Arbor: University of Michigan Press, 1972) is still the best of the scholarly monographs. On Ford's vision, see Nevins and Hill I, 276. The literature on Henry Ford is also vast. For an excellent recent scholarly biography, see Steven Watts, *The People's Tycoon: Henry Ford and the American Century* (New York: Alfred A. Knopf, 2005).

2. Recognition of the opportunity is evident in the early automobile enthusiasts' magazines. For Henry Ford's interest in the five-hundred-dollar car, see "Automobile Notes of Interest," *New York Times*, 3 Feb. 1907, 11. Wik, *Henry Ford and Grass-Roots America*, 233–236, debunks the myth that Ford was the first to envision an affordable car for the masses.

3. "The Automobile Outlook for 1910," *Washington Post*, 31 Oct. 1909, AS1, gives 275 automakers. For a nice graphical depiction of the number of auto manufacturers in the United States, Great Britain, France, and Germany broken down by type of car made for the period 1895–1919, see Gijs Mom, *Electric Vehicle: Technology and Expectations in the Automobile Age* (Baltimore: Johns Hopkins University Press, 2004), 23. On rising prices, see Nevins and Hill I, 276.

4. E. C. Shumard, "Foresight of Henry Ford," letter to the editor, *Scientific American*, 25 Dec. 1915, 557. Shumard was an early Ford dealer (1903) in Cincinnati. Writing after Ford's great success and nearly a decade after the conversation in question, his recollection of Ford's comments in 1906 must be accepted with appropriate caution. Donald B. Dodd, comp., *Historical Statistics of the States of the United States: Two Centuries of the Census, 1790–1990* (Westport, CT: Greenwood, 1993), 104. On how well Ford understood his customers' needs, see Nevins and Hill I, 495. On Ford's feelings about the Model T's popularity with rural Americans, see ibid., 396–397.

5. Nevins and Hill I, 388–393. See esp. Watts, *People's Tycoon*, 112–118, on Henry Ford's personal involvement.

6. See Stern, *Tin Lizzie*, 52, for a reproduction of the ad, and 53 and 58 for the public response. The twenty-four lower-priced models are mentioned by Wik, *Henry Ford and Grass-Roots America*, 236. On the letters, see Nevins and Hill I, 401, 626n, citing the original source, *Ford Times*, 15 Oct. 1908, 10. On being an immediate sensation, see "The Ford Headquarters," *Washington Post*, 25 Oct. 1908, AU6; and "Model T Ford Meeting with Success," *Washington Post*, 20 Dec. 1908, S3. Mom, *Electric Vehicle*, 107–112, also suggests that the crisis of 1907 checked the growth in the market for large, expensive automobiles as well as marked a turning point in automotive history. No doubt this pause in the trend helped Ford gain some important early traction in its intended market niche.

7. One can trace the dealer efforts and the support the company provided for these efforts in local newspapers. See, e.g., notices about Charles E. Miller and Brother in the automobile pages of the *Washington Post*, 1908–1912. On the Seattle race, see Nevins and Hill I, 405–406. On other

competitions, see "Sixteen Cars Finish," *Washington Post*, 16 May 1909, S3; "Cars Now Number Twenty-Seven for Auto Race," *Chicago Daily Tribune*, 5 June 1909, 11; "Forty Cars Are Assured in Run from Savannah," *Atlanta Constitution*, 31 Oct. 1909, A5; "Big Hill Climb on Saturday," *Atlanta Constitution*, 25 Mar. 1910, 11; "Hats Off to the Ford!" *Atlanta Constitution*, 12 June 1910, B2; "In the Auto World," *Washington Post*, 14 July 1910, 9; "In the Auto World," *Washington Post*, 27 July 1910, 9; "Auto Notes," *Washington Post*, 31 July 1910, ES3; "Model T Ford Does Fine Work," *Christian Science Monitor*, 20 Aug. 1910, 13; "1,530-Mile Trip Ends," *Washington Post*, 28 Aug. 1910, ES3; "Ford Victories of Recent Date," *Atlanta Constitution*, 18 Sept. 1910, A5; "Ford Feature of Syracuse Meet," *Atlanta Constitution*, 27 Sept. 1910, 10; "Thirty-Two Autos Entered to Contest Today at Fairmount," *Christian Science Monitor*, 8 Oct. 1910, 15; "Takes Grit on Desert," *Los Angeles Times*, 9 Oct. 1910, II:AS3; "South Carolina in Fall Meet," *Atlanta Constitution*, 16 Oct. 1910, A3; "Woman First in Auto Race," *Washington Post*, 13 Nov. 1910, A2; "Ford Feature of Reliability Run and Hill Climb," *Christian Science Monitor*, 15 July 1911, 12; "Gossip of the Automobilists and Notes of the Trade," *New York Times*, 3 Sept. 1911, C8; "Auto Passes Worst Road," *Washington Post*, 4 Oct. 1911, 4; "Victory for Ford Car," *Washington Post*, 10 Dec. 1911, R3. On stock cars, see "Stock Cars in Racing," *Washington Post*, 5 Mar. 1911, E10. On durability, see "Ford Model T Taxicab Makes Mileage Record," *Christian Science Monitor*, 15 Apr. 1911, 18. On freeing horses, see "In the Automobile World," *Washington Post*, 12 May 1911, 14. For "famous car," see "Ford Agents Now Have Dissected Auto Chassis," *Christian Science Monitor*, 8 Mar. 1911, 9.

8. Ibid., 9. For Hawkins's efforts, see Watts, *People's Tycoon*, 128–132. K. P. Drysdale, "The Novice and His First Automobile," *Harper's Weekly*, 6 Jan. 1912, 26.

9. On the importance of price, see Gerald Stanley Lee, "Is Ford an Inspired Millionaire?" *Harper's Weekly*, 14 Mar. 1914, 10; J. George Frederick, "Automobiles by the Million," *Review of Reviews*, October 1915, 459; and Albert L. Clough, "Democratizing the Motor Car," *Independent*, 1 May 1916, 178. On the existence and nature of two markets, see "The Motor Rage in America," *Living Age*, 17 Sept. 1910, 754; "The Motor Car and Country Life," *Craftsman*, May 1911, 227; and Thaddeus S. Dayton, "The Poor Man's Motor Car," *Harper's Weekly*, 11 Jan. 1913, 10. For contemporaneous discussions of the appeal of automobiles and the Model T, especially to farmers and rural markets, see G. F. Carter, "Automobiles for Average Incomes," *Outing Magazine*, January 1910, 416; Walter Langford, "What the Motor Vehicle Is Doing for the Farmer," *Scientific American*, 15 Jan. 1910, 51; Edward S. Martin, "The Motor Craze," *Outlook*, 19 Nov. 1910, 629; Charles Moreau Harger, "Automobiles for Country Use," *Independent*, 1 June 1911, 1207; "What the Auto Stands For," editorial, *Independent*, 24 Aug. 1911, 440; Dayton, "Poor Man's Motor Car," 11; and Theodore M. Von Keler, "The Farmer and the Motor Car," *Collier's*, 9 Jan. 1915, 34. See, too, Nevins and Hill I, 493–494.

10. Henry Ford, as quoted by Harry Franklin Porter, "What's Behind Ford's $1,000,000 a Week?" *System*, December 1916, 605. This figure for the Model T's fuel economy was widely used. See also Wik, *Henry Ford and Grass-Roots America*, 76.

11. On ease of fixing, see Porter, "What's Behind Ford's $1,000,000 a Week?" 600. Sears Roebuck is from E. B. White, *Farewell to Model T: From Sea to Shining Sea* (New York: Little Bookroom, 2003), 9. Wik, *Henry Ford and Grass-Roots America*, 2.

12. For contemporaneous accounts that discuss the appeal of the automobile to this market, see Herbert Ladd Towle, "The Every Day Automobile," *Harper's Weekly*, 4 Feb. 1911, 11; Harger,

"Automobiles for Country Use," 1207; Dayton, "Poor Man's Motor Car," 10; Frederick, "Automobiles by the Million," 457; Anthony M. Rud, "How to Buy a Car of Moderate Pice," *Illustrated World*, February 1917, 923.

13. On favorable context, see Wik, *Henry Ford and Grass-Roots America*, 21. Roscoe Sheller, *Me and the Model T* (Yakima, WA: Franklin, 1965), vii and 58. On the Model T and doctors, see Wik, *Henry Ford and Grass-Roots America*, 22. The point is also very evident in the early automobile enthusiasts' magazines like *Horseless Age*. See also Mom, *Electric Vehicle*, 106–107. The observation about sharing rides was made by Martin, "Motor Craze," 630, and Judson C. Welliver, "The Automobile in Our Country," *Collier's*, 8 Jan. 1916, 72. On the suspension of Model T advertising, see Allan Nevins and Frank Ernest Hill, *Ford: Expansion and Challenge: 1915–1933* (New York: Scribner's, 1957) (hereinafter Nevins and Hill II), 262–263. See Bronson Batchelor, "Motorizing America," *Independent*, 1 Mar. 1915, 320, who argued that the growth of the industry was due more to mass production (and declining prices) than advertising. On the importance of word-of-mouth generally, see Thaddeus S. Dayton, "How to Sell a Car," Harper's Weekly, 7 Jan. 1911, 10. In connection with the Model T, see G. S. Lee, "The Clue to Mr. Ford," *Everybody's Magazine*, January 1916, 91.

14. Survey cited by Wik, *Henry Ford and Grass-Roots America*, 40. See Harger, "Automobiles for Country Use," 1207, for peer pressure as well as Minor Hinman, 5–6 Nov. 1938, "American Life Histories: Manuscripts from the Federal Writers' Project, 1936–1940," Manuscript Division, Library of Congress, accessed at http://memory.loc.gov/ammem/wpaintro/wpahome. Hinman did not reveal the make of the car he bought. On the role of women, see Dayton, "How to Sell a Car," 9; and Frederick, "Automobiles by the Million," 461. Lee, "Is Ford an Inspired Millionaire?" 10.

15. Sheller, *Me and the Model T*, 75–78. The five-hundred-dollar barrier dates are from Watts, *People's Tycoon*, 146; Les Henry, "The Ubiquitous Model T," in Floyd Clymer, *Henry's Wonderful Model T, 1908–1927* (New York: Bonanza, 1955), 118.

16. This tipping-point date is estimated from the year-over-year rates of increase evident in "Appendix III: Total Sales of Ford Cars," Nevins and Hill I, 644. For dealers, see Watts, *People's Tycoon*, 129. For every third car, Lee, "Is Ford an Inspired Millionaire?" 9. For sales to the farm market, see Stern, *Tin Lizzie*, 81. The regional rates of ownership are from Ronald R. Kline, *Consumers in the Country: Technology and Social Change in Rural America* (Baltimore: Johns Hopkins University Press, 2000), 65, whose figures are from a 1920 Bureau of the Census report. Sheller, *Me and the Model T*, 85–92.

17. This lore was fondly recalled by White, *Farewell to Model T*, 18. On complaints and jokes, see Clymer, *Henry's Wonderful Model T*, 167. For Ford clubs, see Wik, *Henry Ford and Grass-Roots America*, 36–37. Daniel J. Boorstin, *Americans: The Democratic Experience* (New York: Random House, 1973), 89–164.

18. The accessory figure is from Clymer, *Henry's Wonderful Model T*, 191.

19. Many people have recognized the obvious importance of timing in the Model T's introduction and eventual success. See, e.g., Wik, *Henry Ford and Grass-Roots America*, 14. Henry Ford, as quoted by Brinkley, *Wheels for the World*, 100.

20. On the Model T's speed and safety, see Carter, "Automobiles for Average Incomes," 413; White, *Farewell to Model T*, 30; and Stern, *Tin Lizzie*, 13. On the top speed, see Clymer, *Henry's Wonderful Model T*, 28; and John C. Gerber, *O Marvelous Model T!* (Iowa City, IA: Maecenas, 1991), 8.

On the acceleration, see White, *Farewell to Model T*, 11. "Ford Accessories at September Show," *Chicago Daily Tribune*, 12 Aug. 1917, D7. William K. Vanderbilt, Jr., and Albert C. Bostwick pioneered speedboat racing. See "Motor Boat for Bostwick," *New York Times*, 9 May 1904, 7; and "New Speed Records by Automobile Boats," *New York Times*, 23 Sept. 1904, 7. Harry Harkness became an important early aviator. Foxhall Keene, on the other hand, retreated to polo and fox-hunting.

21. White, *Farewell to Model T*, 12. Harger, "Automobiles for Country Use," 1208.

22. Watts, *People's Tycoon*, 131, quotes Hawkins. The original source is Norval A. Hawkins, *The Selling Process: A Handbook of Salesmanship Principles* (Detroit, MI: n.p., 1920).

23. For the two sales surges, see H. G. Weaver, "A Study Relating to the Past, Present and Future of the Domestic Passenger Car Market," Public Relations, Consumer Research, General Motors, 12, 22, in Edward Stettinius Papers, Special Collections, University of Virginia Library, Charlottesville, accession 2723, box 18. On the timing of the first surge, see Raymond B. Prescott, "Cars Selling for Less Than $1,000 81.6 per cent of 1923 Output," *Automotive Industries*, 21 Feb. 1924, 372. See Watts, *People's Tycoon*, 135–148, for an excellent account of the Highland Park achievement. The classic analysis of the technological path to mass production at Highland Park remains David Hounshell, *From the American System to Mass Production, 1800–1932: The Development of Manufacturing Technology in the United States* (Baltimore: Johns Hopkins University Press, 1984). For prices, see *Automotive Industries*, 15 Mar. 1950, 76. Income and hours are derived from U.S. Department of Commerce, Bureau of the Census, *Historical Statistics of the United States: Colonial Times to 1970*, pt. 1 (Washington, DC: GPO, 1975), Series F 1-5: Gross National Product, Total and Per Capita, in Current and 1958 Prices: 1869 to 1970, 224; and Series D 765-778: Average Hours and Average Earnings in Manufacturing, in Selected Non-manufacturing Industries, and for "Lower-Skilled" Labor 1890 to 1926, 168.

24. Sheller, *Me and the Model T*, vii. On the 1920s consumer credit revolution, see the complementary Martha L. Olney, *Buy Now, Pay Later: Advertising, Credit, and Consumer Durables in the 1920s* (Chapel Hill: University of North Carolina Press, 1991); and Lendol Calder, *Financing the American Dream: A Cultural History of Consumer Credit* (Princeton, NJ: Princeton University Press, 1999).

25. Figures are from Olney, *Buy Now, Pay Later*, 86; and Calder, *Financing the American Dream*, 18. Calder, *Financing the American Dream*, 17, 19. On auto debt quintupling, see Olney, *Buy Now, Pay Later*, 92. For the historical perspective on Americans and debt, see Calder, *Financing the American Dream*, 22–28.

26. On the roots of financing, see Olney, *Buy Now, Pay Later*, 105–106; Calder, *Financing the American Dream*, 161–181; Alfred P. Sloan, Jr., *My Years with General Motors* (New York: Currency and Doubleday, 1963, 1990), 302–303. For contemporary observers who make this same point, see J. D., "Time Sales of Automobiles Have Reached Enormous Proportions," *Automotive Industries*, 26 July 1923, 160; James Dalton, "The Trade-In Problem Is Eighty Years Old," *Motor*, September 1924, 20; J. A. Estey, "Financing the Sale of Automobiles," *Annals of the American Academy* 116 (1924): 44. Calder, *Financing the American Dream*, 159–161, also argues that credit sales of expensive, technologically complex farm equipment in the nineteenth century also prepared the way for automobiles. Will Rogers, quotations compiled by Bryan B. Sterling and Frances N. Sterling, *Will Rogers Speaks* (New York: M. Evans, 1995), 30. Rogers made this comment in December 1929, but the compilers provide no original source or context.

27. On the Ford reluctance, see Calder, *Financing the American Dream*, 189–191. On bankers, see ibid., 187; and "Why Is an Automobile Finance Company?" *Automotive Industries*, 9 Nov. 1922, 919. On origins, see the indispensable Edwin R. A. Seligman, *The Economics of Instalment Selling: A Study in Consumers' Credit with Special Reference to the Automobile*, 2 vols. (New York: Harper, 1927), 1:43–44; Sloan, *My Years with General Motors*, 303; and Calder, *Financing the American Dream*, 187–188. But see "How Automobile Instalment Sales Are Financed," *Literary Digest*, 11 Sept. 1926, 87, where C. C. Hanch, general manager of the National Association of Finance Companies, is quoted as saying, "About fourteen years ago [ca. 1912] the first so-called automobile finance companies came into existence." Calder produces evidence to show that the practice began as early as 1910 with private individuals interested in selling their existing cars, 185. On GMAC, see Seligman, *Economics of Instalment Selling*, 1:48; Olney, *Buy Now, Pay Later*, 118–127; and Calder, *Financing the American Dream*, 191–192. Olney stresses the important role that the automakers' desire for "production smoothing" played in the creation of consumer financing. Companies like GMAC were created, not to encourage retail sales, which were already booming, but to help dealers pay for taking and holding inventory in off-peak seasons. This argument was presented earlier and with greater salience in Martha L. Olney, "Credit as a Production-Smoothing Device: The Case of Automobiles, 1913–1938," *Journal of Economic History* 49:2 (1989): 377–391. Olney is far more knowledgeable in this area than I am, but I think she underappreciates the larger context of roaring consumer demand, including the demand for credit, that made providing consumer credit an obvious opportunity worth pursuing at the same time as dealer financing. This history is covered by Seligman, *Economics of Instalment Selling*, 1:43–51; and Olney, *Buy Now, Pay Later*, 109, 126–128.

28. On terms, see Seligman, *Economics of Instalment Selling*, 2:434, 441–443. The terms for financed used-car sales were more conservative, typically 40 percent down and the balance in ten monthly payments. Seligman and his researchers found that these terms had been stable since 1921. The percentage range for financed sales is an estimate based on multiple sources. See, e.g., Estey, "Financing the Sale of Automobiles," 44; and Henry G. Hodges, "Financing the Automobile," *Annals of the American Academy* 116 (1924): 50. Seligman concluded that this percentage may have been too high, *Economics of Instalment Selling*, 2:425–426, yet see John J. Raskob, "The Development of Installment Purchasing," 9, 11, paper read to the Academy of Political Science, New York, 17 Nov. 1926, reprinted by General Motors and distributed to shareholders, 11 Dec. 1926, Warshaw Collection (no. 60), Archives, National Museum of American History, Washington, DC, Autos, box 6, folder 15, General Motors. GM executive Raskob clearly had early knowledge of Seligman's conclusions but still used the 75 percent figure in November 1926. For other estimates, see Olney, *Buy Now, Pay Later*, 95–97; and Calder, *Financing the American Dream*, 201, who suggests 60–75 percent by 1930. The closer analysis is Seligman, *Economics of Instalment Selling*, 2:403, 424–426. Figures for financing by category are from ibid., 455. On GM's dominance, see Olney, "Credit as a Production-Smoothing Device," 383. Calder, *Financing the American Dream*, 157.

29. For the articles, see Calder, *Financing the American Dream*, 21, who in turn cites Ernestine Wilder, *Consumer Credit Bibliography* (New York: Prentice-Hall, 1938). Allen Sinsheimer, "Does the Automobile Instalment Plan Hurt Business?" *National Retail Clothier*, 20 Dec. 1923, 50, 52.

30. Calder does a nice job summarizing the broader impact of Seligman's study, *Financing the Ameri-*

can Dream, 237–254. The assertion that "there is no new principle in installment selling" made in the 1929 publication *Recent Economic Changes in the United States*, 1, clearly reflects the impact of Seligman's conclusions. Raskob served on the committee that prepared the report. On the findings, see Seligman, *Economics of Instalment Selling*, generally, but see 2:513–516 and esp. 514n. On GM's losses, see Raskob, "Installment Purchasing," 7. Raskob's remarks are apparently a preview of Seligman's published findings. On the lack of advertising, see Olney, *Buy Now, Pay Later*, 160, 183. Calder, *Financing the American Dream*, 222, argues that department store newspaper advertising did encourage credit buying. On the percentage of families that owned automobiles, see "Two-Car-Family Market Needs Factory Development," *Automobile Industries*, 17 Dec. 1927, 889. Frank Stricker, "Affluence for Whom?—Another Look at Prosperity and the Working Classes in the 1920s," *Labor History* 24:1 (1983): 29–33.

31. Ray W. Sherman, "Take Off Your Hat to the Man Who Buys on Time," *Motor*, October 1925, 27. See also "How Automobile Instalment Sales Are Financed," *Literary Digest*, 11 Sept. 1926, 8-7; and Walter Engard, "The Blessing of Time Sales," *Motor*, April 1928, 40, for much the same argument. Raskob, "Installment Purchasing," 8–9. Charles Coolidge Parlin and Fred Bremier, *Passenger Car Industry: Report of a Survey* (Philadelphia: Commercial Research Division, Curtis, 1932), 11–12. Calder, *Financing the American Dream*, 30, 206–208.

32. *Facts and Figures* (Automobile Manufacturers Association, 1930), 82–83. See Alexander Johnston, "Motor Car Industry as a Consumer of American Products," *Motor*, December 1922, 28–29, 76, for similar statistics from the early 1920s. G. Rex Henrickson, "Trends in the Geographic Distribution of Suppliers of Some Basically Important Materials Used at the Buick Motor Division, Flint, Michigan," Institute for Human Adjustment, Horace H. Rackham School of Graduate Studies, University of Michigan, March 1951, 1. Buick in 1950 had thirteen hundred suppliers in forty-two of the forty-eight states, as well as southern Canada. Most were located within a 650-mile radius of Flint.

33. David A. Walker, *Iron Frontier: The Discovery and Early Development of Minnesota's Three Ranges* (n.p.: Minnesota Historical Society, 1979), 230, 258.

34. On automobile tourism in the 1920s, see Warren J. Belasco, *Americans on the Road: From Autocamp to Motel, 1910–1945* (Cambridge, MA: MIT Press, 1979). On the connection with the wilderness movement, see Paul S. Sutter, *Driven Wild: How the Fight against Automobiles Launched the Modern Wilderness Movement* (Seattle: University of Washington Press, 2002).

35. Wik, *Henry Ford and Grass-Roots America*, 76–77, mentions the letters. For expressions of concern, see "New Fuels," *Scientific American*, 13 Apr. 1918, 339; "Are We Wasting Our Petroleum?" *Scientific American*, 20 Apr. 1918, 372; "The Declining Supply of Motor Fuel," *Scientific American*, 8 Mar. 1919, 220; C. F. Kettering, "Combustion of Fuels in Internal-Combustion Engines," *Journal of the Society of Automotive Engineers* 7:3 (1920): 224. Gasoline production is from T. A. Boyd, "Motor Fuel from Vegetation," *Journal of Industrial and Engineering Chemistry* 13:9 (1921): 836. Boyd's article provides a nice overview of the problem and an analysis of the potential solutions. The depletion estimate is from D. H. Killeffer, "Seek New Fuel Supply to Replace Gasoline," *New York Times*, 4 Jan. 1925, A3. Killeffer, then New York secretary of the American Chemical Society, provides a nice overview of the fuel problem in the mid-1920s.

36. William M. Burton, "Manufacture of Gasolene," U.S. patent no. 1,049,667, issued 7 Jan. 1913. "Amazing Chemical Discoveries," *New York Times*, 14 Mar. 1915, SM15. The yield is from Killeffer, "Seek New Fuel Supply to Replace Gasoline," A3.

37. "Declining Supply of Motor Fuel," 220. For Ford's interest in alcohol, see Wik, *Henry Ford and Grass-Roots America*, 143–145.

38. For the date of Kettering's interest in the "fuel problem," see Kettering, "Combustion of Fuels in Internal-Combustion Engines," 224. The discovery that tetraethyl lead used as a gasoline additive eliminated engine "knock" and the implications of using the compound for this purpose have received considerable scholarly attention. However, the story has been told from two very different perspectives. Biographers, business historians, and historians of science and technology have largely celebrated the discovery as a triumph of systematic corporate-funded scientific research successfully solving a pressing technological, business, and national economic problem. Examples in this group include T. A. Boyd, *Professional Amateur: The Biography of Charles Franklin Kettering* (New York: E. P. Dutton, 1957), 98–106, 142–158; Joseph C. Robert, *Ethyl: A History of the Corporation and the People Who Made It* (Charlottesville: University of Virginia Press, 1983), 93–127; Stuart W. Leslie, *Boss Kettering* (New York: Columbia University Press, 1983), esp. 149–180; and Thomas P. Hughes, *American Genesis: A Century of Invention and Technological Enthusiasm* (New York: Penguin, 1989), 223–225. A second group of scholars, writing more recently, has called attention to the issues of human health, environmental impact, and corporate self-interest involved in the decision to use tetraethyl lead. In this group are Gerald Markowitz and David Rosner, who covered this ground in "A 'Gift of God'? The Public Health Controversy over Leaded Gasoline during the 1920s," *American Journal of Public Health* 75:4 (1985): 344–352; and *Deceit and Denial: The Deadly Politics of Industrial Pollution* (Berkeley, CA: Milbank Memorial Fund, 2002), 12–35; William Kovarik, "Charles F. Kettering and the 1921 Discovery of Tetraethyl Lead in the Context of Technological Alternatives," paper originally presented to the Society of Automotive Engineers Fuels and Lubricants Conference, Baltimore, 1994, revised in 1999, and available online at www.radford.edu/%7Ewkovarik/papers/kettering.html; "Henry Ford, Charles Kettering and 'the Fuel of the Future,'" *Automotive History Review*, Spring 1998, 7–27; and "Ethyl: The 1920s Environmental Conflict over Leaded Gasoline and Alternative Fuels," paper presented at the American Society for Environmental History Annual Conference, 26–30 Mar. 2003, Providence, RI, copy in possession of the author; Alan P. Loeb, "Birth of the Kettering Doctrine: Fordism, Sloanism and the Discovery of Tetraethyl Lead," *Business and Economic History* 24 (1995): 72–87; Christian Warren, *Brush with Death: A Social History of Lead Poisoning* (Baltimore: Johns Hopkins University Press, 2000), esp. 116–133, 203–223; and Jamie Lincoln Kitman, "The Secret History of Lead," *Nation*, 20 Mar. 2000, 11–44. This section is based on the works just cited and a review of the principal primary sources used by these authors. Kovarik's interpretation, while still probably correct, is usefully tempered by Loeb's cautionary points. The author thanks Loeb for sharing his analysis. On the viability of alcohol, see "Alcohol as an Automobile Fuel," *Scientific American*, 6 July 1918, 4. On the favorable view of GM's researchers toward alcohol as late as 1921, see Boyd, "Motor Fuel from Vegetation," 839–841.

39. Robert, *Ethyl*, 115. "Ethyl Gasoline for New Engine," *Wall Street Journal*, 3 Nov. 1924, 9. For the number of service stations and sales volume, see "No Ethyl Gasoline Sold in New York," *Wall Street Journal*, 1 Nov. 1924, 8. For the number of states, see "Shift Ethyl Inquiry to Surgeon General," *New York Times*, 21 May 1925, 7.

40. Hamilton was the most important figure in the field of occupational health in the United States in the first half of the twentieth century. She almost single-handedly created the field, especially through her studies of lead. Most of her early findings were published by the federal government

when she was employed as a special investigator by the Department of Labor. These studies were the basis for Alice Hamilton, *Industrial Poisons in the United States* (New York: Macmillan, 1925). The Midgley-Henderson exchange is from Warren, *Brush with Death*, 122.

41. See, e.g., "Stopping Sale of Gasoline Containing Lead Is Urged," *Washington Post*, 29 Oct. 1924, 1. Henderson quoted in "Third Victim Dies from Poison Gas," *New York Times*, 29 Oct. 1924, 23. For the state bans, see "Grand Jury to Get Ethyl Poison Case," *New York Times*, 2 Nov. 1924, 22. Henderson quoted in "Sees Deadly Gas a Peril in Streets," *New York Times*, 22 Apr. 1925, 25.

42. Markowitz and Rosner, *Deceit and Denial*, 25–33, provide a nice summary of the conference. Rosner and Markowitz first popularized Howard's telling bon mot. See Rosner and Markowitz, "'Gift of God'?" 344–352. The *New York Times* quoted Howard as saying "a gift from Heaven." See "Shift Ethyl Inquiry to Surgeon General," *New York Times*, 21 May 1925, 7.

43. "Report No Danger in Ethyl Gasoline," *New York Times*, 20 Jan. 1926, 13. See also Kitman, "Secret History of Lead," 30; and Markowitz and Rosner, *Deceit*, 34.

44. Boyd is quoted in "Tells of New Type of Auto and Fuel," *New York Times*, 7 Aug. 1925, 4. David Farber in *Sloan Rules: Alfred P. Sloan and the Triumph of General Motors* (Chicago: University of Chicago Press, 2002), 81–86, concludes that GM president Alfred P. Sloan knew about the problems but moved forward because of the profits involved. Robert, *Ethyl*, 134, reports rumors among Ethyl employees that Sloan placed Earle W. Webb in the Ethyl presidency in 1925 to oversee shutting down the company. The 1940 figure is from ibid., 168. For the tetraethyl lead produced, see Warren, *Brush with Death*, 127–128. GM profits are from Farber, *Sloan Rules*, 86. Drucker is quoted by Kitman, "Secret History of Lead," 25.

45. This perspective is most prominent in Kovarik, "Charles F. Kettering." For an interpretation that is based principally on Kovarik's view but goes farther in this direction, see Kitman, "Secret History of Lead." For Loeb's analysis, personal communication. For examples of published reservations about alcohol, see "New Fuels," *Scientific American*, 13 Apr. 1918, 339; and Boyd, "Motor Fuel from Vegetation," 840–841. On iron carbonyl, see Robert, *Ethyl*, 139–141; "New German Chemical for Engine Knocking," *New York Times*, 28 Mar. 1926, 10:15; "Motor Mileage Gain Promised by New Fuels," *Christian Science Monitor*, 10 Sept. 1926, 1; "Rubber Plantation Methods Unlikely to Undergo Change," *Washington Post*, 11 Sept. 1926, 3; and "Nation Solves Dye Problems," *Los Angeles Times*, 11 Sept. 1926, 4. Ellis is quoted in "Research in Ethyl Gas Step for Improved Fuels," *New York Times*, 9 Nov. 1924, 10:13.

46. "How Henry Ford Rode Out Storm without a Loan," *Christian Science Monitor*, 7 March 1921, 1.

47. Frederick Lewis Allen, *Only Yesterday: An Informal History of the Nineteen-Twenties* (New York: Harper, 1931), 118.

48. "A Blind Leader of the Blind," *Freeman*, 26 Dec. 1923, 365. Joseph A. Schumpeter, *Capitalism, Socialism and Democracy* (New York: Harper Torchbooks, 1942; reprint ed., 1976), 83.

4. AN INDUSTRIAL EPIC

1. *Ford Industries* (n.p.: Ford Motor Company, 1924), 35.

2. Cutting out middlemen and charging a single profit appeared in many sources. See ibid., 89.

3. On flow, see Allan Nevins and Frank Ernest Hill, *Ford: Expansion and Challenge: 1915–1933* (New York: Scribner's) (hereinafter Nevins and Hill II), 202–203.

4. On the percentage of iron and steel in a Model T, see William F. Verner, notebook, Small Accessions, accession 521, Ford Motor Company Archives (hereinafter FMC), Benson Ford Research Center, The Henry Ford (formerly the Henry Ford Museum and Greenfield Village). In the early to-mid-1950s, the FMC conducted a series of tape-recorded oral history interviews with individuals who had worked with Henry Ford or at the Ford Motor Company during Ford's lifetime. These interviews were subsequently transcribed, typed, and bound. They are now part of the FMC. Hereinafter they are designated FOH. Unless otherwise indicated, the interviews date from 1952 to 1956 and are part of accession 65. On steel, see E. G. Liebold, FOH, 3:178; A. M. Wibel, FOH, 144–145; and "Independent Steel Production Denied as Reason for Expansion of Ford Plant," *New York Times,* 27 Feb. 1934, 27, which indicated that to this point Ford had never produced more than 5 percent of its steel needs. On Ford's suspicions, Charles Fearn Sutherland, Jr., "Consumption and Marketing of Forest Products in the Automobile Industry: A Case Study of the Ford Motor Company" (Ph.D. diss., University of Michigan, 1961), 7.

5. On coal use, see Ford News Bureau, press release, 29 July 1949, FMC, acc. 536, box 36. This figure was comprehensive, including among other items the use of coal by suppliers and in transporting raw materials. For the deposit estimate, see "Ford Controls Many Raw Materials," *New York Times,* 9 Sept. 1934, 10:14.

6. Although many people have written about the Rouge, we lack a definitive scholarly treatment of the great complex. The best place to begin remains Nevins and Hill II, esp. 200–216, 279–299. This should be supplemented with Lindy Biggs, *The Rational Factory: Architecture, Technology, and Work in America's Age of Mass Production* (Baltimore: Johns Hopkins University Press, 1996). Biggs provides the best analysis of the Rouge as the preeminent American plant engineering accomplishment of the first portion of the twentieth century. See also Tom McCarthy, "Henry Ford, Industrial Ecologist or Industrial Conservationist? Waste Reduction and Recycling at the Rouge," *Michigan Historical Review* 27:2 (2001): 53–88. On planning, see E. G. Liebold, FOH, 3:178–179; A. M. Wibel, FOH, 146–148; William F. Verner, FOH, 10–11, 19; and R. J. Walker, FOH, 48, 53. See, however, George R. Thompson, FOH, 15, and Dana Mayo, FOH, 1, for possible contradictions of this statement. Allan Nevins and Frank Ernest Hill, *Ford: Decline and Rebirth, 1933–1962* (New York: Scribner's, 1963) (hereinafter Nevins and Hill III), 9, argues that even in the late 1920s and early 1930s GM made a much larger percentage of its own parts than Ford. GM, however, did not integrate backward into raw materials.

7. *Ford Industries,* 11. For the number of guides, see "Dearborn Is the Home of the Ford Motor Company," press release, n.d. [early 1930s], FMC, acc. 450, box 3, Dearborn Plant Story. Most of these facts are from "Suggested Visitors Trip," 13 Mar. 1939, FMC, acc. 629, box 3, Releases. Others are recurring points from Ford's Rouge publicity literature. On the guides' use of the facts, see also Ric Bohy, "Ford Plans to End Tours at the Rouge," *Detroit News,* 17 June 1980, 1-A. Charles Sorenson, FOH, Rouge—Final, 3–4.

8. "Dearborn Is the Home of the Ford Motor Company." For the figure of eight million, see Bohy, "Ford Plans to End Tours at the Rouge," 1-A, 4-A. The last tour was given on 27 June 1980. They resumed several years ago. "Rotunda Nears Completion at Dearborn," *Ford News,* February 1936, n.p. "An Industrial Epic," proof for an advertisement that ran in the *Saturday Evening Post,* 28 June 1924, FMC, acc. 19, box 151, Institutional Ads—1924. Free publicity is from Nevins and Hill II, 262. On Rivera, Linda Downs, "The Rouge in 1932: The Detroit Industry Frescoes by Diego Rivera," in *The Rouge: The Image of Industry in the Art of Charles Sheeler and Diego Rivera*

(Detroit, MI: Detroit Institute of Arts, 1978), 49. The museum's director, William Valentiner, sold the idea of commissioning murals for the museum to Edsel Ford, the president of the museum's governing body, the Detroit Arts Commission. The younger Ford agreed to put up ten thousand dollars of his money to pay for them. Rivera's original commission from the museum merely suggested murals on the general theme of Detroit history and the development of industry, but Rivera became so fascinated by the Rouge that he convinced the museum to commission him to paint all the walls in the Garden Courtyard with the Rouge as the central theme of the work.

9. "Distribution of 'The Ford Industries,'" cover memo, Advertising Department, 2 Jan. 1925, FMC, acc. 78, box 51. The films included *A Tour through the Rouge Plant* (1926), *The Ford Rouge Plant* (1937), *Symphony in F* (1940), *River Rouge Plant* (1952), and *The Rouge* (1962). More than 1.5 million feet of film were donated by the Ford Motor Company to the National Archives of the United States as the Ford Historical Film Collection in 1963. It is now housed at the National Archives in College Park, MD, where videotapes of the films are available for viewing and copying. Raw material flows to the Rouge were a recurring theme in the films. See "New Ford Film Pays Tribute to Raw Material Producers," *Automotive Industries*, 15 June 1956, 182. On expositions, see "Ford Industrial Exposition" and "Visit the Ford Industrial Exposition," circulars dated 1928, Ford Motor Company, books 561 and 562, N. W. Ayer Collection, Advertising History Collection, Archives Center, National Museum of American History, Smithsonian Institution, collection 59, box 330, folder 1, marked "Book 561 Ford, 1927–1928."

10. The "Industrial Epic" ads ran in the *Saturday Evening Post*, 28 June–27 Dec. 1924. They also appeared in *Country Gentleman* and leading newspapers. All proofs for the ads discussed in this chapter are found in FMC, acc. 19, box 151, Institutional Ads–1924. Nevins and Hill II, 264, provide a paragraph of background context that relates this series of ads to the company's larger advertising policy during the mid-1920s.

11. For a good introduction to the field of industrial ecology, see Thomas E. Graedel and Braden R. Allenby, *Industrial Ecology* (Englewood Cliffs, NJ: Prentice Hall, 1995). Other scholars who pioneered this field include Robert H. Socolow, Robert Frosch, and Robert U. Ayres. For an application of the basic concepts of industrial ecology to the automobile and the automobile industry, see Thomas E. Graedel and Braden R. Allenby, *Industrial Ecology and the Automobile.* (Upper Saddle River, NJ: Prentice Hall, 1998). Frosch worked for General Motors.

12. Ovid M. Butler, "Henry Ford's Forest," *American Forestry*, December 1922, 725. For a thorough discussion of the northern Michigan and Wisconsin lumber industry from 1860 to 1890, see William Cronon, *Nature's Metropolis: Chicago and the Great West* (New York: W. W. Norton, 1991), 148–206. Cronon's focus is on white pine. Lumbermen turned to hardwoods after the white pine was depleted and the extension of the railroad system made it more economical to move hardwoods—which did not float—to market.

13. "Ford Company Buys Railroad, Big Land Tract," *Chicago Daily Tribune*, 11 July 1920, 8. For background on Ford's forests, see Butler, "Henry Ford's Forest"; and "Reforestation," typescript, FMC, acc. 476, box 6, R-S 1922. An especially good description of the Iron Mountain operation is chap. 11, "Saving the Timber," in Henry Ford, *Today and Tomorrow* (New York: Doubleday, Page, 1926), 120–134. See also W. G. Nelson, "Waste-Wood Utilization by the Badger-Stafford Process," *Industrial and Engineering Chemistry* 23:4 (1930): 312–315. Nelson managed the plant. See letter to Albert E. Dunford, 18 Feb. 1924, FMC, acc. 572, box 23–11.24, Raw Materials, which indicates that there was one ton of water in every thousand board feet of timber.

14. "New Sawmill Methods Show 35 to 50 Percent Saving," *Ford News*, 1 Apr. 1924, 1. "By-Product Business of Ford Is Enormous," *Washington Post*, 18 Jan. 1925, AU5.

15. "Ford Leads in Conserving of Lumber Supply," press release, 3 July 1923, FMC, acc. 545, box 1, Publicity—1923, 1. "Timber Conservation Is Aided by Ford Methods," *Ford News*, 1 Mar. 1924, 1. "Ford Restoring Michigan Forests," *Wall Street Journal*, 24 Sept. 1923, 8.

16. "Ford Harvesting Forest," *New York Times*, 13 Nov. 1921, 21; "Ford Restoring Michigan Forests," *Wall Street Journal*, 24 Sept. 1923, 8; "Ford Developing Lumber Industry," *Christian Science Monitor*, 3 June 1924, 17. "Ford Upsets Precedents in Interest of Lumber Conservation," *Automotive Industries*, 17 Apr. 1924, 866–867; "Ford Does It Again," *Automotive Industries*, 24 Apr. 1924, 936–937; "Minimize Lumber Waste," *Automotive Industries*, 5 June 1924, 1241.

17. On lumber from elsewhere, letter to Albert E. Dunford; and Eric Stromquist, FOH, 8, where he claimed that these loggers did follow Ford practice. On Ford clear-cutting, see Stromquist, FOH, 6; Robert Edwards, FOH, 9; and Walter Doehler, FOH, 44, who all make the same claim. There is evidence that contractors who cut timber on Ford land were forced to practice selective logging. See Robert Edwards, FOH, 11. The quotation is from Stromquist, FOH, 6. On the Sidnaw-area operation, see Gust Erickson, FOH, 9; Leo Brennan, FOH, 5–6; Robert Edwards, FOH, 9; P. A. Alexander, FOH, 18; and Carl G. Miller, FOH, 7–8. The *New York Times* piece, "Ford Harvesting Forest," indicated that the Sidnaw camp was in operation by November 1921. Erickson recalled selective logging at Sidnaw in 1924. Edwards remembered the company doing selective logging for two to three years, though he thought it started in 1925 or 1926. See W. DuB. Brookings, letter to E. G. Kingsford, 25 Aug. 1927, and E. G. Kingsford, letter to Chamber of Commerce of the United States, 30 Aug. 1927, FMC, acc. 476, box 1, 476-C-Misc. For more candid assessments by Kingsford, see Kingsford, letter to James D. Lacey and Co., 15 Feb. 1929 in the same place; and Kingsford, letter to W. A. Holt, 5 Dec. 1929, FMC, acc. 476, box 5, Mr. Kingsford. The letter to Holt explains the higher costs and practical difficulties with burning the slash.

18. On losses, see W. E. Carnegie, letter to Edsel Ford et al., 11 Nov. 1925, FMC, acc. 572, box 14, 11.12.5. On industry use in 1923, see Chrysler Corporation, "Wood and Steel Trends, 1921–1948," reproduced as pl. 2 in Bernard Gibson Cline, Jr., "Wood in the Automobile Industry" (M.S. thesis, Yale School of Forestry, 1948). The Model A figure is from Geo. A. Moss, "Log and Lumber Conditions as of May 1, 1929," memorandum dated 3 June 1929, FMC, acc. 370, box 1, Logs 1927–1929, and the 1937 figure is from *Ford News*, October 1937, 204. With the exception of the Pierce-Arrow, Cline, "Wood in the Automobile Industry," 7, claims that the American automobile was 100 percent metal by 1937.

19. Ford News Department, approval draft press release, 19 Oct. 1953, 1, FMC, acc. 837, box 1. "Most pollution" is obviously a broad generalization, to which numerous exceptions could be made, some of them important, but they are not worth pursuing here.

20. On Highland Park, see George E. Hagemann, "$15,000,000 Salvaged by Ford Plants," *Management and Administration* 9:6 (1925): 557; and William F. Verner, FOH, 17. Nevins and Hill II, 201, mentions "the inadequacy of Highland Park" as one of Ford's chief motives for building the Rouge but does not mention the waste problems at the older plant specifically.

21. Hanson, FOH, 118.

22. Verner, FOH, 17–18.

23. William B. Mayo, letter to the Ford Motor Company, 13 Nov. 1916, recorded in the Ford Motor

Company Minutes Books, as quoted in appendix 1, from "William B. Mayo and Henry Ford, 1912–1932," 27, FMC, acc. 819.

24. *Ford Industries,* 94.

25. On beginning, see Hagemann, "$15,000,000 Salvaged by Ford Plants," 557. That the Rouge represented the quintessence of the modern, rational factory is the central argument of Biggs's *Rational Factory,* although she says little about waste reduction. Ford, *Today and Tomorrow,* 110. L. D. Middleton, "Ford Salvage Practice," 15 Feb. 1939, 5, FMC, acc. 629, box 3, Salvage.

26. Ford's descriptions of its waste reduction activities at the time usually employed these two terms. Much this same distinction and terminology were used by outside waste reduction experts. See, e.g., Ray M. Hudson, "The Six Ways to Eliminate Waste," *Management and Administration* 10:2 (1925): 71–72. A third waste reduction category was energy efficiency. A fourth category, postconsumer recycling of automobiles, is discussed in the next chapter. On benzol, see "How Henry Ford Rode Out Storm without a Loan," *Christian Science Monitor,* 7 Mar. 1921, 1. Revenue is from "By-Product Business of Ford Is Enormous," AU5. On dealer resentment, see Nevins and Hill II, 262.

27. On process, see *Ford Industries,* 57; and Hanson, FOH, 117. On capacity, see "By-Product Business of Ford Is Enormous," AU5. On use, see Hanson FOH; Ford Motor Company—Cement, 1947, FMC, acc. 453, box 1, Historical Data—Materials; and *Ford Industries,* 57.

28. On salvage sales, see "By-Product Business of Ford Is Enormous," AU5. On most important, as well as other metals, see Middleton, "Ford Salvage Practice," 5. On "home" scrap volume, see *Ford Industries,* American Legion ed. (Dearborn, MI: Ford Motor Company, 1931), 45, Rouge vertical file, Dearborn Historical Museum.

29. Early-twentieth-century trade and engineering journals make the point for the United States. For a survey of German efforts, see Theodor Koller, *Utilization of Waste Products* (London: Scott, Greenwood, 1918). For a similar work that deals primarily with British efforts, see John B. C. Kershaw, *Recovery and Use of Industrial and Other Wastes* (London: Ernest Benn, 1928). Both Koller and Kershaw demonstrate the older roots of waste reduction and recycling while highlighting the substantial recent contributions of chemists and engineers. On this traditional interest in the profitable uses of waste, see Susan Strasser, *Waste and Want* (New York: Metropolitan, 1999); and Carl Zimring, *Cash for Your Trash* (New Brunswick, NJ: Rutgers University Press, 2005).

30. The classic treatment of the conservation movement remains Samuel P. Hays, *Conservation and the Gospel of Efficiency* (Pittsburgh, PA: University of Pittsburgh Press, 1959, 1999). The analysis here, while dealing with a somewhat later period than that treated by Hays in his book, is consistent with his conclusions. On the role of Taylorite engineers, see Edwin T. Layton, *The Revolt of the Engineers* (Cleveland, OH: Case Western Reserve University Press, 1971), 194–195, 212–213. The rise and fall of the interest of engineers in industrial conservation is well charted in Layton. On Hoover, see Kendrick A. Clements, *Hoover, Conservation, and Consumerism: Engineering the Good Life* (Lawrence: University of Kansas Press, 2000), esp. 44–47, 60–61, 75–76.

31. See "Amazing Chemical Discoveries," *New York Times,* 14 Mar. 1915, SM15, where the author notes that nearly all coke ovens in Europe were by-product coke ovens, whereas in the United States only about 25 percent were. William Crawford Hirsch, "Profiting from the Scrap Heap in the Automotive Factory," *Automotive Industries,* 3 June 1920, 1270. For specific automaker examples, see "Salvaging Scrap in Cadillac Plant," *Iron Age,* 13 Dec. 1923, 1571–1574; "Future

Profits May Come from Scrap Pile," *Automotive Industries*, 30 Oct. 1924, 761–764; H. V. Kimble, "Saving Steam: Studebaker's Salvage Division Recovers Every Bit of Waste," *Forbes*, 15 Apr. 1926, 24–25, 42; "Scrap Iron Briquettes for Cupola Use Effect Saving for Dodge," *Automotive Industries*, 22 Jan. 1927, 80–81; "Saving $3,000,000 from Waste," *Brass World*, March 1927, 83–84; John Younger, "Production Progress—a Glimpse into Future Possibilities," *Automotive Industries*, 1 Oct. 1927, 472; and "How Dodge Bros. Salvage Waste Materials," *Iron Trade Review*, 21 June 1928, 1595–1597.

32. Nevins and Hill II, 203, 283, 290, makes the general point about Ford's unwillingness to settle for "best practice," although not with specific reference to waste reduction activities. For pride, see Frank C. Riecks, FOH, 38. On the use of the Badger-Stafford process at Iron Mountain, see Nelson, "Waste-Wood Utilization by the Badger-Stafford Process," 312–315. On cement, see "Plant at Rouge Converts Slag into Cement," *Ford News*, 15 June 1924, 1, 6; and "Cement Plant Doubles Output in Three Years," *Ford News*, 15 Aug. 1927, 4. On paper, see "New Paper Mill Gives Value to Waste," *Ford News*, 15 Apr. 1924, 1. On pulverized coal, see "Rouge Boilers World's Largest," *Ford News*, 1 Nov. 1920, 5.

33. John H. Van Deventer, "What Ford Is Doing and How He Does It," *Industrial Management* 64:3 (1922): 130. George E. Hagemann, "$15,000,000 Salvaged by Ford Plants," *Management and Administration* 9:6 (1925): 557–560, 10:1 (1925), 33–36. For examples of such articles, see "Iron Mt. Plant Makes Power from Wood Waste," *Power Plant Engineering*, 1 Dec. 1925, 1186–1194; "Scrapping Ships at Ford Plant," *Iron Age*, 4 Nov. 1926, 1271–1273; and "Reclaiming Sixty Tons Battery Lead Daily," *Brass World*, October 1927, 339. On Ford assistance, see Fay Leon Faurote, "Henry Ford Still on the Job with Renewed Vigor," *Industrial Management*, 54:4 (1927), 200.

34. Hard-headed, clear-eyed engineers were by no means immune to the mystique. Indeed, some contributed to it. See Faurote, "Henry Ford Still on the Job with Renewed Vigor." For a debunking assessment from an engineer, see Halbert P. Gillette, "Ford's Business Philosophy," *Engineering and Contracting*, March 1928, 137–140. *Ford Industries*, 99. On the bottom line, see, e.g., Ray M. Hudson, "Cutting down the Waste Pile: Costs and Benefits of Industrial Waste Elimination," *Management and Administration* 9:5 (1925): 414. There were exceptions to this generalization, especially when the war still dominated discussions of waste reduction. For one, see W. Rockwood Conover, "Salvaging Miscellaneous Wastes," *Industrial Management* 57:1 (1919): 16. H. E. Howe, "Possibilities in Saving and Utilizing Industrial Wastes," *Industrial Management* 57:2 (1919): 93. For examples where the point about added cost is made, see "It Pays to Reclaim Scrap," *Factory and Industrial Management*, November 1928, 944; and Milton Wright, "A Use for Everything," *Scientific American*, February 1929, 120. Stuart Chase, "The Challenge of Waste to Existing Industrial Creeds," *Nation*, 23 Feb. 1921, 284.

35. On lack of records, see Nevins and Hill II, 296. Hagemann, "$15,000,000 Salvaged by Ford Plants," 557. See "By-Product Business of Ford in Enormous," AU5, for similar figures, as well as the four million dollars in sales for waste materials in 1924. Henry Ford, *Today and Tomorrow*, 91. The author has found no corroboration for Ford's claim. On the four-to-five-million-dollar profit, see "Converting Waste into Millions"; "Salvage Takes Its Place," 1:3; "This is being released," FMC, acc. 545, box 10, Salvage, 1. The figure does not include profit from by-product sales. The vagueness of the 1930s figures and the fact that they did not vary in publicity releases over the decade argues for skepticism here as well.

36. Harold N. Denny, "Henry Ford Plans for a New Adventure," *New York Times*, 26 Mar. 1933, SM16. The best oral histories for Henry Ford waste stories are those of A. M. Wibel, Ernest G. Liebold, Frank Hadas, Philip E. Haglund, Charles E. Sorenson, Harry Hanson, John L. McCloud, Frank C. Riecks, George R. Thompson, and John W. Thompson. Ernest G. Liebold, FOH, 7:610. A. M. Wibel, FOH, 323.

37. See, e.g., Frank C. Riecks, FOH, 36–38, 57–58, and John L. McCloud, FOH, 311–313. Howard P. Segal, "'Little Plants in the Country': Henry Ford's Village Industries and the Beginning of Decentralized Technology in Modern America," *Prospects: An Annual of American Cultural Studies* 13 (1988): 200, notes opposition by Ford managers to Henry Ford's village industries project as well. Frank Hadas, FOH, 259–264. The project was the automobile disassembly line that is discussed in chapter 5. John W. Thompson, FOH, 71.

38. On Ford's love of birds, see Harvey Whipple, "Protecting and Breeding Wild Birds," *Craftsman*, December 1911, 270–281; "Henry Ford, Friend of the Birds," *American Game Protective Association*, 1 Dec. 1913, 5, 8; and John B. Gorgan, "The Triumph of Henry Ford's Ideals," *National Magazine*, January 1915, 623. Henry Ford, *Today and Tomorrow*, 46–47. Press Release, June 1930, 2, FMC, acc. 545, box 10, Salvage. John W. Thompson, FOH, 79. These efforts, while effective to a degree, were not very successful by post–World War II environmental standards.

39. Denny, "Henry Ford Plans for a New Adventure," SM7.

40. The no justification point was made by Faurote, "Henry Ford Still on the Job with Renewed Vigor," 197. Nevins and Hill II, 204, also underscores it.

41. Quotation from interview with Henry Ford by Geo. Sylvester Viereck, *New York American*, 5 Aug. 1928, found in FMC, acc. 511, box 1, Small Brown Notebook. "My Dutch mother is in my workshops. She is in my workshops to this extent—it is impossible for me to tolerate disorder or uncleanliness anywhere."

42. This observation is attributed to George Orwell. For a later advertisement with smokestacks, see "Take a Big Look at the Big New Ford," advertisement proof sheet dated 4 Oct. 1940, Ford Motor Company, N. W. Ayer Collection, collection 59, Archives, National Museum of American History, box 202, folder 1, marked "Ford Motor Company New and Used Cars 1940." On the decline of muckraking and reform, see Frederick Lewis Allen, *Only Yesterday* (New York: Harper, 1931), 235. See also Reynold M. Wik, *Henry Ford and Grass-Roots America* (Ann Arbor: University of Michigan Press, 1972), 6–8, who attributes Ford's favorable public image in part to the favorable timing of rising to prominence after Progressive Era "muckraking" and before the loss of authority that American business suffered during the Great Depression.

5. THE DEATH AND AFTERLIFE OF AUTOMOBILES

1. The epigraph is from Carleton B. Case, comp., *Ford Smiles: All the Best Current Jokes about a Rattling Good Car* (Chicago: Shrewsbury, 1917), 43. On Ned Cobb, see Theodore Rosengarten, *All God's Dangers: The Life of Nate Shaw* (New York: Alfred A. Knopf, 1974), 249–251, 287. Rosengarten changed Ned Cobb's name to Nate Shaw. I thank Pete Daniel for bringing Cobb to my attention. Automobile ownership by African Americans in the 1920s and 1930s was a sensitive subject. George Milburn used it as a pretext for a lynching in his novel *Catalogue* (New York: Harcourt, Brace, 1936).

2. The representative nature of the Cobbs is fully evident in those biographies where subjects discussed automobiles in "American Life Histories: Manuscripts from the Federal Writers' Project, 1936–1940," Manuscript Division, Library of Congress, accessed at http://memory.loc.gov/ammem/wpaintro/wpahome.

3. On concerns, see Virginia Scharff, *Taking the Wheel: Women and the Coming of the Motor Age* (New York: Free Press, 1991), 138 141; and Beth L. Bailey, *From Front Porch to Back Seat: Courtship in Twentieth-Century America* (Baltimore: Johns Hopkins University Press, 1988).

4. This joke was in neither Case, *Ford Smiles*, nor George Milburn, ed., *A Book of the Best Ford Jokes* (Girard, KS: Little Blue Books, [c. 1927]). I thank whoever brought it to my attention.

5. "The Growing Motor Replacement Market," *Literary Digest*, 30 Aug. 1930, 44. The changing nature of Ford jokes is based on an analysis of 217 Ford jokes current in 1917 or earlier from Case with 78 jokes current in 1927 from Milburn, ed., *Book of the Best Ford Jokes*. Many of the jokes in these compilations were about cars generally. A few were over my head. Jokes in these two categories were not included in my analysis. There was little duplication in the two collections. For horns, see Case, *Ford Smiles*, 76. On disposition, see Milburn, *Book of the Best Ford Jokes*, 6.

6. C. E. Griffin, "The Life History of Automobiles," *Michigan Business Studies* 1:1 (1926), second printing 1928, in the Warshaw Collection of Business Americana, collection 60, Archives Center, National Museum of American History, Washington, DC. For mentions of the study, see "Actuarial Computation Fixes Average Life of Car at 7.04 Years," *Automotive Industries*, 8 Apr. 1926, 612–614; "Car's Life Seven Years," *Los Angeles Times*, 30 May 1926, G9; "Life of Cars Compared," *Ford News*, 15 Dec. 1926, 1, 3; "The Life of an Automobile," *Literary Digest*, 8 Jan. 1927, 67; "The Future of the Automobile," *Executive's Magazine*, May 1927, 12–13; and "How Long a Car Lives," *Motor*, December 1927, 56. See chart in Griffin, "Life History of Automobiles," 10. On the two kinds of obsolescence, see James Dalton, "What Factory Men Say about the Used Car," *Motor*, October 1926, 122.

7. On this point, see "Style—an Eternal Force in Car Merchandising," *Automotive Industries*, 12 June 1924, 1261; "In Choosing the Color Scheme of Your New Car Harness the Rainbow," *Motor*, June 1923, 33; "Bodies Have Become Most Important Factor in Making Sales," *Automotive Industries*, 5 July 1923, 4; Norman G. Shidle, "Has Ford Lost His Big 'Sales Punch' in the Low-Priced Field?" *Automotive Industries*, 29 Oct. 1925, 727; and "Color as a Sales Appeal," *Automotive Industries*, 13 May 1926, 825.

8. On the early importance of styling, see R. Damon, "Depreciation Owing to Advances in the Art and Fashion," *Horseless Age*, 11 Feb. 1903, 237; and "Style in Automobiles," *Country Life in America*, August 1905, 449. On steel and closed cars, see Charles K. Hyde, *Riding the Roller Coaster: A History of the Chrysler Corporation* (Detroit, MI: Wayne State University Press, 2003), 51–52, 62–63. On the popularity of closed cars, see Alexander Johnston, "The Balance Swings to the Closed Car," *Motor*, September 1922, 23, where Edward S. Jordan is quoted to this effect; H. H. Buggie, "The Open Season for Closed Cars," *Motor*, January 1922, 59, 266; Alfred P. Sloan, Jr., *My Years with General Motors* (New York: Currency, Doubleday, 1963; reprint ed., 1990), 152; and Charles Coolidge Parlin and Fred Bremier, *Passenger Car Industry: Report of a Survey* (Philadelphia: Commercial Research Division, Curtis, 1932).

9. Sloan, *My Years with General Motors*, 235; "Why Paint Them Black?" *Automotive Industries*, 26 Apr. 1923, 938.

10. On Duco, see "Exceptional Durability Is Claimed for New Body Finish," *Automotive Industries*, 26 July 1923, 158; and Sloan, *My Years with General Motors*, 236. The use of Duco on GM cars later became a point in the GM–Du Pont Anti-Trust Case brought by the Justice Department. It was alleged that Du Pont forced GM to use Duco. See, however, Charles Stewart Mott, letter to Pierre S. du Pont, 12 Jan. 1953, Mott Papers, 77-7.4-2A, Kettering/GMI Alumni Foundation Collection of Industrial History, Kettering University, Flint, MI, where Mott takes responsibility for forcing Duco on several GM divisions on his own initiative. On sharing with the industry, see Duco ad, *Automotive Industries*, 30 Oct. 1924, 7. For a thousand colors, see "Body Production — Color Viewed as Potent Factor in Promoting Car Sales," *Automotive Industries*, 4 Feb. 1926, 182. The best biography of Sloan is David Farber, *Sloan Rules: Alfred P. Sloan and the Triumph of General Motors* (Chicago: University of Chicago Press, 2002). The author's characterization of Sloan rests chiefly on this book.

11. "Attractive Body Designs," *Automotive Industries*, 13 Nov. 1924, 861. On recent successes, see "Art in Bodies," *Automotive Industries*, 29 Jan. 1925, 201. "Ford Changes Body Lines," *Motor*, October 1923, 52.

12. Harry Tipper, "What Factors Will Determine Sales Volume in the Future?" *Automotive Industries*, 1 May 1924, 962–963.

13. Of all the historians who have written about General Motors and styling in the 1920s, Sally Clarke is perhaps alone in recognizing that this business opportunity also presented a substantial business risk. See Clarke, "Managing Design: The Art and Colour Section at General Motors, 1927–1941," *Journal of Design History* 12:1 (1999): 65–79. "Art in Bodies," 201.

14. As quoted in Farber, *Sloan Rules*, 59. On Sloan's fear, see ibid., 63. Pierre S. du Pont shared Sloan's fear; see Allan Nevins and Frank Ernest Hill, *Ford: Expansion and Challenge: 1915–1933* (New York: Scribner's, 1957) (hereinafter Nevins and Hill II), 389. However, see Shidle, "Has Ford Lost His Big 'Sales Punch' in the Low-Priced Field?" 727, for the argument that quality, service, and comfort, not styling, were more obvious alternatives to price competition.

15. On the meeting, see Sloan, *My Years with General Motors*, 163–168. On the limits to commitment, see Beverly Rae Kimes and Robert C. Ackerson, *Chevrolet: A History from 1911* (Kutztown, PA: Automobile Quarterly, 1986), 45. Sloan, *My Years with General Motors*, 167.

16. The well-known Chevrolet story recounted here is based on Kimes and Ackerson, *Chevrolet*, 28–68; Ed Cray, *Chrome Colossus: GM and Its Times* (New York: McGraw-Hill, 1980), esp. chap. 10; Farber, *Sloan Rules*, 96–99; Richard S. Tedlow, *New and Improved: The Story of Mass Marketing in America* (Boston: Harvard Business School Press, 1996), 172–73; and Douglas Brinkley, *Wheels for the World: Henry Ford, His Company, and a Century of Progress* (New York: Viking, 2003), 340–341, 360–361, 385–386. Knudsen, as quoted by Farber, *Sloan Rules*, 97. On up-to-date features, see Kimes and Ackerson, *Chevrolet*, 40; and Brinkley, *Wheels for the World*, 340–341.

17. These differences were a part of the epic struggle between Ford and GM for dominance of the American automobile industry, perhaps the most famous and oft told in American business history. For a selection that ranges from the scholarly to the popular, see Nevins and Hill II; Tedlow, *New and Improved*; Cray, *Chrome Colossus*; Brinkley, *Wheels for the World*; Kimes and Ackerson, *Chevrolet*; and Alfred D. Chandler, *Giant Enterprise: Ford, General Motors, and the Automobile Industry* (New York: Harcourt Brace, 1964); as well as Anthony Patrick O'Brien, "How to Succeed in Business: Lessons from the Struggle between Ford and General Motors during the 1920s and

1930s," *Business and Economic History* 18 (1989): 79–87. On mechanical improvements in GM cars, see Nevins and Hill II, 416–417. On profits from parts, see ibid., 262–264, 265. Griffin, "Life History of Automobiles," 34–35.

18. E. C. Barringer, *Iron Trade Review*, as excerpted in "Scrapping the Ancient Auto," *Literary Digest*, 17 Apr. 1926, 21.

19 "'Rosey,' the Junkman, Has Made a Fortune Out of Old Motor Cars," *American Magazine*, September 1924, 69–70. The best history of the American junk men and the scrap industry is Carl A. Zimring, *Cash for Your Trash: Scrap Recycling in America* (New Brunswick, NJ: Rutgers University Press, 2005). "The Cars of Yesteryear," *Literary Digest*, 26 Mar. 1927, 67.

20. On the stripping process, see Barringer, *Iron Trade Review*, 21–22. Charles H. Lipsett, *Industrial Wastes and Salvage: Conservation and Utilization* (New York: Atlas, 1963), 159. On scrap value, see ibid., 143. On the severity of pollution from steelmaking, see Andrew Hurley, *Environmental Inequalities: Class, Race, and Industrial Pollution in Gary, Indiana, 1945–1980* (Chapel Hill: University of North Carolina Press, 1995), esp. 15–45.

21. On siting, see "For Metal Only," as told to Myron M. Stearns, *Saturday Evening Post*, 8 Sept. 1928, 46. On space needed, see "The Wanderings and Later Ends of Some Old Cars," *Literary Digest*, 15 Oct. 1921, 52; and Al C. Faeh, "The Junk Yard Idea Grows," *Motor*, February 1928, 51. On unsightliness and advertising, see Stearns, *Saturday Evening Post*, 46. For critics, see "Ways of Controlling Automobile 'Graveyards,'" *American City*, October 1929, 171.

22. On the alternative fates, see Barringer, *Iron Trade Review*, 21. For a fascinating description of picking over cars, see Stearns, *Saturday Evening Post*, 19, 43–44, 46. See also "Ways of Controlling Automobile 'Graveyards,'" 171. "What to Do with Old Automobile Bodies," *American City*, July 1927, 95. Of course, private operators looking to dispose of a car also used landfills. For one anecdote, see Stearns, *Saturday Evening Post*, 44. On building foundations, see "Where Can Good Automobiles Go When They Die?" *Literary Digest*, 22 Feb. 1930, 61. On farm use, see Barringer, *Iron Trade Review*, 21; and "The Cars of Yesteryear," *Literary Digest*, 26 Mar. 1927, 67. For the fate of most, see ibid.

23. Barringer, *Iron Trade Review*, 21–22. See also "Reducing Scrapped Cars," editorial, *Automotive Industries*, 9 Sept. 1926, 427.

24. Henry Ford, *My Life and Work*, as quoted in Nevins and Hill II, 412.

25. Farber, *Sloan Rules*, persuasively argues that amorality was a cardinal virtue of the professional management style that Alfred P. Sloan, Jr., brought to GM.

26. See Steven Watts, *The People's Tycoon: Henry Ford and the American Century* (New York: Alfred A. Knopf, 2005), 366–375, for a succinct treatment of this transition.

27. The classic overview of advertising in the 1920s is Roland Marchand, *Advertising the American Dream: Making Way for Modernity, 1920–1940* (Berkeley: University of California Press, 1985). For a brief overview of automobile advertising to the late 1920s, see Frank Presbrey, *The History and Development of Advertising* (New York: Greenwood, 1929; reprint ed., 1968), 556–564; see also Martha L. Olney, *Buy Now, Pay Later: Advertising, Credit, and Consumer Durables in the 1920s* (Chapel Hill: University of North Carolina Press, 1991), 153, 157–158. For Ford, see Watts, *People's Tycoon*, 125–128; and Nevins and Hill II, 262–264.

28. On creating a positive corporate image, see Roland Marchand, "The Corporation Nobody Knew: Bruce Barton, Alfred Sloan, and the Founding of the General Motors 'Family,'" *Business His-*

tory Review 65 (1991): 825–875. The dollar volume is from Marchand, *Advertising the American Dream*, 6, who cites Daniel Pope. On the new advertising, see generally ibid. On GM advertising, see Farber, *Sloan Rules*, 68–71, 103. For quotations from Grant and Presbrey, see Presbrey, *History and Development of Advertising*, 561.

29. Paul M. Mazur, *American Prosperity: Its Causes and Consequences* (New York: Viking, 1928), 95. Ray W. Sherman, "The Public Wants to Spend Millions for Style," *Motor*, June 1926, 27.

30. Mazur, *American Prosperity*, 93, 55.

31. On GM's commitment to better understanding consumers, see Sally Clarke, "Consumers, Information, and Marketing Efficiency at GM, 1921–1940," *Business and Economic History* 25:1 (1996): 186–195. Sloan, *My Years with General Motors*, 149–150. "Art in Bodies," 201.

32. Harley J. Earl was born in Los Angeles (or Hollywood) on 22 November 1893, the son of carriage maker Jacob W. Earl and his wife, Abbie Taft. He worked as a designer for his father's firm, Earl Automobile Works, 1918–19, and then as the general manager of the Don Lee Coach and Body Corporation until 1926. Don Lee bought Earl's father's firm largely to gain the car body design services of the younger Earl. Harley Earl died in Florida in 1969. See his entry in *Who's Who in America*, vol. 30 (1958–59), 802. For further detail, see Bill Henry, "Alumni of Automobile Row," *Los Angeles Times*, 30 Mar. 1930, F5; Sydney Oxberry, "The Appropriate Car," *Motor*, December 1920, 55, for notice of Arbuckle's car without a mention of Earl by name; "Pastel Shades All the Go This Spring," *Los Angeles Times*, 26 Mar. 1922, VI:4; "Harley Earl Has Concluded Tests," *Los Angeles Times*, 11 June 1922, VI:8; "Modern Methods Cut Paint Time," *Los Angeles Times*, 29 July 1923, VI:6; and "Years Spent in Perfecting of Auto Finishes," *Los Angeles Times*, 1 June 1924, F3. The Earl story has been told by many authors. See Sloan, *My Years with General Motors*, 266–70; Stephen Bayley, *Harley Earl and the Dream Machine* (New York: Alfred A. Knopf, 1983), 36–47; David Halberstam, *The Fifties* (New York: Villard, 1993), 123–125; Cray, *Chrome Colossus*, 243–245; David Gartman, *Auto Opium: A Social History of American Automobile Design* (New York: Routledge, 1994), 68–92; Vincent Curcio, *Chrysler: The Life and Times of an Automotive Genius* (New York: Oxford University Press, 2000), 350–353; and Farber, *Sloan Rules*, 100–102.

33. Sloan, *My Years with General Motors*, 269. For fifty million, see Bayley, *Harley Earl and the Dream Machine*, 12. For further background on Earl and GM styling, see Gartman, *Auto Opium*; and Clarke, "Managing Design."

34. See Thomas Hine, *Populuxe* (New York: Alfred A. Knopf, 1986), 93. In his discussion of the 1950s automobile market, Hine recognized correctly, in the author's view, the important connection between having a large (perhaps even the largest) market share and styling leadership.

35. *Historical Statistics of the United States: Colonial Times to 1970*, pt. 2, Series Q 148–162: Motor-Vehicle Factory Sales and Registrations, and Motor Fuel Usage: 1900–1970, 716.

36. The best descriptions of the disassembly line are found in: Press Release, June 1930, Ford Motor Company Archives (hereinafter FMC), Benson Ford Research Center, The Henry Ford, Dearborn, MI, acc. 545, box 10, Salvage; "Auto Salvage Operations," memorandum, 31 Oct. 1930, FMC, acc. 479, box 1, vol. no. 3; Charles Sorenson, transcript, oral history, FMC, acc. 65, Engineering—Final, 53–54 (hereinafter by surname of interviewee with designation FOH); Harold M. Baker, "Ford Reverses the Production Line for Junking Operation," *Automotive Industries*, 28 June 1930, 988–989; "Ford Scraps 375 Automobiles Each Sixteen Hr.; Even the Grease Saved," *Iron Age*, 3 July 1930, 14–15, 60; "Ford Is Salvaging Old Automobiles," *Washington Post*, 6 July 1930,

A4; "Reincarnation of Junk," *New York Times*, 6 July 1930, 109; Edwin P. Norwood, "Walk the Plank," *Review of Reviews*, April 1931, 66–68; "Ford to Scrap Automobiles Almost One Every Minute," *Iron Age*, 5 Nov. 1931, 1200.

37. On the Ford refurbishment program, see Jim Schild, "Let's Junk All the Old Cars," *Old Cars and Marketplace*, 5 Aug. 1993, 6. The idea had been floated earlier. See Ralph C. Rognon, "Re-Manufacture Is Our Next Job," *Motor*, September 1924, 46 47, 100. But dealers were not enthusiastic: Walter Davenport, "Old Cars for New," *Collier's*, 12 Jan. 1929, 43. On the Ford disassembly line, see Sorenson, FOH, 53; and Frank Hadas, FOH, 259. For the take-back suggestion, see "Where Can Good Automobiles Go When They Die?" 59–60. For the benefits, see Press Release, June 1930; and "Derelict Cars Salvaged at Rouge Plant," *Ford News*, 1 Aug. 1930, 170.

38. For the junker price, see "Ford Is Sifting Cash from the Junk Pile," *Business Week*, 19 Mar. 1930, 13. For the scrap percentage, see Edwin P. Norwood, "Where Motor Cars Walk the Plank," *Review of Reviews*, April 1931, 66.

39. "Auto Salvage Operations," 2. In stripping the junkers of reusable parts and materials, Ford was following the practice of junkmen and the scrap or wrecking cooperatives that many auto dealers set up and ran in the late 1920s to cut the junk men out of the business. For descriptions of salvage practices at an automobile wrecking yard and a dealer cooperative, see E. C. Barringer, "Scrap Heap Claims Thousands of Automobiles Annually," *Iron Trade Review*, 11 Mar. 1926, 631–634; and Robert E. Lee, "A Graveyard for Motorized Junk," *National Safety News*, May 1929, 15. These ventures did not employ moving conveyor lines for disassembly purposes as Ford did. On uses for reclaimed material, see "Ford Salvages More Than Thirty Thousand Old Cars," Press Release, n.d. [but 1930], FMC, acc. 545, box 10, Salvage. On the baling press, see Sorenson, FOH, 53; and "Cars Scrapped in Giant Baling Press, Melted in Huge Open-Hearth," *Iron Age*, 2 Mar. 1930, 352. This article is also useful for distinguishing between the first and second disassembly lines at the Rouge. Disassembly time is from Baker, "Ford Reverses the Production Line for Junking Operation," 988. For shift size and processing capacity, see Baker, "Ford Reverses the Production Line for Junking Operation," 1988; "Ford Scraps 375 Automobiles Each Sixteen Hr.," 14. Baker indicates that 110 men were employed. On spectacular job, see Sorenson, FOH, 53. For eighteen thousand, see Press Release, June 1930; "Ford's 'Demolition Line' Makes Junking a Fast, Thorough Job," *Business Week*, 2 July 1930, 13; "Derelict Cars Salvaged at Rouge Plant," *Ford News*, 1 Aug. 1930, 170. On "practical," see Press Release, June 1930; and "Derelict Cars Salvaged at Rouge Plant," 170.

40. The difficulty is explained by Sorenson, FOH, 54. "Ford to Junk Five Thousand Cars Daily," *Wall Street Journal*, 13 Apr. 1931, 1.

41. See "Frank Dow Spends Years Saving Metal for Company," *Ford Rouge News*, 3 Aug. 1956, 2, in which Dow, the salvage car supervisor in 1932, indicated that the disassembly line was in operation for "nearly 14 years." Sorenson, FOH, 53–54. Frank Hadas, FOH, 263.

42. This attachment is fully evident in the American Life Histories narratives. Will Rogers, quotation from a compilation by Bryan B. Sterling and Frances N. Sterling, *Will Rogers Speaks* (New York: M. Evans, 1995), 229. Rogers made the comment on 18 Oct. 1931, but the compilers do not provide the original source or context. Olney, *Buy Now, Pay Later*, 187–189. Olney's argument is summarized by Peter Temin, "The Great Depression," in Stanley L. Engerman and Robert E. Gallman, eds., *Cambridge Economic History of the United States*, vol. 3 (New York: Cambridge University Press, 2001), 310.

43. These figures and those that immediately follow are from *Historical Statistics of the United States: Colonial Times to 1970*, pt. 2, Series Q 148–162: Motor-Vehicle Factory Sales and Registrations, and Motor Fuel Usage: 1900–1970, 716, and Series Q 199–207: Miles of Travel by Motor Vehicles: 1921 to 1970, 718. Figures for miles traveled and gasoline consumption are for all motor vehicles, not just passenger cars. "Ghosts of the Gas Age," *Journal of American Insurance*, July 1964, 68, clipping in "Scrappage" vertical file at the National Automotive History Collection, Detroit Public Library.

44. Robert S. Lynd and Helen Merrell Lynd, *Middletown in Transition: A Study in Cultural Conflicts* (New York: Harcourt Brace, 1937), 265, 492, 245, 266.

6. CADILLACS AND COMMUNITY

1. The material on Elvis Presley and the Cadillac in this and the following paragraphs is drawn principally from two sources: Jerry Osborne, ed., *Elvis: Word for Word* (New York: Harmony, 1999); and Peter Guralnick, *Last Train to Memphis: The Rise of Elvis Presley* (Boston: Little, Brown, 1994). For specific references to Cadillacs in Osborne through 1958, see 8, 17, 37, 41, 42, 63–64, 75, 82, 84–85, 100, 105, 120. For the more revealing Cadillac and automobile references in Guralnick, see 16, 45, 52, 148, 179, 184–185, 196, 317, 462, 474, 483.

2. Thomas Hine, *Populuxe* (New York: Alfred A. Knopf, 1986), 23. For an analysis of this market that deftly and succinctly combines economic and psychological analysis, see Avner Offer, *The Challenge of Affluence: Self-Control and Well-Being in the United States and Britain since 1950* (New York: Oxford University Press, 2006), 193–209.

3. Hine, *Populuxe*, does a nice job in capturing the importance of 1955 in Americans' relationship with the automobile; see 12, 90–106. On market share, see Robert Sheehan, "A Big Year for Small Cars," *Fortune*, August 1957, 106. For spending on styling, "Autos—'War' of the Giants," *Newsweek*, 10 June 1957, 89. On new Chevrolets, see Hine, *Populuxe*, 90. On wild animals, see "Battle Report—Detroit," *Newsweek*, 21 Mar. 1955, 81. On car spending, see David Halberstam, *The Fifties* (New York: Villard, 1993), 495. The GNP figure is from *Historical Statistics of the United States*, pt. 1 (Washington, DC: GPO, 1975), Series F 1-5: Gross National Product, 224.

4. For "desperate hunger," see Halberstam, *The Fifties*, 118. The best description of the postwar conversion to consumer-driven economic prosperity is Lizabeth Cohen, *A Consumer's Republic: The Politics of Mass Consumption in Postwar America* (New York: Alfred A. Knopf, 2003), esp. 112–133. "The Fortune Survey," *Fortune*, December 1943, 10, 16. Table VM-201A: Annual Vehicle Distance Traveled in Miles and Related Data, 1936–1995 1/ By Vehicle Type, in *Highway Statistics Summary to 1995* (Washington, DC: Federal Highway Administration, 1996), V-15, V-16.

5. On strategy, see Halberstam, *The Fifties*, 120. According to Halberstam, what Wilson actually said was, "We at General Motors have always felt that what was good for the country was good for General Motors as well." For this quotation, see 118, where Halberstam draws on Ed Cray, *Chrome Colossus: General Motors and Its Times* (New York: McGraw-Hill, 1980), 7, as his source.

6. Unless otherwise indicated, this section on Cadillac draws primarily on William H. Whyte, Jr., "The Cadillac Phenomenon," *Fortune*, February 1955, 106–109, 174–184.

7. On Cadillac as the model, see Hine, *Populuxe*, 101. On Cole, Cadillac, and Chevrolet, see Halberstam, *The Fifties*, 490–495. Driveway is from David E. Cole, interview with the author. Halberstam, *The Fifties*, 495. For more on the 1955 Chevrolets, see Hine, *Populuxe*, 87–91.

8. On the Ford crisis, see Allan Nevins and Frank Ernest Hill, *Ford: Decline and Rebirth, 1933–1962* (New York: Scribner's, 1962, 1963) (hereinafter Nevins and Hill III), 294–345; and Douglas Brinkley, *Wheels for the World: Henry Ford, His Company, and a Century of Progress* (New York: Viking, 2003), 491–509. Elmo Roper, "A Survey of Attitudes towards and Opinions about Ford Motor Company," October 1944, microform, Ford Motor Company Records (hereinafter FMC), Benson Ford Research Center, The Henry Ford. On the GM preference, see Roper, "Survey," 6, 20B. On GM as model, see the comments of R. E. Roberts in "Ford Official Reveals 1946 Operating Loss," *Automotive Industries*, 15 Dec. 1953, 80. On the 1949 Ford, see "Young Henry's $72,000,000 Gamble," *Business Week*, 14 June 1948, 68. On Ford's advertising, see "Notes on Ford Problem," n.d. [but 1949–1950], Detroit Office, Ford: Correspondence, 1949–1979, Norman Strouse Papers, J. Walter Thompson Company Archives (hereinafter JWT), John W. Hartman Center for Sales, Advertising, and Marketing History, Duke University, Durham, NC, box 1, folder 5; and William J. Griffin, Jr., "Ford Car Advertising: A Time for Communication," 9 Mar. 1956, Office Files and Correspondence, Ford, Correspondence and Memos, 1956, Winfield Taylor Papers, JWT, box 1, folder 8.

9. "Why Do People Buy the Cars They Do?" *Popular Mechanics*, February 1956, 312. Obviously, this poll was not representative of U.S. auto buyers. The SRI quotation is from a summary of the study (commissioned by the *Chicago Tribune*) that appeared as "Why Do People Choose One Type of Car and Not Another?" Ford Motor Company Research Department Letter, 20 Jan. 1954, 2, FMC, acc. 465, box 2—Research Dept., Letters, 1954. The study was titled "Automobiles, What They Mean to Americans."

10. The characterization of Adams is from Whyte, "Cadillac Phenomenon," 109.

11. Guy Shipler, Jr., "The Hidden Reasons Why You Buy a Car," *Popular Science*, Nov. 1957, 92. Shipler indicates that this characterization of Cadillac owners is based on the SRI–*Chicago Tribune* study. "Why Negroes Buy Cadillacs," *Ebony*, September 1949, 34.

12. The quotation is from Eric Larrabee, "The Edsel and How It Got That Way," *Harper's*, September 1957, 72. Understatement is from Paul Fussell, *Class: A Guide through the American Status System* (New York: Touchstone, 1983), 54. "Conspicuous reserve" is from Vance Packard, *Hidden Persuaders* (New York: David McKay, 1957), 53–54.

13. Elaine Tyler May, *Homeward Bound: American Families in the Cold War Era* (New York: Basic, 1988), 3.

14. For "more, new, better" and the Cold War connection, see Cohen, *Consumer's Republic*, 119, 124–125.

15. For strangers, see May, *Homeward Bound*, 24. Hine, *Populuxe*, 36. On coping, see May, *Homeward Bound*, 14. Hine repeatedly stresses the lack of authority; see *Populuxe*, 9, 21, 24, 27, 32, 34, 36. The quotation that follows is his and appears at 32.

16. For average wholesale prices of U.S. automobiles, 1900–1949, see "Fifty Years of Motor Vehicle Production," *Automotive Industries*, 15 Mar. 1950, 76. On Ford's GM-trained management, see Halberstam, *The Fifties*, 118; and Brinkley, *Wheels for the World*, 506–507, 528–529. Both authors make the point that the managers went to Ford with GM's blessing. Griffin, "Ford Car Advertising," 4. Griffin, reviewing Ford's postwar marketing (1946–1956) for JWT staff, indicated that a primary marketing objective was to "sell value rather than price." "Car Makers Stack Their Chips on Luxury," *Business Week*, 23 Feb. 1952, 157–160.

17. On horsepower from 1930 to 1958, see *Automotive Industries*, 15 Mar. 1962, 109. On Cole and the

1955 Chevrolet V-8 engine, see Halberstam, *The Fifties*, 494–495. On the industry's adoption of the V-8, see Joseph Geschelin, "New Highs Established in Horsepower and Torque," *Automotive Industries*, 1 Feb. 1955, 48, and 15 Mar. 1962, 109; and "V-8s Seize the Market," *Business Week*, 21 May 1955, 116.

18. On populuxe, see Hine, *Populuxe*, 91. Hine called particular attention to the 1955 models in this regard. Halberstam, *The Fifties*, 123–127.

19. For "gorp," see Hine, *Populuxe*, 104. Harley J. Earl, "I Dream Automobiles," *Saturday Evening Post*, 7 Aug. 1954, 82. On tailfins, see also Hine, *Populuxe*, 83–106. On the near failure of the tailfin, see Whyte, "Cadillac Phenomenon," 181.

20. On Earl and "dynamic obsolescence," see Hine, *Populuxe*, 85.

21. For Ford's chromium problem at Monroe, see "U.S. Blocks Ford Plan to Dump in Lake Erie," *Detroit News*, 22 June 1977, 4B; "State Officials Are Probing . . . ," *Detroit News*, 23 Jan. 1978, 2A; and "Suit Cites Ford Pollution," *Detroit News*, 2 Feb. 1978, B5.

22. The environmental impacts from automobile manufacturing were fundamentally those of steel manufacturing. See Andrew Hurley, *Environmental Inequalities: Class, Race, and Industrial Pollution in Gary, Indiana, 1945–1980* (Chapel Hill: University of North Carolina Press, 1995), esp. 15–45. The employment figure is from Brinkley, *Wheels for the World*, 501. For the number of stacks, see Approval Draft Press Release, 1 Sept. 1953, 4, FMC, acc. 837, box 1. Many of these stacks were for ventilation purposes only.

23. Bill Wolf, "Running Sores on Our Land," *Sports Afield*, November 1948, 91. This clipping is in the Jack A. Van Coevering Papers (hereinafter VC), Bentley Historical Library, University of Michigan, Ann Arbor, box 16. For orange and acids, see Program Evaluation Division, "A Report to the Deputy Administrator: Pollution Control in the U.S.—Some Examples of Recent Accomplishments," 24 Nov. 1976, 33, in the Russell E. Train Papers, Library of Congress, box 30. On Superfund, see Cheryl Eberwein, "Dingell Fights for Clean Rouge," *Dearborn Press and Guide*, 28 Aug. 1986, 3-A. For "biggest proving ground," see Approval Draft Press Release, 24 Nov. 1953, 2, HFM, acc. 837, box 1.

24. Postwar pollution from the Rouge and its impact on surrounding Dearborn bears comparison with the impact of U.S. Steel's pollution on Gary, Indiana, the subject of Andrew Hurley's pathbreaking *Environmental Inequalities*. In Dearborn protests against pollution began earlier and the class and race dimensions of the protests were less marked. Dearborn had few African-American residents. Class later became a more prominent factor in Dearborn protests. Ford was nowhere near as recalcitrant in dealing with pollution as U.S. Steel, probably because of the greater potential for an impact on sales from negative publicity.

25. This section is congruent with and builds on the argument in Samuel P. Hays with Barbara D. Hays, *Beauty, Health, and Permanence: Environmental Politics in the United States, 1955–1985* (New York: Cambridge University Press, 1987). On the concern of the Michigan United Conservation Clubs, which represented 173 organizations, see "Fight Begins on Pollution," *Detroit Free Press*, n.d. [but late August 1945], VC, box 1. For evidence of ongoing support for the water pollution fight, see the letters printed in "What Readers have to Say on 'Save Our Streams' Articles," *Detroit Free Press*, 15 June 1947, in VC, box 2. Van Coevering's first column on the subject, "Seventy Millions Needed to Check Pollution of Michigan Waters," appeared in the *Detroit Free Press* on 20 May 1945, VC, box 1.

26. Van Coevering's interest culminated in a series of Sunday articles in 1947 entitled "Save Our

Streams" (SOS). These articles began with "Pollution Facts," *Detroit Free Press*, 25 May 1947, VC, box 2. The last article in the series ran on 28 Sept. 1947. For GM's problems with oil, acids, plating wastes, and pickling liquors at its Flint plants, see Jack Van Coevering, "Poor Butterfly: Pollution Hits Saginaw Area," *Detroit Free Press*, 15 June 1947, VC, box 2. For a mention of Ford's Rouge problems, see "Widespread Pollution in County Waters Bared," *Detroit Free Press*, 28 Sept. 1947, VC, box 2. Van Coevering also favorably singled out early antipollution efforts by both General Motors in Flint and Ford at the Rouge. See Jack Van Coevering, "Crusader Strikes Blow for Clean Streams," *Detroit Free Press*, 7 Mar. 1948; "Michigan Industry Eagerly Joins War on Water Pollution," *Detroit Free Press*, 21 Mar. 1948; "Ford Mobilizes Anti-Pollution Army," *Detroit Free Press*, 11 Apr. 1948; and "GM Waging All-Out War to Curb Stream Pollution in Plants throughout U.S.," *Detroit Free Press*, 23 May 1948, VC, box 2.

27. J. K. Hoskins, letter to G. H. Ferguson, 25 June 1945, Records Relating to Water and Hazardous Materials, Records of Predecessor Water Pollution Control Organizations, Correspondence Relating to the International Joint Commission's Pollution Investigation, 1945–57, RG 412, Records of the Environmental Protection Agency, National Archives, College Park, MD, box 1, folder "IJC — Correspondence July-Oct 1945." The International Joint Commission was formed as a result of the 1909 Boundary Waters Treaty between the United States and Canada. Under the terms of the treaty, three American and three Canadian commissioners were charged with examining issues with respect to the waters that formed the international border between the two nations and making recommendations to the two governments. Article 4 of the treaty specifically enumerated transborder water pollution as one of the commission's responsibilities. The IJC's study of pollution in the Detroit River was conducted between July 1946 and December 1949. The commissioners' recommendations were accepted and published by the two governments in 1951.

28. *Report of the International Joint Commission United States and Canada on the Pollution of Boundary Waters* (Washington, DC, and Ottawa. 1951), 63–65. For pollutant levels, see "Public Hearing, Ford Motor Company Rouge Plant with the Water Resources Commission . . . October 28, 1952," Department of Natural Resources, RG 86-77, Michigan State Archives, Lansing, box 16, folder 15, 11–12.

29. On the broad support, see Jack Van Coevering, "Why It's Up to You to Do Job of Ending Pollution of Waters," *Detroit Free Press*, 13 July 1947, 6; and "Aroused Public Pushing Demand for Quick Action," *Detroit Free Press*, 19 Dec. 1948, in VC, box 2. On Michigan's leadership, see Jack van Coevering, "Fight for Clean Streams Spreading over Nation," *Detroit Free Press*, 6 Feb. 1949, VC, box 2. For the award, see "A Gratifying Honor," *Detroit Free Press*, 13 Apr. 1948; and Lyall Smith, "Van Coevering Gets Award for Campaign," *Detroit Free Press*, 21 Jan. 1949, in VC, box 2.

30. Terence Kehoe, *Cleaning Up the Great Lakes: From Cooperation to Confrontation* (DeKalb: Northern Illinois University Press, 1997), 5–7.

31. "City Moves to Halt Smoke Nuisance," *Dearborn Press*, 17 July 1947, 1; "Ask Action on Smoke Control," *Dearborn Guide*, 9 Dec. 1948, 1. See Lynne Page Snyder, "'The Death-Dealing Smog over Donora, Pennsylvania': Industrial Air Pollution, Public Health Policy, and the Politics of Expertise, 1948–1949," *Environmental History Review* 18:1 (1994): 117–139. Dearborn residents may well have known about municipal air pollution control programs in Saint Louis and Los Angeles. One of the housewives' primary complaints was the soot produced by the coal-fired locomotives of the New York Central Railroad, which suggests that they were aware that railroads around the country were then being pushed to do more about this problem. On passage of the law, see "Ask

Action on Smoke Control," *Dearborn Guide*, 9 Dec. 1948, 1. On efforts to control railroad smoke, see Joel A. Tarr, "Railroad Smoke Control: The Regulation of a Mobile Source," in Tarr, *The Search for the Ultimate Sink: Urban Pollution in Historical Perspective* (Akron, OH: University of Akron Press, 1996), 262–283. Cohen, *Consumer's Republic*, 135, has argued that women who exercised greater civic authority in their communities during World War II "found themselves marginalized from public life during reconversion." This case may reflect that or simply a more traditional working-class perspective on the inappropriateness of housewives protesting problems that originated in male workplaces. "Hire Inspector for Smoke Law," *Dearborn Press*, 12 May 1949, 1.

32. "New Dust Collectors Installed in Foundry," *Rouge News*, 22 Apr. 1949, 8.

33. On Ford's worries about sportsmen, see "Meeting on Industrial Wastes Control with Ford Manufacturing Staff," n.d. [but almost certainly the first half of 1949], FMC, acc. 837, box 2. For more detail on the problem with sportsmen, see H. L. Mottl, letter to Mr. Bauerback, 21 Feb. 1949; and D. S. Harder, letter to Michigan United Conservation Clubs, undated letter draft, both in FMC, acc. 837, box 2.

34. Hays and Hays, *Beauty, Health, and Permanence*, 3. For the connection between sportsmen and automobiles, evident as early as the 1920s, see Paul S. Sutter, *Driven Wild: How the Fight against Automobiles Launched the Modern Wilderness Movement* (Seattle: University of Washington Press, 2002), 40–41. The "Treaty of Detroit" was a milestone in American labor history. For a discussion of its significance, see Nelson Lichtenstein, *The Most Dangerous Man in Detroit: Walter Reuther and the Fate of American Labor* (New York: Basic, 1995), 271–298. For its significance in the context of postwar consumer developments, see Cohen, *Consumer's Republic*, 155.

35. Hays and Hays, *Beauty, Health, and Permanence*, 4–5, 22–23, also underscore the importance of protecting the quality of home life as a source of environmental protest. Ibid., 23.

36. On Parkhurst, see "Smoke Signals," *Dearborn Press*, 23 Dec. 1948, 4. "Hartleb Named to Engineer Board," *Dearborn Guide*, 18 June 1953, 11.

37. On twentieth-century California history, see the Americans and the California Dream Series by Kevin Starr (New York: Oxford, 1973–2002); and Starr, *Coast of Dreams: California on the Edge, 1990–2003* (New York: Alfred A. Knopf, 2004).

38. For good roads, see David W. Jones, "California's Freeway Era in Historical Perspective," Institute of Transportation Studies, University of California, Berkeley, June 1989, 93–94, which makes this very important point.

39. On Southern California and air pollution, see James E. Krier and Edmund Ursin, *Pollution and Policy: A Case Essay on California and Federal Experience with Motor Vehicle Air Pollution, 1940–1975* (Berkeley: University of California Press, 1977). See also Marvin Brienes, "The Fight against Smog in Los Angeles, 1943–1957" (Ph.D. diss., University of California, Davis, 1975). More recently, Scott Hamilton Dewey devoted three chapters in his dissertation, "Don't Breathe the Air: Air Pollution and the Evolution of Environmental Policy and Politics in the United States, 1945–1970" (Rice University, 1997), to the Los Angeles air pollution story. Dewey has since followed up with an article, "'The Antitrust Case of the Century': Kenneth F. Hahn and the Fight against Smog," *Southern California Quarterly* 81:3 (1999): 341–376; and *Don't Breathe the Air: Air Pollution and U.S. Environmental Politics, 1945–1970* (College Station: Texas A&M University Press, 2000). Jack Doyle devotes several early chapters to the Southern California smog story in his book *Taken for a Ride: Detroit's Big Three and the Politics of Pollution* (New York: Four Walls

Eight Windows, 2000). Joshua Dunsby, "Localizing Smog," in E. Melanie DuPuis, ed., *Smoke and Mirrors* (New York: New York University, 2004), 170–200, offers a more sociological take. The interpretation offered here blends these sources with additional primary source research. For earlier episodes, see, e.g., Charles L. Senn, letter to George M. Uhl, 30 July 1943, 3, and I. A. Deutch, "Various Aspects of Air Pollution Control," paper delivered before the American Society of Heating and Ventilating Engineers, 8 May 1946, 1, in John Anson Ford Papers (hereinafter JAF), Huntington Library, San Marino, CA, box 25, B III, 5a, aa (2 and 5).

40. Dunsby stresses the appeal of clean air. On Royce, see "Man Who Organized the First Campaign Gives His Answers," *Los Angeles Times*, 19 Nov. 1953, A5. See stenographer's notes, Board of Directors Meetings, Los Angeles County Chamber of Commerce, 14 Sept. 1944, 11, Los Angeles County Chamber of Commerce Records (hereinafter LACCC), Regional History Center, University of Southern California, Los Angeles.

41. John Anson Ford, letter to George L. Kelley, 24 Dec. 1946, JAF, box 25, B III, 5a, aa (5).

42. For the legislation, see *Statutes of California 1947* (Sacramento: California State Printing Office, 1948), chap. 632, 1640–1651. For support, see Ed Ainsworth, "Anti-Smog Bill to Be Explained for Senate Unit," *Los Angeles Times*, 25 May 1947, A1. Dewey, "'The Antitrust Case of the Century,'" 344, touts the caliber of the Los Angeles organization.

43. See, e.g., "Report to the Los Angeles County Board of Supervisors by the L.A. County Smoke and Fumes Commission," 13 Mar. 1944, 3, and H. O. Swartout, letter to Los Angeles County Board of Supervisors, 1 Aug. 1944, 3, both in JAF, box 25, B III, 5a, aa (2); and H. O. Swartout and I. A. Deutch, "The 'Smog' Problem," September 1945, JAF, box 25, B III, 5a, bb (1). See also Harold W. Kennedy, *History, Legal and Administrative Aspects of Air Pollution Control in the County of Los Angeles*, report submitted to the Board of Supervisors of the County of Los Angeles, 9 May 1954, 48. Louis C. McCabe, "National Trends in Air Pollution," paper presented at National Air Pollution Symposium, Pasadena, CA, 10 Nov. 1949, 4, JAF, box 25, B III, 5a, bb (no folder). "Air Pollution Control District–Index" [apparently a series of questions and answers prepared for public relations purposes], stamped 17 July 1950, 8, JAF, box 26, B III, 5a, ff (1).

44. Arnold O. Beckman, interview with Jeffrey L. Sturchio and Arnold Thackray, 23 July 1985, transcript, Chemical Heritage Foundation, Philadelphia. For Haagen-Smit and his involvement with smog research, see Zus Haagen-Smit, interview by Shirley K. Cohen, Pasadena, CA, 16, 20 Mar. 2000, Oral History Project, California Institute of Technology Archives, retrieved June 2004 from the World Wide Web: http://resolver.caltech.edu/CaltechOH:OH_Haagen-Smit_Z (hereinafter Haagen-Smit interview). Zus Haagen-Smit was his wife. During this interview, Haagen-Smit read extensive excerpts from a tribute to her husband by Beckman. Ed Ainsworth, "Cause of Smog's Eye Smarting Told," *Los Angeles Times*, 26 Feb. 1949, 1.

45. Haagen-Smit interview, 24–26. On Haagen-Smit's work for the APCD, see "Supervisors Hire Research Scientist," *Los Angeles Times*, 26 Apr. 1950, 9. See Arthur J. Will, letter to Los Angeles County Board of Supervisors, 15 Mar. 1954, 5, JAF, box 25, B III, 5a, dd, in which Will indicates that Haagen-Smit was paid a thousand dollars a month from 15 May 1950 to 1 July 1951. On causes, see "Caltech Scientist Will Discuss Smog," *Los Angeles Times*, 16 Nov. 1950, A2; and William S. Barton, "Puzzle of Smog Production Solved by Caltech Scientist," *Los Angeles Times*, 20 Nov. 1950, pt. 2, 1. His conclusions were subsequently published in academic journals. See A. J. Haagen-Smit et al., "Investigation on Injury to Plants from Air Pollution in the Los Angeles Area,"

Plant Physiology 27:1 (1952): 18–34; and A. J. Haagen-Smit, "Chemistry and Physiology of Los Angeles Smog," *Industrial and Engineering Chemistry* 44:6 (1952): 1342–1346.

46. On the Los Angeles inversion, see "Climate Found Smog's Cohort in Los Angeles," *Chicago Daily Tribune*, 28 Dec. 1958, 17; and "Second Worst Smog of Year Irritates City," *Los Angeles Times*, 3 Dec. 1946, 1.

47. Haagen-Smit quoted in Barton, "Puzzle of Smog Production Solved by Caltech Scientist," 1.

48. Call is quoted in Air Pollution Foundation, *Final Report, 1961* (San Marino, CA: Air Pollution Foundation, 1961), 8.

49. See Air Pollution Foundation, *Final Report, 1961*, 12–13, for an acknowledgment of Haagen-Smit's assistance, and ibid., 25, for accepting the auto's role. W. L. Faith, N. A. Renzetti, and L. H. Rogers, *Automobile Exhaust and Smog Formation* (San Marino, CA: Air Pollution Foundation, 1957), 15.

50. For evidence of continuing disbelief, see constituent letters throughout the Kenneth Hahn Papers (hereinafter KH) at the Huntington Library. On the origin of the alert regulations, see Beckman interview, 32. For discussion of driving bans, see "Auto Exhaust Pollutants Put at 1100 Tons Daily," *Los Angeles Times*, 16 Oct. 1954, 8. Haagen-Smit often made this claim about the beneficial effect of freeways early on. See William S. Barton, "War with Smog Held 'Over Hump,'" *Los Angeles Times*, 20 Mar. 1953, A1; and "Car Exhausts Blamed for Irritants in Smog," *Los Angeles Times*, 5 Oct. 1953, 11.

51. See Hahn's obituary, Richard Simon, Judith Michaelson, and Bettina Bixall, "People's Politician Hahn Dies," *Los Angeles Times*, 13 Oct. 1997, 1. Hahn kept the heat on the APCD in the late 1940s. See Ed Ainsworth, "State Uses 'Hands Off' Policy in City Battle against Smog," *Los Angeles Times*, 11 Dec. 1948, A1; and "Merited Rebuke," *Los Angeles Times*, 30 Oct. 1949, A4. For these letters, see *Smog: Record of Correspondence between Kenneth Hahn, Los Angeles County Supervisor, and the Presidents of General Motors, Ford, and Chrysler . . .* , 7th ed. (Los Angeles: Los Angeles County Board of Supervisors, 1972). The Ford denial is from Ralph Nader, *Unsafe at Any Speed* (New York: Grossman, 1965), 153.

52. "Automotive Industry Aid Pledged in Fumes Study," *Los Angeles Times*, 3 Feb. 1954, 14. On GM's industrial waste committee, see *Smog*, 6.

53. On task groups, see Induction System Task Group, "Automotive Exhaust Hydrocarbon Reduction during Deceleration by . . . Induction System Devices," *SAE Transactions* 66 (1958): 384. On cross-licensing, see "Smog Control Antitrust Case," *Congressional Record*, vol. 117, pt. 12, 18 May 1971, 15626, 15628. For Heinen's role, see Robert E. Bedingfield, "A Student of Air Pollution Control by Automobiles," *New York Times*, 1 Feb. 1970, F3. For board approval, see "Smog Control Antitrust Case," 15626.

54. On interfirm agreements, see "Smog Control Antitrust Case," 15630. "Automakers Pledge Help in Smog War," *Los Angeles Times*, 9 Jan. 1954, A1; and "Automotive Engineer Group Arrives for Smog Checkup," *Los Angeles Times*, 26 Jan. 1954, A1.

55. On the device, see Kenneth F. Hahn, letter to Harry F. Barr, 30 Mar. 1966, reprinted in *Smog*, 20. Also see "Auto Industry Urged to Speed Antismog Device," *Los Angeles Times*, 19 July 1957, B3. On it being a carburetor, see "Hahn Confers on Smog Issue," *Los Angeles Times*, 20 Jan. 1955, 36. On the two ways, see "Automakers Pledge Help in Smog War"; and "Cities Get Year of Grace for Change," *Los Angeles Times*, 20 Oct. 1954, 1.

56. On Houdry's work, see "Cities Get Year of Grace for Change"; and "Poulson Says Larson . . .

Lied about Their Talk," *Los Angeles Times*, 22 Oct. 1954, 1. On Houdry, see Stacy V. Jones, "Anti-Smog Device Uses Catalyst," *New York Times*, 30 June 1962, 23; and "Arthur Pew, Jr., Sun Oil Aide, 65," *New York Times*, 20 Jan. 1965, 39, as well as more generally the Eugene J. Houdry Papers, Collections of the Manuscript Division, Library of Congress, Washington, DC (hereinafter Houdry-LOC), and the Eugene J. Houdry Collection, Archives, Chemical Heritage Foundation, Philadelphia (hereinafter Houdry-CHF). See also "Eugene Jules Houdry," in *American National Biography*; and "Eugene J. Houdry, Inventor, Was 70," *New York Times*, 19 July 1962, 27. On catalytic cracking, see "Arthur Pew, Jr., Sun Oil Aide, 65," 39. See also Alex G. Oblad, "The Contributions of Eugene J. Houdry to the Development of Catalytic Cracking," in Burtron H. Davis and William P. Hettinger, Jr., eds., *Heterogeneous Catalysis: Selected American Histories* (Washington, DC: American Chemical Society, 1983), 61–75.

57. See "Oxy-Catalyst, Inc.: A Brief History," 1 May 1974; Eugene J. Houdry, "Practical Catalysis and Its Impact on Our Generation," paper given at the International Congress on Catalysis, Philadelphia, 10–14 Sept. 1956, both in Houdry-LOC; and Clarence H. Thayer, "Eugene Jules Houdry and His Catalytic Processing Activities," 8 Feb. 1975, Houdry-CHF. On the earlier work, see the Eldred U.S. patent no. 1,043,580 issued in 1912. On his first patent, see G. Alex Mills, "Eugene J. Houdry—Pioneer in Catalytic Oxidation," presentation to the Division of the History of Chemistry, American Chemical Society, 29 Aug. 1984, Philadelphia, 9, in Houdry-CHF. On platinum, etc., see Houdry, "Practical Catalysis and Its Impact on Our Generation"; and Eugene J. Houdry and Gordon P. Larson, "An Aspect of Catalysis in Modern Life," paper presented to the American Chemical Society, Washington, DC, 26 Mar. 1962, in Houdry-LOC.

58. On the OCM, see Houdry, "World's Biggest Cleaning Job," 175–185; and Houdry and Larson, "An Aspect of Catalysis in Modern Life," 9. Southwest Research Institute, *Report No. 8: Field Evaluation of Houdry Catalytic Exhaust Converters* (Los Angeles: Air Pollution Foundation, June 1955), iii. The author thanks Kathleen C. Taylor for bringing this study to his attention.

59. William G. Agnew, interview with the author.

60. On Nebel and the task group, see Exhaust System Task Group, "Oxidation Seems Best Exhaust Gas Cleanser," *SAE Journal* 65:11 (1957): 120. For GM's conclusions, see J. M. Campbell, memo to C. A. Chayne, 29 Oct. 1954, in Office Files—Smog, Kettering Papers, Kettering/GMI Alumni Foundation Collection of Industrial History, Kettering University, Flint, MI. Agnew did not think there was any catalyst research going on at GM Research when he began in 1952.

61. J. Robert Mondt summarizes the prospects in his book *Clean Cars: The History and Technology of Emission Control since the 1960s* (Warrendale, PA: SAE International, 2000), 84–86. On industry promises, see Nader, *Unsafe at Any Speed*, 155. Houdry's role in the invention and adoption of the catalytic converter has not been widely appreciated. For an exception, although one where Houdry was not credited by name, see George R. Lester, "The Development of Automotive Exhaust Catalysts," in Davis and Hettinger, eds., *Heterogeneous Catalysis*, 417–418. See also Kathleen C. Taylor, interview with the author. The importance of Houdry's work is evident in the more than twenty patents that he filed for and received in this area between 1949 and 1964. See, e.g., U.S. patent no. 2,742,437, filed 29 Sept. 1952.

62. My argument in this section is congruent with Hays and Hays, *Beauty, Health, and Permanence*, esp. 2–5, 34–35, 529.

63. Americans' strong embrace of Detroit's automobiles in the early 1950s is a good example of Daniel J. Boorstin's "communities of consumption." See Boorstin, *The Americans: The Democratic*

Experience (New York: Random House, 1973), 89–164. The connection between consumption and community is also stressed by Hine, *Populuxe,* 12, 17. On emphatic ways and cars, see ibid., 4–5. Whyte, "Cadillac Phenomenon," 108–109. E. Franklin Frazier, *Black Bourgeoisie* (New York: Free Press, 1957), 235; see 233–334 for his acknowledgment that middle-class African Americans embody "in an acute form many of the characteristics of modern bourgeois society, especially in the United States." Hine, *Populuxe,* 4, 106. See, however, Offer, *Challenge of Affluence,* 201–207, which argues that American automobile consumers in the 1950s embraced comfort in the form of more feature-laden low-priced models more so than status.

7. DISENCHANTED WITH DETROIT

1. "Buyers' strike" is from Larry Doyle, as quoted by John Brooks, "Annals of Business: The Edsel," a two-part article that appeared in the *New Yorker,* 26 Nov. 1960, 57–102, and 3 Dec. 1960, 199–224. For a brief economic and psychological analysis of this transition, see Avner Offer, *The Challenge of Affluence: Self-Control and Well-Being in the United States and Britain since 1950* (New York: Oxford University Press, 2006), 210–221.

2. On the Volkswagen Beetle, see Walter Henry Nelson, *Small Wonder: The Amazing Story of the Volkswagen* (Boston: Little, Brown, 1965; reprint ed., 1970); Dan R. Post, *Volkswagen: Nine Lives Later* (Arcadia, CA: Horizon House, 1966); and Phil Patton, *Bug: The Strange Mutations of the World's Most Famous Automobile* (New York: Simon and Schuster, 2002). For brief versions, see Huston Horn, "The Beetle Does Float," *Sports Illustrated,* August 1963, 58–67; David Halberstam, *The Fifties* (New York: Villard, 1993), 636–639; and Douglas Brinkley, *Wheels for the World: Henry Ford, His Company, and a Century of Progress* (New York: Viking, 2003), 544–545, 586–587. For unchanged design, see "Volkswagen: An Auto City in West Germany," *Business Week,* 11 July 1953, 109. Text for commercial is from J. D. Strohm to J. N. Hastings, "Report on 8th Visual Communications Conference," 8 May 1963, "1964 Seminars and Communications Conferences, Speeches, Shows," Kensinger Jones Papers, John W. Hartman Center for Sales, Advertising, and Marketing History, Duke University, Durham, NC, box 4. "Volkswagen Says It Will Not Change Basic Construction," *Automotive Industries,* 15 Jan. 1958, 33; and "No Change in VW," *Automotive Industries,* 1 Aug. 1959, 18.

3. On dealer advertising, see Ed Brown, "Import Discounts Abound in N.Y.," *Automotive News,* 4 Nov. 1957, 67; and Martin Trepp, "Imports Near 7% of Wash. Sales," *Automotive News,* 25 Nov. 1957, 6. For word of mouth and press, see Edgar Snow, "Herr Tin Lizzie: Saga of the Volkswagen," *Nation,* 3 Dec. 1955, 474; and Carl Spielvogel, "It's the Difference That Sells the Volkswagen," *New York Times,* 20 Oct. 1955, 38. Snow, "Herr Tin Lizzie," 476. Conflation is from Ruth M. Eddy, "Foreign-Car Whiz," *Automotive News,* 5 Nov. 1956, 20.

4. For first customers, see Horn, "The Beetle Does Float," 61. For "bugs," see Ed Brown, "She Sells Volkswagens," *Automotive News* [March 1956], n.p. In Malcolm Gladwell's terminology from the *The Tipping Point: How Little Things Can Make a Big Difference* (Boston: Little, Brown, 2000; reprint ed., 2002), these people were "mavens." Dretzin is quoted in Spielvogel, "It's the Difference That Sells the Volkswagen," 28. "Honest car" is from Arthur Railton, *Popular Mechanics,* 1956, as quoted by Halberstam, *The Fifties,* 639.

5. For the German background, see Patton, *Bug,* 7–66. For Hitler's specifications, see Post, *Volkswagen,* 46.

6. On Nordhoff, see Post, *Volkswagen*, 120–125; and "Heinz Nordhoff, Volkswagen President, Dies at 69," *New York Times*, 13 Apr. 1968, 25. On the link to Henry Ford, see Vinton McVicker, "People's Car," *Wall Street Journal*, 20 Mar. 1956, 1. On selling to the U.S. market, see Nelson, *Small Wonder*, 178–193.

7. For 1955 and 1956 import sales, see "Foreign Cars," *Automotive News*, 20 Feb. 1956, 2; and "Foreign-Car Registrations," *Automotive News*, 3 Mar. 1958, 1. For the 1957 backlog, see Maynard M. Gordon, "Continued U.S. Shortage of Foreign Cars Seen," *Automotive News*, 3 June 1957, 4; and B. J. Widick, "Small Cars: If You Can't Lick 'Em . . . ," *New Republic*, 26 Aug. 1957, 9. On used car sales, see William Carroll, "Imports Are Big in Los Angeles," *Automotive News*, 28 Oct. 1957, 2; and Robert M. Lienert, "Breakthrough for Import Cars," *Automotive News*, 2 Dec. 1957, 92.

8. For individualists, see "At Last, GM Invades Small Car Market," *Business Week*, 29 June 1957, 41; and Robert Sheehan, "A Big Year for Small Cars," *Fortune*, August 1957, 105. For surveys of foreign and small car buyers, see, e.g., "GM to Sell Its Europe-Built Vauxhall and Opel Cars in U.S. in Fall through Pontiac, Buick Dealers," *Wall Street Journal*, 21 June 1957, 3; Sheehan, "A Big Year for Small Cars," 108; "The Small Car—The Big Pitch," *Newsweek*, 14 Oct. 1957, 106; Kimmis Hendrick, "Renaults Pass VWs in California Race," *Christian Science Monitor*, 17 Sept. 1958, 10; Bill Carroll, "'Invasion Accomplished,'" *Los Angeles Times*, 20 Mar. 1960, K10; and Horn, "The Beetle Does Float," 62. On the importance of the Southern California and New York City markets, see Robert M. Lienert, "Breakthrough for Import Cars," *Automotive News*, 2 Dec. 1957, 98. For Los Angeles County, see Gladwin Hill, "California Sets Pace in Off-Beat Sales," *New York Times*, 5 Apr. 1959, A12. Carroll, "'Invasion Accomplished,'" K10.

9. On the symbolism, see Lienert, "Breakthrough for Import Cars," 92; "Prestige Sells Foreign Cars," *Automotive News*, 16 Dec. 1957, 6; and "Ford Eyes 5% Import Niche," *Automotive News*, 12 May 1958, 45. For Bing Crosby, see "On the Slow Road," *Time*, 12 May 1958, 88. Horn, "The Beetle Does Float," 62. "Avant-garde" is from Daniel M. Burnham, "Economy Cars," *Wall Street Journal*, 10 Apr. 1957, 1.

10. Bob Garfield, as quoted in Patton, *Bug*, 104.

11. For a sampling, see Eric Larrabee, "The Edsel and How It Got That Way," *Harper's*, Sept. 1957, 70; Raymond Loewy, "Jukeboxes on Wheels," *Atlantic Monthly*, April 1955, 36–38; Edgar Snow, "Herr Tin Lizzie," *Nation*, 3 Dec. 1955, 474–476; David Cort, "You, Too, Can Drive a Juke Box," *Nation*, 22 Dec. 1956, 534–536; and "Osborn Looks at the '57 Cars," *New Republic*, 21 Jan. 1957, 11–13. Loewy designed cars for Studebaker that were critically well received but ignored by consumers.

12. Offer, *Challenge of Affluence*, 215, concludes that the recession was caused by the change in consumer tastes for automobiles. For "out of touch," see "Autos: Lower Targets," *Time*, 23 Dec. 1957, 64; "Slowdown in Detroit," *Time*, 10 Mar. 1958, 82; "Those Big Cars, Small Cars—And Detroit," *Newsweek*, 28 Apr. 1958, 69–70; and Herbert Brean, "'Volkswagen, Go Home!'" *Life*, 28 July 1958, 83–84, 87–88, 90. "On the Slow Road," *Time*, 12 May 1958, 84. "Those Big Cars, Small Cars—And Detroit," 69.

13. John Keats, *The Insolent Chariots* (Philadelphia: Lippincott, 1958), 186, which he indicates was a characterization from a friend. The book was on the *New York Times* best-seller list for five weeks in October and November 1958.

14. "AMA Votes Racing Ban," *Automotive News*, 10 June 1957, 2.

15. On prices, see Keats, *Insolent Chariots*, 165. On styling, see ibid., 39, 48, 54–55, 64. *Insolent Chariots* is riddled with Keats's anticonsumer bias, but see esp. 57–64.

16. On advertising, see Larrabee, "The Edsel and How It Got That Way," 70. Vance Packard, *Hidden Persuaders* (New York: David McKay, 1957). Packard's book was published on 29 Apr. 1957. It reached the *New York Times* best-seller list on 2 June and stayed there for the next year, occupying the number-one position for six of seven weeks between 4 August and 15 September and the number-two position for most of the fall. On motivation research in historical context, see Russell W. Belk, "Studies in the New Consumer Behaviour," in Daniel Miller, ed., *Acknowledging Consumption: A Review of New Studies* (London: Routledge, 1995), 59; and Jackson Lears, *Fables of Abundance: A Cultural History of Advertising in America* (New York: Basic, 1994), 251–254. I thank Shelley Kaplan Nickles, "Object Lessons: Household Appliance Design and the American Middle Class, 1920–1960" (Ph.D. diss., University of Virginia, 1999), 284n, for pointing out these references, as well as for her own analysis of motivation research, 303–318.

17. Perrin Stryker, "'Motivation Research,'" *Fortune*, June 1956, 222. On more scientific, see "Inside the Consumer: The New Debate: Does He Know His Own Mind?" *Newsweek*, 10 Oct. 1955, 89. For automobile examples, see ibid., 89; Stryker, "'Motivation Research,'" 146; "Psychology and the Ads," *Time*, 13 May 1957, 51–52; Packard, *Hidden Persuaders*, 17–18, 52–56, 97–88, 128–131; and Keats, *Insolent Chariots*, 71–75.

18. Timothy D. Wilson, *Strangers to Ourselves: Discovering the Adaptive Unconscious* (Cambridge, MA: Belknap Press of Harvard University Press, 2002), 82–83, 184–188. For commissioned studies that involved U.S. automakers, see "Inside the Consumer" (Chrysler); Stryker, "'Motivation Research,'" 146 (Chevrolet), 228 (Buick), 230; and Larrabee, "The Edsel and How It Got That Way," 70–73 (Edsel). For an internal document that mentions the application of motivation research, see "1958 [Ford] Car Advertising Program," 25 July 1957, 1, Office Files and Correspondence, Ford, Advertising Plans (1958 Model) 1957, Winfield Taylor Papers (hereinafter WT), J. Walter Thompson Company Archives, John W. Hartman Center for Sales, Advertising, and Marketing History, box 3, folder 10.

19. Halberstam, *The Fifties*, 629–635. Thomas Frank, *Conquest of Cool: Business Culture, Counterculture, and the Rise of Hip Consumerism* (Chicago: University of Chicago Press, 1997).

20. Packard, *Hidden Persuaders*, 52–56, 69.

21. Vance Packard, "The Ad and the Id," *Reader's Digest*, November 1957, 118–121. Packard discusses Dichter's study in *Hidden Persuaders*, 87–88.

22. Keats, *Insolent Chariots*, 71–73.

23. Ibid., 45. Halberstam, *The Fifties*, 488.

24. Keats, *Insolent Chariots*, 56; see 26, 86, for his automobile virtues.

25. "Third of Market Seen for Ford," *Automotive News*, 15 Oct. 1956, 67.

26. The story of the Edsel is recounted by Larrabee, "The Edsel and How It Got That Way," 67–73; William Carroll, "Is There a Silver Lining in Cloud over Edsel?" *Automotive News*, 27 Jan. 1958, 6, 41, 43; "The Edsel Story," *Consumer Reports*, Apr. 1958, 216–218; Brooks, "Annals of Business"; and C. Gayle Warnock, *The Edsel Affair . . . What Went Wrong?* (Paradise Valley, AZ: Pro West, 1980). Warnock was the public relations director for Ford's Edsel Division. He suggests that Ford destroyed most of its corporate files on the Edsel after the debacle. My analysis rests heavily on Brooks and Warnock. For a brief recounting of the Edsel disaster, see Brinkley, *Wheels for the*

World, 575–581. A more recent account of the debacle is Thomas E. Bonsall, *Disaster in Dearborn: The Story of the Edsel* (Stanford, CA: Stanford University Press, 2002). On advertising cost, see Larrabee, "The Edsel and How It Got That Way," 73.

27. For scapegoat, see Eric Larrabee, "Detroit's Great Debate: 'Where Did We Go Wrong?'" *Reporter*, 17 Apr. 1958, 19. Keats, *Insolent Chariots*, 82–113. Arthur W. Baum, as quoted in Warnock, *Edsel Affair*, 179. The original article was Arthur W. Baum, "I Rode in Detroit's Mystery Car," *Saturday Evening Post*, 31 Aug. 1957, 18–19, 77–79. On liberties, see Warnock, *Edsel Affair*, 248.

28. On not different enough, see Ben Plegar, "Back in 1925, Auto Buyer Had Choice of Fifty-Three Names," *Washington Post*, 29 Nov. 1959, 61. On motivation research, see Larrabee, "The Edsel and How It Got That Way," 73. The dealer is quoted in "Captives Test Small-Car Market," *Automotive News*, 10 Mar. 1958, 55.

29. Warnock, *Edsel Affair*, 242–244, 245, summarizes the case for high comparative price being the problem. Brooks, "Annals of Business," 3 Dec. 1960, 205. See Warnock, *Edsel Affair*, 208–209, 215, 244–246, for more on quality.

30. Keats, *Insolent Chariots*, 83; Brooks, "Annals of Business," 3 Dec. 1960, 207–208, 217; and Warnock, *Edsel Affair*, 241, all make the point about Sputnik's effect on the Edsel's market reception. Republican U.S. senator Styles Bridges actually made the connection between tailfins and Cold War unpreparedness; see Halberstam, *The Fifties*, 702. On rockets and small cars, see "Fear of World Linked to Rise on Small-Car Sales," *Automotive News*, 16 Dec. 1957, 4. *Consumer Reports*, January 1958, 30–33. Warnock, *Edsel Affair*, 179–180, called this review "the most searing condemnation." Carroll, "Is There a Silver Lining in Cloud over Edsel?" 41. Kenyon and Eckhardt is quoted in Warnock, *Edsel Affair*, 233–234.

31. "The Small Car—The Big Pitch," *Newsweek*, 14 Oct. 1957, 107; "Auto Prestige: Conspicuous Consumption Is Waning," *Time*, 31 Mar. 1958, 76; and Brean, "'Volkswagen, Go Home!'" 83–84, 87–88, 90.

32. "The Perennial Rebellion," *Fortune*, May 1958, 104.

33. On awareness, see Dan Cordtz and George Melloan, "Detroit's Future," *Wall Street Journal*, 21 Aug. 1958, 1. Keats, *Insolent Chariots*, 46, 48.

34. On Romney, see James Playsted Wood, "George Romney," *Journal of Marketing* 27:3 (1963): 77–78; and T. George Harris, *Romney's Way: A Man and an Idea* (Englewood Cliffs, NJ: Prentice-Hall, 1967), 145. James P. Thurber, Jr., "The Little Car: Search for a Modern Model T Is Chronicle of Recurring Defeat," *Wall Street Journal*, 31 Dec. 1954, 1.

35. On corporate suicide, inspiration, and the pay cut, see Harris, *Romney's Way*, 135, 157–158, 144. Ernest Breech is quoted by Warnock, *Edsel Affair*, 241.

36. "Colbert Says Chrysler's Small Car Emphasis in '53 Cut Sales Sharply," *Wall Street Journal*, 7 Feb. 1958, 7. Thurber, "The Little Car," 1, 8.

37. The quotation about Romney is from Dan Cordtz, "Big Three Auto Firms Unlikely to Build Really Small Car in Next Eighteen Months," *Wall Street Journal*, 20 May 1958, 26.

38. On small imports, see "At Last, GM Invades Small Car Market," 41; and Sheehan, "A Big Year for Small Cars," 105. For Detroit's thinking on small cars, see "Flivver Again," *Newsweek*, 20 Jan. 1947, 71–72; "How a Typical 'Small' Car Has Grown Up: The Chevrolet through the Years," *Business Week*, 7 June 1947, 20–21; "'N.x.i.': Nash Formula for a $1,000 Car," *Business Week*, 7 Jan. 1950, 25; Thurber, "The Little Car," 1, 8; and Sheehan, "A Big Year for Small Cars," 108. For a review of the American automobile industry's post–World War II attitudes and activities around small

cars, see Lawrence J. White, "American Automobile Industry and the Small Car, 1945–70," *Journal of Industrial Economics* 20:2 (1972): 179–192. See also Brinkley, *Wheels for the World,* 534–535; and the testimony of Harlow H. Curtice, 30 Jan. 1958, before the Kefauver Senate antitrust subcommittee, as quoted by Ed Cray in *Chrome Colossus: General Motors and Its Times* (New York: McGraw-Hill, 1980), 405. The larger Ambassador comprised a portion of American Motors sales.

39. On the term *compact,* see "Detroit 'Missionary' at Large . . . ," *Newsweek,* 24 Feb. 1958, 84. On the Plymouth Valiant, see Charles K. Hyde, *Riding the Roller Coaster: A History of the Chrysler Corporation* (Detroit, MI: Wayne State University Press, 2003), 181–183. Dan Cordtz, "Compact Car Clash," *Wall Street Journal,* 23 June 1959, 1. For Falcon sales, Offer, *Challenge of Affluence,* 214. This figure, which included Mercury Comet sales, is apparently for calendar year 1960. For 1961 compact production, see Allen F. Jung, "Impact of the Compact Cars on New-Car Prices: A Reappraisal," *Journal of Business* 35:1 (1962): 70.

40. For the 1959 Volkswagen figure, see "'American Language' Absorbs Foreign Cars," *Christian Science Monitor,* 19 Feb. 1960, 17. For the total imports, see *Motor Vehicle Facts and Figures* (1973/74), 20; and Joseph C. Ingraham, "Shift to Compact Sweeps Car Field," *New York Times,* 25 Apr. 1960, 43. On Volkswagen advertising, see Horn, "The Beetle Does Float," 65–67; Nelson, *Small Wonder,* 213–229; Frank, *Conquest of Cool,* 53–73; and Patton, *Bug,* 91–105.

41. Market research available in the Ford Motor Company Collection at the Benson Ford Research Center at The Henry Ford and the J. Walter Thompson Company Archives at Duke University suggests a persistent undercurrent of dissatisfaction on the part of consumers with the quality of automobiles. For a good discussion of the importance of quality, see "New Car Purchase Motivation Study," Ford Division Marketing Research, June 1958, Office Files and Correspondence, Ford, WT, box 6, folder 8. On Chrysler's quality problems, see Hyde, *Riding the Roller Coaster,* 176. Quality was not a criticism that made it into the mainstream news media, however. For examples where it did, see "On the Slow Road," 85; and Joseph C. Ingraham, "Luxury Small Cars Added to 1961 Lines," *New York Times,* 19 June 1960, 70.

42. Halberstam, *The Fifties,* 635–636.

8. IF WE CAN PUT A MAN ON THE MOON . . .

1. "Car Population Now Equal to Family Total," *Automotive News,* 10 Sept. 1956, 2. *Historical Statistics of the United States: Colonial Times to 1970,* pt. 2, Series Q 175–186, "Percent Distribution of Automobile Ownership, and Financing: 1947 to 1970," 717. Gordon M. Buehrig, "Automobiles for 1970," speech to the Annual Technical Convention, American Society of Body Engineers, 27 Oct. 1954, 3, in Kettering Papers (hereinafter KP), Kettering/GMI Alumni Foundation Collection of Industrial History, Kettering University, Flint, MI, 12/63.

2. "Percent Distribution of Automobile Ownership, and Financing," 717.

3. "Preliminary Views on the 1950 Budget for Ford Cars and Ford Trucks," n.d. [1 of 2], Detroit Office, Ford, Client Presentations 1948–1950, Norman Strouse Papers, J. Walter Thompson Company Archives (hereinafter JWT), John W. Hartman Center for Sales, Advertising, and Marketing History, Duke University, Durham, NC, box 1, folder 9. Thompson called it a "hidden market" that had "not as yet reappeared." There had been a discernible although minor trend toward second car ownership in the 1920s that had been arrested and partially reversed by the Depression and

World War II, but it showed no sign of reviving. For anecdotes about 1920s two-car families, see "Two-Car Smith and the Saturation Point," *Automotive Industries,* 9 Oct. 1924, 651; and Fraser Nairn, "How Many Cars Do You Need?" *Country Life,* January 1925, 35–39. For examples about the saturation point and second car prospects generally, see "Two-Car Families, as Mr. Ayres Sees Them," *Automotive Industries,* 15 Jan. 1925, 99; K. W. Stillman, "Replacement Market Alone May Soon Absorb 2,900,00 Cars a Year," *Automotive Industries,* 29 Apr. 1926, 725; John B. Kennedy, "You Can't Afford to Walk," *Collier's,* 16 July 1927, 46; "Two-Car-Family Market Needs Factory Development," *Automotive Industries,* 17 Dec. 1927, 889–890; and "General Motors Attacks the Used Car Problem," *Motor,* June 1928, 46. Data from *Automobile Facts and Figures* (1951), 46, suggests that the figure for two-car ownership may have fallen as low as 2 percent during World War II. See "Percent Distribution of Automobile Ownership, and Financing," 717, where families owning an automobile jumped from 54 percent in 1948 to 70 percent in 1955.

4. On touting two cars, see William J. Griffin, Jr., "Ford Car Advertising: A Time for Communication," 9 Mar. 1956, Office Files and Correspondence, Ford, Correspondence and Memos, 1956, Winfield Taylor Papers, JWT, box 1, folder 8. For examples of Ford's second car advertising, see "Fords by the Pair" and "Twice the fun . . . ," JWT Advertisements, Ford: 1955–1956, Wallace W. Elton Papers, JWT, oversize (20 × 16) box 4; Ford 1957 General, JWT/Domestic Ads, JWT, box 20.

5. "Stranded in Suburbia . . . ," JWT Co., 1956, Domestic Advertisement Collection, JWT Co., 1953–1958, JWT, box 8.

6. On success, see "'55 Hardtops, Wagons Spark 'Revolution' in Body Style," *Automotive News,* 5 Mar. 1956, 4; and Maynard M. Gordon, "Cinderellas of the Auto Industry," *Automotive News,* 3 Dec. 1956, 42. On 1949 sales, see *Automotive Industries,* 15 Mar. 1950, 77. On the market through 1957, see John W. Conley, "Station Wagon Information," memorandum to Paul E. Urschalitz, 26 Dec. 1957, Office Files and Correspondence, Ford, Correspondence and Memoranda, 1957 Sep–Dec, box 1, folder 12; Craig Barker, "1958 Ford Car," memorandum to distribution, 30 Apr. 1958, 2, Office Files and Correspondence, Ford, Model data (1958 Model) 1958, box 4, folder 17; and John W. Conley, memorandum to distribution, 2 Apr. 1958, Office Files and Correspondence, Ford, Correspondence and Memos, 1958 Apr–Jun, box 1, folder 14, all in Winfield Taylor Papers, JWT. On second car station wagon sales, see "Mainline—Customline Ranch Wagon Mix," memorandum, n.d. [but probably 1954], 1, Ford Motor Company Archives, the Benson Ford Research Center, The Henry Ford, acc. 695, box 8, Survey—Station Wagons—1954. On multiple car ownership, see "Percent Distribution of Automobile Ownership, and Financing," 717; and *Pocket Guide to Transportation, 1998* (Washington, DC: Bureau of Transportation Statistics, U.S. Department of Transportation, December 1998), 5.

7. See Bradford C. Snell, "American Ground Transport: A Proposal for Restructuring Automobile, Truck, Bus, and Rail Industries," in Part 4A of Hearings in S. 1167, The Industrial Reorganization Act, before the Subcommittee on Antitrust and Monopoly of the Committee of the Judiciary, U.S. Senate, 93rd Cong., 2nd Sess. (Washington, DC: GPO, 1974). Scott L. Bottles, *Los Angeles and the Automobile: The Making of the Modern City* (Berkeley: University of California Press, 1987). The author paraphrases an advertisement that he saw while reviewing issues of *Horseless Age, Automobile, Motor,* and *Motor Age* published through 1910.

8. Many authors have rightly recognized that the late 1950s opened the door to personal cars and

new forms of market segmentation based on this assumption. See, e.g., Thomas Hine, *Populuxe* (New York: Alfred A. Knopf, 1986), 91–92.

9. This data is from table VM-201A: Annual Vehicle Distance Traveled in Miles and Related Data, 1936–1995 1/ By Vehicle Type, in *Highway Statistics Summary to 1995* (Washington, DC: Federal Highway Administration, 1996), V-15, V-16.

10. The figure for junkyard owners is from Charles H. Lipsett, *Industrial Wastes and Salvage: Conservation and Utilization* (New York: Atlas, 1963), 147, 154, and that for scrap dealers from Albert R. Karr, "Sagging Scrap: New Steel Technology Helps Cut Waste Metal Sales to Half '56 Level," *Wall Street Journal*, 31 Jan. 1962, 1. For an overview of the mid-twentieth-century scrap industry, see *Monopoly and Technological Problems in the Scrap Steel Industry, Report of the Select Committee on Small Business, United States Senate*, 16 Oct. 1959, Report no. 1013 (Washington, DC: GPO, 1959), 1–34 (hereinafter *Select Committee*); "In the Matter of Luria Brothers and Company, Inc. et al.," *Federal Trade Commission Decisions*, vol. 62 (Washington, DC: GPO, 1967), esp. 272–281 (hereinafter Luria-FTC); and Carl A. Zimring, *Cash for Your Trash: Scrap Recycling in America* (New Brunswick, NJ: Rutgers University Press, 2005). On burning the hulks, see Lipsett, *Industrial Wastes and Salvage*, 153, 313–314; and "Ghosts of the Gas Age," *Journal of American Insurance*, July 1964, 24.

11. For the 1956 volume sales figure, see Lipsett, *Industrial Wastes and Salvage*, 147. For the 1956 dollar sales figure, see Lee Berton, "Robot's Diet—Old Autos," *Detroit News*, 28 June 1964, n.p. The open-hearth and basic oxygen furnace charge percentages are from *Select Committee*, 3, 6. For the 1963 price, see "Where Do Old Autos Go to Die?" *Newsweek*, 20 July 1964, 66.

12. On the crisis, see *Select Committee*, 6–8; and Charles E. Dole, "Rising Junk Cars Evoke Crisis," *Christian Science Monitor*, 2 Aug. 1966, 11. The New York figures are from "Where Do Old Autos Go to Die?" 68; Roger B. May, "Scrap-Steel Shredding Units, Law Changes Could Solve Cities' Derelict Cars Problem," *Wall Street Journal*, 9 May 1967, 32; and Monty Hoyt, "Junk-Automobile Blight Spreads throughout U.S. . . . and Startled States Act," *Christian Science Monitor*, 2 Dec. 1970, 9. The Detroit figures are from "A Junker's Graveyard," *Detroit Free Press*, 27 Feb. 1967, n.p. For the 1966 estimate, see Bob Thomas, "Scrap Industry Relegating Old Methods to the Junk Heap," *Los Angeles Times*, 24 July 1966, M1.

13. This figure is from John A. Volpe, secretary of transportation, "Abandoned Autos: A National Dilemma," *Phoenix Quarterly* 1:1 (1969): 3, a publication of the Institute of Scrap Iron and Steel, Inc. in the National Automotive History Collection (hereinafter NAHC), Detroit Public Library, "Scrappage" vertical file.

14. On location, see Lipsett, *Industrial Wastes and Salvage*, 159; and Zimring, *Cash for Your Trash*, 104, 107, 124.

15. For Lady Bird Johnson's involvement and analysis of the Highway Beautification Act in general, see Lewis L. Gould, *Lady Bird Johnson: Our Environmental First Lady* (Lawrence: University of Kansas Press, 1988; reprint ed., 1999), 90–108; and Zimring, *Cash for Your Trash*, 124–128. Federal financing for screening is from Volpe, "Abandoned Autos," 4.

16. On the Prolers, see Lipsett, *Industrial Wastes and Salvage*, 313–314; Robert Lindsey, "Shredders Slowly Reducing Heaps of Junked Cars," *New York Times*, 26 Mar. 1972, 60; and Calvin Lieberman, "Creative Destruction," *Invention and Technology*, Fall 2000, 58. On Newell, see Zimring, *Cash for Your Trash*, 116.

17. See Lieberman, "Creative Destruction," for a good present-day description of the automobile shredding process.

18. On the Proler-Neu venture, see "Where Do Old Autos Go to Die?" *Newsweek*, 20 July 1964, 67; and Thomas, "Scrap Industry Relegating Old Methods to the Junk Heap," M1.

19. See "Machine Gobbles Up 1,400 Autos a Day," *Detroit News*, 25 June 1969, 20-C; and Dan Fisher, "There May Be a Lot of Recycled Detroit in That Japanese Import," *Los Angeles Times*, 23 Apr. 1972, n.p.

20. On the Ford-Luria agreement, see "Ford Foundry to Use 250,000 Junked Cars," *Dearborn Guide*, 16 Dec. 1965, 1; "Using Junked Cars," editorial, *Detroit Free Press*, 30 Dec. 1965, n.p.; James L. Kerwin, "Plant to Hammer Junkers into Usable Steel," *Detroit Free Press*, 13 May 1966, 14-D; Tom Kleene, "Firm Thrives on 'Klunkers,'" *Detroit Free Press*, 8 Aug. 1967, n.p.; "Turning Junk and Trash into a Resource," *Business Week*, 10 Oct. 1970, reprint, 6; and Zimring, *Cash for Your Trash*, 102. General Motors sensed this opportunity as well and bankrolled a pilot project in the Traverse City, Michigan, area in 1970 to experiment with facilitating the movement of abandoned cars into the scrap metal recycling stream. See, e.g., "GM to Begin Project to Retrieve Junked and Abandoned Cars," *Wall Street Journal*, 15 June 1970, 1, NAHC, "Scrappage" vertical file. For the 1973 Ford figure, see Terry Zintl, "Metal Demand Gives Old Cars New Role," *Detroit Free Press*, 2 May 1975, n.p. On ferrous castings and steel, see L. R. Mahoney and J. J. Harwood, "The Automobile as a Renewable Resource," *Resources Policy* 1:5 (1975): 254.

21. "Car Wreckers 'Out Produce' Junked Autos," *Chicago Tribune*, 31 Oct. 1969, A3. On two hundred large operations, see L. R. Mahoney, J. Braslaw, and J. J. Harwood, "Effect of Changing Automobile Materials on the Junk Car of the Future," SAE Technical Paper Series, no. 790299, 1979, 1. The transition to automobile shredders was abetted by the consolidation of the American scrap industry in the 1950s and 1960s. See Luria-FTC and Zimring, *Cash for Your Trash*, 102–130.

22. On Ford's concession, see, e.g., "Ford Motor Company Progress Report Air Pollution Control Rouge Area, 1965," copy on file in the City of Dearborn Archives, Dearborn, MI. For the duck kills, see George S. Hunt and Archibald B. Cowan, "Causes of Deaths of Waterfowl on the Lower Detroit River—Winter 1960," 1; and "Water Pollution Information Relating to Detroit, Michigan," n.d. [but before 11 Dec. 1961], 6, Records Pertaining to Regulation of Interstate Bodies of Water (hereinafter RPRIBW), 1948–1975, Office of Enforcement and General Council, Records of the Environmental Protection Agency (hereinafter EPA), RG 412, National Archives, College Park, MD, box 82, folder "Detroit River Reports #1." For the Water Resources Commission staff view, see R. W. Purdy, "The Water Situation Today—Pollution and Its Control," remarks made before and published in "Water Resources Conference, Lansing Civic Center, December 4, 1962," 9–14, in George W. Romney Papers (hereinafter GWR), Bentley Historical Library, University of Michigan, box 26, folder "Miscellaneous Reports—1963."

23. On Dingell's efforts, see John Dingell to ?Philip A. Hart, notes of telephone conversation, 25 Jan. 1960, and Milton P. Adams, statement before the U.S. Senate Committee on Public Works, 8 May 1961, 6–7, in Philip A. Hart Papers (hereinafter PAH), Bentley Historical Library, box 79, folder "Poll—Water Legislation." For the relevant part of the statute, see *United States Statutes at Large . . . 1961*, vol. 75 (Washington, DC: GPO, 1961), Public Law 87-88-July 20, 1961, Section 7, amending Section 8 of the Federal Water Pollution Control Act, 207–208. On the support of the

senators and Congressman John Dingell, see Milton P. Adams, memorandum to Members of Water Resources Commission and Others, 29 Nov. 1961, 3, RPRIBW, 1948–1975, Office of Enforcement and General Council, EPA, box 82, folder "Detroit River Reports #1."

24. U.S. Department of Health, Education, and Welfare, Public Health Service, "Conference in the matter of Pollution of the navigable waters of the Detroit River and Lake Eerie and their Tributaries in the State of Michigan. Proceedings, Volume 1. Second Session. June 15–18, 1965," 376–377.

25. For Ford's reaction, see reported remarks by environmental manager Frank Kallin from an undated (but certainly June 1965) editorial in an unnamed Detroit newspaper clipping in GWR, box 337, folder "Conservation—Water Resources Standards." See, too, William C. Treon, "Polluters Gripe, UAW Cheers Report," *Cleveland Plain Dealer*, 18 June 1965, 9. For the expected compliance steps, see Michigan Water Resources Commission, Transcripts of Conferences held on Pollution of Detroit River and Michigan Waters of Lake Erie, RPRIBW, 1948–1975, Office of Enforcement and General Counsel, EPA, box 85, folder "Detroit River Reports #4." In June 1965 the Michigan legislature amended the 1949 water pollution control act to remove language that forced the WRC to prove that polluters had "willfully" intended to pollute the waters in question and that the pollution had done actual harm to public health, fish and wildlife, or lawful occupations, tests so difficult to meet that there had not been a single prosecution under the law from 1949 to 1965. See Hugh McDonald, "New Law Puts Teeth in State's War on Pollution," *Grand Rapids Press*, 2 July 1965; and "Pollution Bill Rolls Despite Dem's Attack," *Journal*, 25 June 1965, in GWR, box 430, folder "Conservation—1965." See also Michigan Water Resources, *Quarterly Bulletin*, 30 June 1965, 4, in GWR, box 336, folder "Water Pollution of Lake Erie . . . 6/65." On Ford's reluctance to commit, see "Water Pollution Hearing Ordered for City, Firms," *Detroit News*, 1 Apr. 1966, 8A; James L. Kerwin, "State Sets Sights on Polluters," *Detroit News*, n.p.; and State of Michigan Water Resources Commission, Proceedings against Ford Motor Company . . . for Abatement of Pollution of the Rouge River, Notice of Determination and Hearing, 31 Mar. 1966, RPRIBW, 1948–1975, Office of Enforcement and General Counsel, EPA, box 80, folder "Detroit River Litigation #1." For some sources of public pressure, see "Water Resources Commission, Program Review, Evaluation and Recommendations Submitted in Response to General Departmental Communication No. 12, June 1964," 2–3, in GWR, box 73, "Water Resources Commission—Loring Oeming File 1964"; and Laurence M. Braun, letter to Murray Stein, 7 Feb. 1965, RPRIBW, 1948–1975, Office of Enforcement and General Counsel, EPA, box 80, folder "Detroit River Correspondence #3." On the UAW interest, see "Statement of Olga Madar," in Public Health Service, "Conference in the Matter of Pollution of the Navigable Waters of the Detroit River and Lake Eerie and their Tributaries in the State of Michigan. Proceedings, Volume 1. Second Session. June 15–18, 1965," U.S. Department of Health, Education, and Welfare, Public Health Service, 1493–1505; Walter P. Reuther, letter to George W. Romney, 11 Mar. 1966; Albert H. Wilson, Recording Secretary, Ford Local 600, UAW, letter to George W. Romney, 22 Apr. 1966; George W. Romney, letter to Albert H. Wilson, 9 May 1966, all in GWR, box 139, folders "Water Resources—M-R 1966" and "Water Resources—S-Z 1966." For Romney's views, "Water Pollution Abatement in Michigan: The Romney Record, 1963–66," 3, in GWR, box 336, folder "Water—Romney Record." On Ford's capitulation, see Minutes, Water Resources Commission, 28–29 Apr. 1966, 8, Department of Natural Resources, RG 74-13, lot 45, series 1, boxes

1–3 (hereinafter DNR), Michigan State Archives, Lansing; and "Four Accept Water Rule," *Grand Rapids Press*, 28 Apr. 1966, 44, in EPA.

26. On pickling acid, see Minutes, Water Resources Commission, 15–16 May 1969, 16, DNR. On glass and coke oven wastes, see various status reports in RPRIBW, 1948–1975, Office of Enforcement and General Counsel, EPA, box 79, folder "Detroit River Compliance #1"; and Minutes, Water Resources Commission, 15–16 Jan. 1969, 11–12, DNR. On contradiction, see James L. Kerwin, "State Unit Put on Spot in Pollution War," *Detroit News*, 28 Apr. 1967, 7-B, in GWR, box 436, folder "Pollution." On the decrease by 1970, see "Activity Report, Michigan Water Resources Commission, November 1970," 1, in William Milliken Papers (hereinafter WM), Bentley Historical Library, box 758, folder "Env—WRC Activities, 1971."

27. For the announcement, see "Ford to Spend $30 Million to Expand Rouge Foundry," *Dearborn Guide*, 1 Apr. 1965, 3. Background on Dearborn's South End is from Paul J. Hardwick, interview with the author.

28. On the smell, Hardwick interview; and Robert Selwa, "Foul Air in Area Turns White Snow into Dirty Grey," *Dearborn Guide*, 12 Jan. 1967, 1. On homes, see Dan Hall, "Ford Promises Cleaner Air," *Dearborn Press*, 22 Dec. 1966, 1. For cars, school, and steel particles, see Robert Selwa, "South End School Uses Filter System to Stop Black Soot," *Dearborn Guide*, 19 Jan. 1967, 1–2. Hardwick interview.

29. On Hart, see "Ford Air Pollution Hit," *Dearborn Guide*, 18 Nov. 1965, 1, 2; "Ford Air Pollution Rapped," *Dearborn Press*, 9 Dec. 1965, 1; and "Council, Ford Battle over Air Pollution," *Dearborn Guide*, 9 Dec. 1965, 1–2. "Council, Ford Battle over Air Pollution," 1–2. Lombardi quoted in "Ford Air Pollution Rapped," 1. For Parkhurst's dig, see "Council, Ford Battle over Air Pollution," 2.

30. For Bruno's foundry employment, see Shelley Eichenhorn, "A Return to the 'Fires of Hell,'" *Detroit News*, 19 Oct. 1977, 6-A. For Hubbard and the dinner, see Robert Selwa, "Air Pollution in South End Seen As 'Hopeless Situation,'" *Dearborn Guide*, 18 May 1967, 1–2. Hardwick interview.

31. "Ban 'Pollution' Float from Memorial Parade," *Dearborn Guide*, 1 June 1967, 1–2; Hardwick interview.

32. The figure is from "Ford to Spend $3.5 Million to Reduce Rouge Pollution," *Dearborn Guide*, 29 June 1967, 1. Accounts of the meeting, including the quotations that follow below, are from "Ford Says It's 'Not Blowing Smoke' in Air Cleanup," *Dearborn Press*, 10 Aug. 1967, 1; "Southeast Area Residents Discuss Air Pollution with Ford Officials," *Dearborn Guide*, 10 Aug. 1967, 1; and Hardwick interview.

33. "Ford to Spend $80 Million on Pollution," *Dearborn Press*, 9 May 1968, 1. On the electrostatic precipitators, see "Air Pollution Control in the Rouge Area. Progress Report for 1968," 4 Mar. 1969, 1, City of Dearborn Archives. On closing the foundry, see "Ford Announces Area Expansion Plans," *Dearborn Times-Herald*, 5 Mar. 1969, 1.

34. See J. Cusumano, "City, FoMoCo Sued for Air Pollution," *Dearborn Press*, 23 Mar. 1972, n.p., which indicates that countywide jurisdiction began in late 1968. On progress, see Mark Lett and Charlie Cain, "Strike Won't Lessen Rouge Pollution Much," *Detroit News*, 26, 27 Sept. 1976, 13-A, 1-B. Hardwick interview.

35. Los Angeles County did not install a countywide system of smog sensors until 1954. The worst smog day in history probably occurred on 24 September 1953, when an ozone reading of 1.0 ppm

was recorded. On 13 September 1955 rates of 0.90 ppm and 0.85 ppm were reached in Vernon and Los Angeles, respectively. See "Less Driving, Burning Urged in Smog Fight," *Los Angeles Times*, 13 May 1955, 1; Ted Sell, "Smog Declared Worst since '55," *Los Angeles Times*, 27 Nov. 1958, 2; and Lee Austin, "All-Out War on Smog Proposed," *Los Angeles Times*, 22 Oct. 1967, SG-B1. "Hahn Asks Auto Firms for Pollution Control," *Los Angeles Times*, 19 Dec. 1958, 16.

36. On the pressure, see "City and County Join Forces to Seek Car Smog Controls," *Los Angeles Times*, 6 Jan. 1959, C1; and "Supervisors Demand Auto Exhaust Law," *Los Angeles Times*, 24 Oct. 1959, 1. Al Thrasher, "Brown Names State Smog Control Board," *Los Angeles Times*, 27 May 1960, B1. On the potential market, see Alexander B. Hammer, "California Smog Spurs a Big Rush," *New York Times*, 14 Aug. 1960, F1. For the thirty applications, see Robert L. Bartley, "Research Partners," *Wall Street Journal*, 18 June 1964, 1.

37. On the Houdry-GM joint program, see "Chronological List of General Motors Contributions to Air Pollution Studies," 3, in Office Files—Smog, in KP; Eugene J. Houdry, "Facts and Figures Pertaining to Air Pollution," n.d. [but 1958 or later], in Eugene J. Houdry Papers, Manuscript Division, Library of Congress (hereinafter Houdry—LOC); and M. F. Homfeld et al., "The General Motors Catalytic Converter," presentation at the March 1962 SAE meeting, Detroit, MI. The Ethyl request is covered in Warren, *Brush with Death*, 205. "Facts You Should Know about Oxy-Catalyst Corp.," ad, *New York Times*, 1 Oct. 1961, F7. On March 1962 report, see Homfeld et al., "The General Motors Catalytic Converter," 11; "Chronological List of General Motors Contributions to Air Pollution Studies," 5; and Eugene J. Houdry and Gordon P. Larson, "An Aspect of Catalysis in Modern Life," 10, paper delivered before the American Chemical Society on 26 Mar. 1962, subsequently published as a report by the society, in Houdry—LOC. On Houdry's patent, see Stacy V. Jones, "Anti-Smog Device Uses Catalyst," *New York Times*, 30 June 1962, 23.

38. "Negotiations On to Acquire Ethyl," *New York Times*, 18 Sept. 1962, 53; and "Albemarle Takes Ethyl Corp. Title," *New York Times*, 1 Dec. 1962, 44. On the constraint, Richard L. Klimisch interview with the author; and Kathleen C. Taylor, personal communication.

39. Klimisch interview.

40. For certification, see "Air Standard Elusive for Car Smog Devices," *Los Angeles Times*, 11 May 1964, A1; Bartley, "Research Partners," 1; Duane Allison, "U.S. Urbanites Sputter over Auto Fumes, *Christian Science Monitor*, 2 Dec. 1964, 17; and Ralph Nader, *Unsafe at Any Speed: The Designed-In Dangers of the American Automobile* (New York: Grossman, 1965), 160. On the bluff, see "Smog Control Antitrust Case," *Congressional Record*, vol. 117, pt. 12, 18 May 1971, 15635. On the shortfall, see J. Robert Mondt, *Cleaner Cars: The History and Technology of Emission Control since the 1960s* (Warrendale, PA: SAE International, 2000), 87; and George R. Lester, "The Development of Automotive Exhaust Catalysts," in Burtron H. Davis and William P. Hettinger, Jr., eds., *Heterogeneous Catalysis: Selected American Histories* (Washington, DC: American Chemical Society, 1983), 420.

41. For the automakers' response, see "Smog Control Antitrust Case," 15634, 15637; and "Chronological List of General Motors Contributions to Air Pollution Studies," 7. For a description of the approach, see Fred W. Bowditch, "Statement of the Automobile Manufacturers Association to Scientific Committee of the Los Angeles County Air Pollution Control District," 19 Jan. 1967, in Kenneth Hahn Papers (hereinafter KH), Huntington Library, San Marino, CA, box 3, Air Quality, folder 3.6.1.13. See also Mondt, *Cleaner Cars*, 87–89. On the failure, see Tom Goff, "Hahn De-

mands Recall of '66 Cars," *Los Angeles Times,* 9 Feb. 1967, B1; Bob Thomas, "Imports Offer Some
New Ways to Combat Smog," *Los Angeles Times,* 16 July 1967, J10; and Spencer Rich, "Detroit:
Cleaning Up," *Washington Post,* 29 Aug. 1971, B1.

42. "Car Makers Hit for Delay on Smog Controls," *Los Angeles Times,* 19 Aug. 1964, 8A. George
Getze, "Griswold Calls for U.S. Smog Action," *Los Angeles Times,* 3.

43. The origins of the smog antitrust case are covered in detail by Morton Mintz, "Smog Fighter
Inspired Auto Industry Lawsuit," *Washington Post,* 26 Jan. 1969, 35. Secondary accounts of the
case include Scott H. Dewey, "'The Antitrust Case of the Century': Kenneth F. Hahn and the
Fight against Smog," *Southern California Quarterly* 81:3 (1999): 341–376; John C. Esposito and
Larry J. Silverman, *Vanishing Air: The Ralph Nader Study Group Report on Air Pollution* (New
York: Grossman, 1970), 40–47; James E. Krier and Edmund Ursin, *Pollution and Policy: A Case
Essay on California and Federal Experience with Motor Vehicle Air Pollution, 1940–1975* (Berke-
ley: University of California Press, 1977), 135–176; and Jack Doyle, *Taken for a Ride: Detroit's Big
Three and the Politics of Pollution* (New York: Four Walls Eight Windows, 2000), 29–49. The only
available primary source on the actual conspiracy is "Smog Control Antitrust Case," 15626–15637.
For Nader's view, see Nader, *Unsafe at Any Speed,* 169. For background on Nader's perspective,
see David P. Riley, "Influence, Picket Signs and the Public Good," *Potomac* magazine, *Washington Post,* 15 Mar. 1970, 14–16, 18–20, 22, 27, 29, 37.

44. On subpoenas, see "Delay in Smog Devices to be Studied by U.S.," *Los Angeles Times,* 20 Jan.
1965, A8; Morton Mintz, "U.S. Probing Auto Firms in Air Pollution," *Washington Post,* 28 Jan.
1965, 1, 21; and Ronald J. Ostrow, "Justice Aide Hits Plea for Antitrust Immunity," *Los Angeles
Times,* 7 Apr. 1966, 22. On investigation, see David Wise, "Auto Industry's Headaches Multiply—
Mr. Hahn Keeps on Writing Letters," *New York Herald Tribune,* 10 Apr. 1966, n.p., clipping in
KH, box 3, Air Quality, folder 3.6.1.12. Wise cites unnamed sources in the Justice Department to
the effect that the Antitrust Division began its investigation based on information that came from
"another agency of the Federal government." Paul Beck, "Auto Smog Probe," *Los Angeles Times,*
22 June 1966, 1; and "Four Major Auto Makers Get Subpoenas in Smog-Device Antitrust Investigation," *Wall Street Journal,* 22 June 1966, 3. For the imminent criminal indictment, see Drew
Pearson and Jack Anderson, "Smog Control Drags," *Washington Post,* 12 Apr. 1968, B15; Mintz,
"Smog Fighter Inspired Auto Industry Lawsuit," 35; "County Wants to 'Ungag' Foreman of Smog
Jury," *Los Angeles Herald-Examiner,* 24 Sept. 1969, n.p., clipping in KH, box 4, Air Quality, folder
3.6.1.19; Morton Mintz, "Justice Dept. Had '68 Memo on Auto Plot," *Washington Post,* 19 May
1971, A2; and "Smog Control Antitrust Case," 15632–15633. For Turner's decision, see Donald F.
Turner, Assistant Attorney General, Antitrust Division, letter to Board of Supervisors, 12 Apr. 1968,
in KH, box 3, Air Quality, folder 3.6.1.15.

45. On Flatow's case, see "Smog Control Antitrust Case," 15631–15633; and summarized by Mintz,
"Justice Dept. Had '68 Memo on Auto Plot," A2. For Flatow's counsel, see Riley, "Influence,
Picket Signs and the Public Good,"18–19.

46. For Turner's rationale, see Mintz, "Smog Fighter Inspired Auto Industry Lawsuit," 35; Riley, "Influence, Picket Signs and the Public Good,"18–19; and Mintz, "Justice Dept. Had '68 Memo on
Auto Plot," A2. For Flatow's dejection, see Riley, "Influence, Picket Signs and the Public Good,"
19. "Antitrust Effort on G.M. Reported," *New York Times,* 1 Mar. 1974, 41.

47. *United States v. Automobile Manufacturers Association, Inc. et al.,* Civil no. 69-75-JWC, filed in

U.S. District Court for the Central District of California. On California efforts, see "U.S. Accused of Blocking Smog Fight," *Los Angeles Times*, 5 Mar. 1969, A20; and Richard W. McLaren, Assistant Attorney General, Antitrust Division, letter to Kenneth Hahn, n.d. [but between 7 and 17 Mar. 1969]; and John D. Maharg, County Counsel, letter to Board of Supervisors, 8 Aug. 1969, both in KH, box 4, Air Quality, folders 3.6.1.18, 3.6.1.16, respectively. On Cutler, see Drew Pearson and Jack Anderson, "Past Poisoning," *Washington Post*, 6 Aug. 1969, B15; and Riley, "Influence, Picket Signs and the Public Good," 14–16, 18–20, 22, 27, 29, 37. Fred P. Graham, "U.S. Settles Suit on Smog Devices," *New York Times*, 12 Sept. 1969, 1.

48. "City Loses Appeal on Auto Pollution Antitrust Case," *New York Times*, 17 Mar. 1970, 28. For the number of state suits, see "Eight More States Sue Car Manufacturers in Air Pollution Case," *New York Times*, 16 Mar. 1971, 74. On the decision, see "Court Bars Trust Laws in Pollution Fight," *Washington Post*, 27 Nov. 1973, A2; "Auto Makers Win Ruling on Pollution," *New York Times*, 27 Nov. 1973, 22; and "Auto Antitrust Suits over Pollution Woes Dismissed by Judge," *Wall Street Journal*, 27 Nov. 1973, 2. This decision was upheld on appeal.

49. Caplan is quoted in "Smog Control Antitrust Case," 15637. Flatow concluded his memo to Turner with Caplan's statement. For the relevant chapter, see Nader, *Unsafe at Any Speed*, chap. 4, 147–169.

50. For taps, see Steven V. Roberts, "Survival Theme of Fair on Coast," *New York Times*, 21 Feb. 1970, 48.

51. This section on Muskie's anti-pollution efforts and the Clean Air Act amendments of 1970 is based on author interviews with staffers Leon G. Billings and Tom Jorling. See also Theo Lippman, Jr., and Donald C. Hansen, *Muskie* (New York: W. W. Norton, 1971), 142–170. See also Doyle, *Taken for a Ride*, 58–75.

52. "Air Quality Act of 1967," Public Law 90-148, *United States Statutes at Large*, vol. 81 (Washington, DC: GPO, 1968), 485–507. William M. Blair, "House Approves a Clean Air Bill," *New York Times*, 3 Nov. 1967, 1. Billings interview.

53. On Nixon and the environment, see John C. Whitaker, *Striking a Balance: Environment and Natural Resources Policy in the Nixon-Ford Years* (Washington, DC: American Enterprise Institute, 1976), 93–105; J. Brooks Flippen, *Nixon and the Environment* (Albuquerque: University of New Mexico Press, 2000); and Russell E. Train, *Politics, Pandas and Pollution: An Environmental Memoir* (Washington, DC: Island Press, 2003). On Nixon's maneuvering with the Democrats, see John Osborne, "Games with Muskie," *New Republic*, 20 Feb. 1971, 15–16; and Flippen, *Nixon and the Environment*, 106–110, 115–116. On little enthusiasm, see Russell E. Train, "The Environmental Record of the Nixon Administration," in the Russell E. Train Papers, Library of Congress (hereinafter RET), box 3, folder "RET Article on 'The . . . .'" On denying the Democrats an issue, see Whitaker, *Striking a Balance*, 95. The quotation is Russell E. Train, interview with the author.

54. Richard Nixon, "Message on Environment," 10 Feb. 1970, House of Representatives, 91st Cong., 2nd Sess., Document no. 91–225, in RET, box 43. See also Flippen, *Nixon and the Environment*, 62–64.

55. On Muskie and the Nader report, Billings interview; and David Vogel, *Fluctuating Fortunes: The Political Power of Business in America* (New York: Basic, 1989), 73–74. On Big Three lobbying, see ibid., 74–75. Russell E. Train, memorandum for the President, 1 Oct. 1970, in RET, box 20, folder "Memoranda to the White House, 1970." Billings interview for Muskie's exclusion.

56. On earlier shortcomings, see Thomas Jorling, "The Federal Law of Air Pollution Control," in Erica L. Dolgin and Thomas G. P. Gilbert, eds., *Federal Environmental Law* (Saint Paul, MN: West, 1974), 1061. On the Muskie, Baker, and Eagleton contributions, Billings interview.

57. Billings interview. Jorling, "The Federal Law of Air Pollution Control," 1144. Vogel, *Fluctuating Fortunes*, 69, has called the 1970 Clean Air Act amendments "one of the strictest, most controversial, and bitterly fought pieces of regulatory legislation enacted by the federal government."

58. "Study Finds Editorials Stress the Environment," *New York Times*, 17 Jan. 1972, clipping in RET, box 43. Public Law 92-500, *United States Statutes at Large*, vol. 86 (Washington, DC: GPO, 1973), 816–904.

59. Russell E. Train, "Year-End Report to the Members of the EPA," 31 Dec. 1974, 4, in RET, box 21.

9. THE ONE WHO GOT IT

1. Tom Jorling, interview with the author. The January 1965 date is suggested in "Smog Control Antitrust Case," *Congressional Record*, vol. 117, pt. 12, 18 May 1971, 15627. For the statute, see Air Quality Act of 1967, Public Law 90-148, *United States Statutes at Large*, vol. 81 (Washington, DC: GPO, 1968), 485–507. See Ralph Nader, *Unsafe at Any Speed* (New York: Grossman, 1965), 168, on the 1968 models. On the fuel economy penalty, see George R. Lester, "The Development of Automotive Exhaust Catalysts," in Burtron H. Davis and William P. Hettinger, Jr., eds., *Heterogeneous Catalysis: Selected American Histories* (Washington, DC: American Chemical Society, 1983), 420; and J. Robert Mondt, *Cleaner Cars: The History and Technology of Emission Control since the 1960s* (Warrendale, PA: SAE, 2000), 115–116. For 1967 spending, see "The Economic Impact of the Federal Environmental Program," EPA, Nov. 1974, VI-2, in Russell E. Train Papers (hereinafter RET), Manuscript Division, Library of Congress, box 43, folder "Domestic Council." On intensification, see David E. Cole, interview with the author, and Jorling interview. See also Jerry M. Flint, "Pollutant Fight in Autos Hailed," *New York Times*, 12 Jan. 1971, 17. On 1967 spending, see Thomas Jorling, "The Federal Law of Air Pollution Control," in Erica L. Dolgin and Thomas G. P. Gilbert, eds., *Federal Environmental Law* (Saint Paul, MN: West, 1974), 1145.

2. Ford's somewhat earlier embrace of the catalytic converter is from author interviews with James C. Gagliardi and Richard L. Klimisch. For Houdry's impact on Keith, see Tim Palucka, "Doing the Impossible," *Invention and Technology* (Winter 2004): 24. On the Engelhard-Ford relationship, see "John Mooney, Co-inventor of the Catalytic Converter, to Receive Distinguished Alumni Medal from New Jersey Institute of Technology," press release, New Jersey Institute of Technology, 29 Sept. 2005, accessed at www.njit.edu.

3. On Kummer, Kathleen C. Taylor, interview with the author. On Weaver and Lassen, see James H. Jones et al., "Selective Catalytic Reaction of Hydrogen with Nitric Oxide in the Presence of Oxygen," *Environmental Science and Technology* 5:9 (1971): 791; J. T. Kummer, "Catalysts for Automobile Emission Control," *Progress in Energy and Combustion Science* 6 (1980): 194; and Gagliardi interview.

4. On the race, Gagliardi interview; Gladwin Hill, "First 'Clean' Cars Cross West Coast Finish Line," *New York Times*, 31 Aug. 1970, 43; "Ford Capri Wins 'Clean Car' Race," *New York Times*, 3 Sept. 1970, 66; and the Mooney NJIT press release. Arnold O. Beckman, transcript of an interview conducted by Jeffrey L. Sturchio and Arnold Thackray, 23 July 1985, 32, Oral History Program, Chemical Heritage Foundation, Philadelphia. Gagliardi interview.

5. See Bruce E. Seely, "Edward Nicholas Cole," in *American National Biography*, vol. 5, 207. The original quotation appeared in "People of the Week," *U.S. News and World Report*, 13 Nov. 1967, 19. On Cole's early knowledge, see Minutes of the General Technical Committee Meeting, 28 Dec. 1954, Office Files—Smog, Kettering Papers, Kettering/GMI Alumni Foundation Collection of Industrial History, Kettering University, Flint, MI, 87-11.2-307.

6. Klimisch interview. On the world's biggest, see "Smog Control Antitrust Case," 15630; and Cole interview. Cole worked in the Fuels and Lubricants Department summers while in graduate school in the early 1960s. Half the research budget is from Joseph M. Colucci, interview with the author. On the tests, see Lester, "Development of Automotive Exhaust Catalysts," 421.

7. On confidence, Klimisch interview; and Lester, "Development of Automotive Exhaust Catalysts," 422. On Cole's motivation, Klimisch interview. See also Jerry M. Flint, "G.M. Shows Off a Variety of Experimental Cars," *New York Times*, 8 May 1969. On the task force, see Mondt, *Cleaner Cars*, 88–89, 231–233.

8. For one company, William D. Smith, "Gulf Oil Planning to Pare Outlays," *New York Times*, 11 Feb. 1970, 86. On SAE papers, see Charles B. Camp, "Smog Control Feud," *Wall Street Journal*, 30 Jan. 1969, 1. For Cole and task force, see Mondt, *Cleaner Cars*, 89, 92–93.

9. The Cole-Hartley collaboration is from the Colucci interview; and Joseph M. Colucci, "Fuel Quality—An Essential Element in Vehicle Emission Control," paper published in *Proceedings of ICEF04: ASME Internal Combustion Engine Division, 2004 Fall Technical Conference, Long Beach, CA, USA, October 24–27, 2004*. See Warren Weaver, Jr., "Senate Hearing Held," *New York Times*, 6 Feb. 1969, 19, for the quotation; and "Head of Union Oil Denies Remarks on Loss of Birds in Slick," *New York Times*, 19 Feb., 32, for the *Times*'s retraction and apology.

10. Jerry M. Flint, "G.M. Sees Autos Fume-Free by '80," *New York Times*, 15 Jan. 1970, 1. Cole's concern about the risk of doing both at the same time is from the Colucci interview. Eric O. Stork, interview with the author. On prelude, see [no title], *New York Times*, 12 Feb. 1970, 41.

11. On conversion cost, see John J. Abele, "Auto-Pollution Fight Takes Familiar, Costly Turn," *New York Times*, 15 Mar. 1970, 148. Cole's lobbying is from Colucci interview. On reduced compression ratios, see Jerry M. Flint, "G.M. Redesigning Auto Engines for Operation on Unleaded Fuel," *New York Times*, 14 Feb. 1970, 1; and Jerry M. Flint, "G.M. Shuns Performance for Cleaner Atmosphere," *New York Times*, 22 Feb. 1970, 90. For loss in fuel economy, see "The Economic Impact of the Federal Environmental Program," EPA, November 1974, I-16, in RET, box 43, folder "Domestic Council" (3 percent); and Russell E. Train, "Decision of the Administrator: In re: Applications for Suspension of 1977 Motor Vehicle Exhaust Emission Standards," 5 Mar. 1975, 43, in RET, box 26 (5 percent). On cooperation, see "Company Plans," *New York Times*, 17 Sept. 1970, 36.

12. For the first EPA regulation, see "Gas Stations Told to Offer a Lead-Free Fuel by 1974," *New York Times*, 28 Dec. 1972, 62; and *Federal Register* 38 (10 Jan. 1973), 1254; and Jorling, "Federal Law of Air Pollution Control," 1076, 1123–1124. Final regulations followed in December 1973. On the second regulation, see "EPA Requires Phase-Out of Lead in All Grades of Gasoline," EPA press release, 28 Nov. 1973, accessed at the EPA's website, www.epa.gov/history/topics/lead/03.htm.

13. For the larger public health fight over leaded gasoline, see Samuel P. Hays, "The Role of Values in Science and Policy: The Case of Lead," in Herbert L. Needleman, ed., *Human Lead Exposure* (Boca Raton, FL: CRC, 1992), 267–283; and Christian Warren, *Brush with Death: A Social History of Lead Poisoning* (Baltimore: Johns Hopkins University Press, 2000), 203–223. Clair C.

Patterson, "Contaminated and Natural Lead Environments of Man," *Archives of Environmental Health* 11 (September 1965): 344–360. For comments by Patterson, see Walter Sullivan, "Lead Pollution of Air 'Alarming,'" *New York Times*, 8 Sept. 1965, 49. The 90 percent figure is from John Quarles, *Cleaning Up America: An Insider's View of the Environmental Protection Agency* (Boston: Houghton Mifflin, 1976), 121.

14. On the health impact of leaded gasoline, see V. M. Thomas, "The Elimination of Lead in Gasoline," *Annual Review of Energy and the Environment* 20 (1995): 301–324; and Philip Shabecoff, "E.P.A. Orders 90 Percent Cut in Lead Content of Gasoline by 1986," *New York Times*, 5 Mar. 1985, A20. On the EPA's leaded gasoline fight, see Quarles, *Cleaning Up America*, 119–141; and Samuel B. Hays, "The Role of Values in Science and Policy: The Case of Lead," in Needleman, ed., *Human Lead Exposure*, 267–286. "Nationwide detoxification" is from Warren, *Brush with Death*, 222.

15. On lobbying Muskie, Leon Billings, interview with the author. For Cole's concession, Jorling interview. Cole's promise is from Jerry M. Flint, "G.M. Cars to Get Pollution Device," *New York Times*, 18 Sept. 1970, 25.

16. Jorling interview.

17. "Clean Air Act Amendments of 1970," Public Law 91-604, *United States Statutes at Large, 1970–1971*, vol. 84, pt. 2 (Washington, DC: GPO, 1971), 1676–1713. On the origin of the reduction percentages, see Delbert S. Barth, "Federal Motor Vehicle Emission Goals for CO, HC and NOx Based on Desired Air Quality Levels," *APCA Journal* 20:8 (1970): 518–523. D. S. Barth, "Light Duty Motor Vehicle Exhaust Emission Standards," memorandum to Eric O. Stork, 19 Oct. 1971, copy in possession of author courtesy of Stork. Advance word is from the Stork interview. Klimisch interview.

18. Jorling interview.

19. Colucci and Gagliardi interviews.

20. William D. Ruckelshaus, interview with the author.

21. On Austin and Hellman and help from catalyst makers, see Stork interview. The Newark police officers story is from the Mooney NJIT press release. On compelling testimony, Ruckelshaus interview; and Dan Fisher, "Chrysler Charges EPA Chief with Meddling," *Los Angeles Times*, 14 Apr. 1973, III:9. For tipped off, Stork interview. Ruckelshaus interview.

22. On Honda, interview with William D. Ruckelshaus by Michael Gorn, published as *EPA History Program, Oral History Interview—1: U.S. Environmental Protection Agency Administrator William D. Ruckelshaus*, EPA, January 1993, 17. Time is from Ruckelshaus and Stork interviews with the author. On Hawkins's persuasiveness, Ruckelshaus interview. NRDC background is from David Hawkins, interview with the author. For decision, see E. W. Kenworthy, "U.S. Agency Bars Delay on Cutting Auto Pollutants," *New York Times*, 13 May 1972, 1. Ruckelshaus interview.

23. For Cole's view, see Robert L. Sansom, *New American Dream Machine: Toward a Simpler Lifestyle in an Environmental Age* (Garden City, NY: Anchor, 1976), 75–76. On worries, Cole and Klimisch interviews. On base metal catalysts, Klimisch and Taylor interviews; Mondt, *Cleaner Cars*, 96–97; Kummer, "Catalysts for Automobile Emission Control," 182; and Lester, "Development of Automotive Exhaust Catalysts," 424. For platinum supply concerns, Taylor interview. Projected platinum needs are from Michael L. Church et al., "Catalyst Formations 1960 to Present," *SAE 1989 Transactions: Journal of Fuels and Lubricants, Section 4—Volume 98*, 459. The details on finding the precious metals are from Mondt, *Cleaner Cars*, 105. For the financial arrangements

with Impala, Klimisch interview; and "Metal Demand Aids South Africans," *New York Times*, 23 Sept. 1972, 37.

24. On Ford-Engelhard, see Kenworthy, "U.S. Agency Bars Delay on Cutting Auto Pollutants," 1; and "Ford in Pact to Buy Engelhard Devices," *New York Times*, 11 July 1972, 51. On Ford-Rustenburg, see "Auto-Pollution Curbs May Spur Use of South Africa's Platinum," *New York Times*, 17 July 1972, 43; and Gagliardi interview. "Lousy bet" is from John F. Lawrence, "Cost vs. Benefit Debate Clouds National Fight for Cleaner Air," *Los Angeles Times*, 13 Mar. 1973, I:6.

25. On Cole rebuffing the White House but failing to dissuade Gerstenberg, see Sansom, *New American Dream Machine*, 39–40.

26. On the court order, see Jorling, "Federal Law of Air Pollution Control," 1117. For the Starkman-GM position, see E. W. Kenworthy, "G.M. Asks Delay on '75 Standards," *New York Times*, 13 Mar. 1973, 1. The Big Three's arguments are recapitulated in "Motor Vehicle Pollution Control Suspension Granted: Decision of the Administrator," *Federal Register*, 26 Apr. 1973, 10317–10330. Hawkins and Klimisch interviews. For the UAW, see "Woodcock Would Delay Auto Smog Curbs," *Los Angeles Times*, 19 Mar. 1973, I:2.

27. On Ruckelshaus's views, "Motor Vehicle Pollution Control Suspension Granted: Decision of the Administrator," 10317–10330; and Ruckelshaus interview. For Chrysler, Dan Fisher, "Stiff '75 Car Smog Rule Set for State," *Los Angeles Times*, 12 Apr. 1973, I:1. On replacement, see Jorling, "Federal Law of Air Pollution Control," 1121.

28. "Car Firms End Fight on '75 Smog Curbs," *Los Angeles Times*, 11 May 1973, I:1. Stork and Klimisch interviews.

29. On the development of GM's first production catalytic converter, Taylor and Klimisch interviews and Mondt, *Cleaner Cars*, 100–105. The weight is from Kummer, "Catalysts for Automobile Emission Control," 181. For background, see Church et al., "Catalyst Formations 1960 to Present," 457. For $300 million, see Lester, "Development of Automotive Exhaust Catalysts," 428.

30. Weight from Kummer, "Catalysts for Automobile Emission Control," 181. See Palucka for detail on Corning's contributions. On Chrysler, see Fisher, "Chrysler Charges EPA Chief with Meddling," III:9; Dan Fisher, "Chrysler Making Last-Ditch Bid to Block Catalyst Plan," *Los Angeles Times*, 16 Aug. 1973, III:10; and "Motor Vehicle Pollution Control Suspension Granted," 10327–10330. Ratio from Lester, "Development of Automotive Exhaust Catalysts," 425.

31. "Cars, Service, Gasoline—All to Cost More," *Los Angeles Times*, 12 Apr. 1973, I:22; "Car Firms End Fight on '75 Smog Curbs," I:1; and "Use of Catalytic Converters for Cars Hit," *Los Angeles Times*, 5 July 1973, I:2. William Kease, letter to the editor, *Los Angeles Times*, 17 Apr. 1973, II:6.

32. On this controversy, see Eric O. Stork, "Background Information on Sulfuric Acid Emissions from Cars . . . : memorandum to the Administrator," 2 June 1975; and E. O. Stork, "Transcript of Bergman Interview on Catalyst/Sulfuric Acid Issues," 4 June 1975, both in RET, box 49, folder "ABC-TV Interview by Jules Bergman"; as well as Eric O. Stork, "The United States Experience with Imposing Automobile Emission Standards," paper delivered to the Australian Society of Automotive Engineers, Perth, 22 Sept. 1976, copy provided to the author by Stork; also Train and Stork interviews. Stanley M. Greenfield, "Assuring That Mobile Source Controls Do Not Adversely Effect Public Health: The Oxidation Catalytic Converter Issue," memorandum, 9 Oct. 1973, in Train Papers, box 48, folder "Sen. Pub. Works Comm. on Clean Air Act . . . November 6, 1973."

33. "Honorable Russell E. Train, Administrator, Environmental Protection Agency, before the Committee on Public Works, United States Senate, November 6, 1973," typewritten transcript in RET,

box 48. For more serious problem, see Stork, "United States Experience with Imposing Automobile Emission Standards," 12–13. Klimisch interview. On the GM Proving Ground tests, Klimisch and Stork interviews; and Stork, "United States Experience with Imposing Automobile Emission Standards,"14–16. Russell E. Train, interview with the author.

34. For percentage, see Russell E. Train, "Decision of the Administrator: In re: Applications for Suspension of 1977 Motor Vehicle Exhaust Emission Standards," 5 Mar. 1975, 3, in RET, box 26. For cost, see "The Economic Impact of the Federal Environmental Program," EPA, November 1974, VI-4, in RET, box 43, folder "Domestic Council." At fifty thousand miles, see Train, "Decision of the Administrator," 7. Fuel economy figure is from Train, "Opening Statement on 1977 Suspension Decision." Big Three spending is from "The Economic Impact of the Federal Environmental Program," EPA, November 1974, VI-1, in RET, box 43, folder "Domestic Council." For performance of first catalytic converters, see Russell E. Train, "Statement . . . before the Subcommittee on Environmental Pollution, Committee on Public Works, United States Senate, May 21, 1975," 3, in RET, box 49. Leonard is quoted in Jack Doyle, *Taken for a Ride: Detroit's Big Three and the Politics of Pollution* (New York: Four Walls Eight Windows, 2000), 81. See Keith Bradsher, "Plan Allows Big Vehicles to Skirt Rules on Pollutants," *New York Times*, 19 Dec. 1997, A26, for Leonard's responsibilities at GM.

35. On nitrogen oxide problems, see Mondt, *Cleaner Cars*, 69–74; and Taylor interview. On the promising approaches, see "The Economic Impact of the Federal Environmental Program," EPA, November 1974, VIII-18–21, in RET, box 43, folder "Domestic Council."

36. For Engelhard-Volvo, see Mooney NJIT press release. On rhodium, see Kummer, "Catalysts for Automobile Emission Control," 194. Klimisch interview.

37. On the three-way catalyst, see Kummer, "Catalysts for Automobile Emission Control," 192–193; and Church et al., "Catalyst Formations 1960 to Present," 460–461. For the oxygen sensor, see Grunde T. Engh and Stephan Wallman, "Development of the Volvo Lambda-Sond System," in *SAE Transactions*, sec. 2, vol. 86 (Warrendale, PA: Society of Automotive Engineers, 1978), 1393–1408. On the GM switch, see Klimisch interview; and Kathleen C. Taylor, personal communication.

38. Taylor and Gagliardi interviews.

39. Colucci interview; Mondt, *Cleaner Cars*, 199–202.

40. On the Clean Air Act, see Gagliardi interview. On Cole, see Stork, Ruckelshaus, Klimisch, Colucci, and Gagliardi interviews.

10. OUT OF MY DEAD HANDS

1. On state responsibilities, see Thomas Jorling, "The Federal Law of Air Pollution Control," in Erica L. Dolgin and Thomas G. P. Gilbert, eds., *Federal Environmental Law* (Saint Paul, MN: West, 1974), 1083.

2. "National Primary and Secondary Ambient Air Quality Standards," *Federal Register* 36 (30 Apr. 1971): 8186. See Paul B. Downing and Gordon Brady, "Implementing the Clean Air Act: A Case Study of Oxidant Control in Los Angeles," *Natural Resources Journal* 18:2 (1978): 247–254, for a critical assessment of how this standard was set. On the contribution of automobiles to carbon monoxide, hydrocarbon and nitrogen oxide pollution, see Gladwin Hill, "A Murky Future for Smog Control," *New York Times*, 30 Jan. 1977, AUTO1. The other figures are from "Testimony of

John R. Quarles, Jr. . . . before the Subcommittee on Urban Mass Transit, House Committee on Banking and Currency, July 30, 1973," reprinted in *Transportation Controls under the Clean Air Act* (Washington, DC: GPO, 1973), 2.

3. For critics, see J. M. Heuss, G. J. Nebel, and J. M. Colucci, "National Air Quality Standards for Automotive Pollutants—A Critical Review," and the rejoinder by Delbert S. Barth et al., "Discussion," both in *Journal of the Air Pollution Control Association* 21:9 (1971): 535–548. On transportation and land use controls, see, in addition to *Transportation Controls under the Clean Air Act;* and Downing and Brady, "Implementing the Clean Air Act," 237–283, the following: James E. Krier and Edmund Ursin, *Pollution and Policy: A Case Essay on California and Federal Experience with Motor Vehicle Air Pollution, 1940–1975* (Berkeley: University of California Press, 1977), 212–247; John Quarles, *Cleaning Up America: An Insider's View of the Environmental Protection Agency* (Boston: Houghton Mifflin, 1976), 196–215; Robert L. Sansom, *New American Dream Machine: Toward a Simpler Lifestyle in an Environmental Age* (Garden City, NY: Anchor, 1976), 161–168; Eli Chernow, "Implementing the Clean Air Act in Los Angeles: The Duty to Achieve the Impossible," *Ecology Law Quarterly* 4:3 (1975): 537–581; John Quarles, "The Transportation Control Plans—Federal Regulation's Collision with Reality," *Harvard Environmental Law Review* 2 (1977): 241–263; R. Shep Melnick, *Regulation and the Courts: The Case of the Clean Air Act* (Washington, DC: Brookings Institution, 1983); and Dennis Williams, *The Guardian: EPA's Formative Years, 1970–1973,* EPA pamphlet, September 1993, 20–24. For additional tools, Tom Jorling, interview with the author. The possible need for less driving was in the air in 1970. See Robert Bendiner, "Of Time and the Flivver," *New York Times,* 24 Aug. 1970, 32.

4. For the EPA's involvement in Southern California's air pollution problem, see Eugene Yee Leong, "Air Pollution Control in California from 1970 to 1974: Some Comments on the Implementation Process" (Ph.D. diss., University of California, Los Angeles, 1974). Leong participated in the implementation process he describes. This ground is also covered in Krier and Ursin, *Pollution and Policy,* 199–247. The standard of 0.08 ppm for ozone was an especially tough standard for the Southern Californians to meet. The local air pollution control districts retained primary responsibility for controlling stationary sources of pollution, although under the direction of ARB. Leong, "Air Pollution Control in California," 23, makes the point about background oxidant levels. See TRW Transportation and Environmental Operations, *Transportation Control Strategy Development for the Metropolitan Los Angeles Region,* report prepared for the EPA (Contract no: 68-02-0048), December 1972 (hereinafter TRW), copy available at the Historical Center, Environmental Protection Agency, Washington, DC, 37, for 600–700 percent. The 241 days is from Chernow, "Implementing the Clean Air Act in Los Angeles," 544, 546n.

5. On the unwillingness, see "U.S. Air Cleanup Plan for L.A. Called Unrealistic by State," *Los Angeles Times,* 6 July 1973, I:2. For the Ruckelshaus perspective, William D. Ruckelshaus, interview with the author. "U.S. Plan to Implement State Air Standards Due by Jan. 15," *Los Angeles Times,* 7 Nov. 1972, I:3; and "Judge Rules EPA Head Failed to Fulfill Duties under Clean Air Act," *Wall Street Journal,* 8 Nov. 1972, 16. On reluctance to implement and the lawsuit, see Leong, "Air Pollution Control in California," 50, 53–54.

6. Coordination is from David G. Hawkins, interview with the author. For the court's decision, see Melnick, *Regulation and the Courts,* 311–313; Jorling, "Federal Law of Air Pollution Control," 1088–1089, 1091; and Dan Fisher, "Air Cleanup Delays Rescinded by Court," *Los Angeles Times,* 2 Feb. 1973, I:1. John E. Bonine, e-mail to the author, 4 Mar. 2006.

7. John F. Lawrence, "Clean Air Enforcers Caught in a Dilemma," *Los Angeles Times*, 27 Nov. 1972, I:1. John Dreyfuss, "U.S. Seeks Public Advice on Environmental Decisions," *Los Angeles Times*, 11 Dec. 1972, II:1.

8. Hawkins interview. Billings is quoted in Lawrence,"Clean Air Enforcers Caught in a Dilemma," I:24.

9. Figures from TRW, 98, 55, 168. For the Ruckelshaus quotation, see Hawkins and Ruckelshaus interviews. Although he does not recall the name of the person who told him this story, Hawkins indicated that his source was in the room when Ruckelshaus responded in this manner. Ruckelshaus does not remember this response but, laughing, indicated that the response sounded like what he probably said. John Dreyfuss, "Clean Air Plan Would Force Rationing of Gas," *Los Angeles Times*, 8 Dec. 1972, I:3. Smashing glasses is from E. F. Ball, letter to the editor, *Los Angeles Times*, 6 Jan. 1973, II:4. "The Proposal to Ration Gasoline," editorial, *Los Angeles Times*, 12 Dec. 1972, II:6. Quarles, *Cleaning Up America*, 201.

10. On the press conference, see "Press Conference of the Honorable William D. Ruckelshaus," Los Angeles, 15 Jan. 1973, Transcripts of Press Conferences and Other Related Records, 1971–1973, Office of Public Affairs, in Records of the EPA (hereinafter EPA), RG 412, box 2, National Archives, College Park, MD; Gladwin Hill, "Strict Auto Curb on Coast Doubted by Ruckelshaus," *New York Times*, 16 Jan. 1973, 81; Chernow, "Implementing the Clean Air Act in Los Angeles," 550–551; and John Dreyfuss, "Gasoline Rationing Plan Seen as Move to Pressure Congress," *Los Angeles Times*, 16 Jan. 1973, I:1. Ruckelshaus as quoted in Sansom, *New American Dream Machine*, 163; and Melnick, *Regulation and the Courts*, 321–322. Ruckelshaus used this same quip in his interview with the author without prompting.

11. Ruckelshaus and Hawkins interviews; Leon G. Billings, interview with the author.

12. For the comments by Reagan and Haagen-Smit, see Hill, "Strict Auto Curb on Coast Doubted by Ruckelshaus," 81. Leong, "Air Pollution Control in California," 57. For secretary's comments, see John A. Jones, "Blow to Economy Seen If Gasoline Is Rationed," *Los Angeles Times*, 16 Jan. 1973, I:3. "Curbs on Automakers Backed in Smog Poll," *Los Angeles Times*, 5 Mar. 1973, I:2. The executive is quoted in "This Way, They'd Be Cheaper to Run," *Los Angeles Times*, 18 Jan. 1973, III:15.

13. On opposition, see John Dreyfuss, "Stringent Control over Car Use on Smoggy Days Only Proposed," *Los Angeles Times*, 6 Mar. 1972, II:1. Press release, 19 Jan. 1973; and Kenneth Hahn, "Statement before the Environmental Protection Agency," 5 Mar. 1973, both in Hahn Papers (hereinafter KH), Huntington Library, San Marino, CA, box 7, Air Quality, folders 3.6.1.35, 3.6.1.36b, respectively. For chamber of commerce, see Gladwin Hill, "Hearings to Open on Alternative Solutions to Smog Problem in Los Angeles," *New York Times*, 4 Mar. 1973, I:50. "Comments to the Environmental Protection Agency," 15 June 1973, 3, Speech Collection, in Automobile Club of Southern California Archives, Los Angeles, box 2, no. 4457. Chass is quoted in Leroy F. Aarons, "Rationed Gas Proposed for Autos in L.A.," *Washington Post*, 16 Jan. 1973, A6. Ruckelshaus Press Conference, 29.

14. For no plan approved, see Jorling, "Federal Law of Air Pollution Control," 1070. On plans of 15 June, see Dan Fisher, "U.S. Proposal Could Ban Autos in L.A.," *Los Angeles Times*, 16 June 1973, I:1. "Opening Statement, Robert W. Fri," 15 June 1973, 12, Transcripts of Press Conferences and Other Related Records, 1971–1973, Office of Public Affairs, EPA, box 3. Fri elaborated on the plans in a second press conference on 27 July 1973. Bonine e-mail.

15. Quarles, *Cleaning Up America*, 204. Tunney is quoted in Gladwin Hill, "Plan to Cut Auto Travel in Los Angeles Basin Scored at Hearing," *New York Times*, 6 Mar. 1973, 35. For the report, see Jorling, "Federal Law of Air Pollution Control," 1095. See excerpts from "Floor Debate—Senate Bill," reprinted in *Transportation Controls under the Clean Air Act*, 54–55.

16. Jorling interview. On comprehensive approach, see Jorling, "Federal Law of Air Pollution Control," 1066.

17. TRW, 34. On state and local officials, John Dreyfuss, "Public Will Sacrifice to Fight Smog, Panel Told," *Los Angeles Times*, 7 Mar. 1973, I:26. For the EPA, see Burt Schorr, "How Badly Do You Want Clean Air?" *Wall Street Journal*, 18 May 1973, 10. For a short summary of the revised plan for Los Angeles, see "Tougher Times Ahead," *Auto Club News Pictorial*, Feb. 1974, 2–3. For a more technical analysis of its feasibility, see Robert G. Lunche, letter to Paul DeFalco, Jr., with appendixes, 1 Aug. 1973, in KH, box 7, Air Quality, folder 3.6.1.36. See also Larry Pryor, "U.S. Control of Land Use Looms," *Los Angeles Times*, 27 Aug. 1973, II:1.

18. On parking and development, see Hawkins interview. See also Melnick, *Regulation and the Courts*, 313–317. Train is quoted in Pryor, "EPA Attempts for First Time to Regulate Local Land Use," I:1. "Remarks by the Honorable Russell E. Train, Administrator, Environmental Protection Agency, before the National Press Club, Washington, DC—September 18, 1973," in Russell E. Train Papers (hereinafter RET), Manuscripts Division, Library of Congress, box 48.

19. Train's views are from author interview. On impractical steps, see "Press Conference of Russell E. Train, EPA Administrator," 13 Sept. 1973, typewritten transcript, in RET, box 26, folder "EPA Press Conference." On reaction, see Larry Pryor, "Proposals on Land Use Stirring Controversy," *Los Angeles Times*, 30 Oct. 1973, II:1; Ray Herbert, "Business, Labor Join to Fight U.S. Parking Space Surcharge," *Los Angeles Times*, 21 Nov. 1973, II:1; and "Energy and the Environment," *Los Angeles Times*, 6 Dec. 1973, I:2.

20. On challenges, see Dick Main, "L.A. Joins in Legal Challenge of U.S. Rules to Cut Auto Use," *Los Angeles Times*, 29 Nov. 1973, I:3; and Leslie Berkman and Doug Shuit, "U.S. May Drop Plan for Parking Surcharge," *Los Angeles Times*, 7 Dec. 1973, II:1. On surprise, see Hawkins interview. For EPA backtracking, see Berkman and Shuit, "U.S. May Drop Plan," II:1; and "Parking Reprieve," *Los Angeles Times*, 8 Dec. 1973, I:1. "Remarks by the Honorable Russell E. Train . . . before Town Hall of California, December 18, 1973," 8, in RET, box 48.

21. "Energy Supply and Environmental Coordination Act of 1974," Public Law 93-319, 22 June 1974, *United States Statutes at Large*, vol. 88 (Washington, DC: GPO, 1976), esp. 256–258. On the National Academy of Science reports, see "The Price of Delay," editorial, *New York Times*, 17 Sept. 1974, 34; and Gladwin Hill, "Air Pollution Drive Lags, But Some Gains Are Made," *New York Times*, 31 May 1975, 57. Mass transit efficiency figures are from Russell E. Train, "Remarks . . . before the American Automobile Association, Chicago . . . September 17, 1974," in RET, box 21. On the Boston effort, see "Concerns in Cities Warned by EPA to Cut Parking," *Wall Street Journal*, 13 Aug. 1974, 2; and "EPA Backs Down from a Plan to Make Firms Cut Parking," *Wall Street Journal*, 6 Feb. 1975, 7.

22. On cases lost, see Melnick, *Regulation and the Courts*, 332–334; and Krier and Ursin, *Pollution and Policy*, 232–233. Philip Shabecoff, "E.P.A.'s Authority over States Is Left in Doubt by Court," *New York Times*, 3 May 1977, 14. Bonine e-mail.

23. "Clean Air Act Amendments of 1977," Public Law 95-95, 7 Aug. 1977, *United States Statutes at Large*, vol. 91 (Washington, DC: GPO, 1980), esp. 693–697. Gladwin Hill, "E.P.A. Sources Say

Agency Ended Efforts to Cut Auto Smog in Cities," *New York Times*, 14 Jan. 1977, 41. Melnick, *Regulation and the Courts*, 300.

24. Train interview. John Quarles, "Runaway Regulation? Blame Congress," *Washington Post*, 20 May 1979, B8. Ruckelshaus interview.

25. Ruckelshaus quoted by Gladwin Hill, "Hearings to Open on Alternative Solutions to Smog Problem in Los Angeles," *New York Times*, 4 Mar. 1973, I:50. Marsha Fullmer, letter to the editor, *Los Angeles Times*, 24 Dec. 1973, II:2. "Quotation of the Day," *New York Times*, 2 Dec. 1973, 95.

26. A status report on the options is provided in Downing and Brady, "Implementing the Clean Air Act," 270–278.

27. On the retrofit issue, see Gladwin Hill, "Car Exhaust Law Repealed in Vote," *New York Times*, 14 Apr. 1975, 34. Opposition to nitrogen oxide devices is described in Krier and Ursin, *Pollution and Policy*, 240–247. Chernow, "Implementing the Clean Air Act in Los Angeles," 561–568, discusses opposition to retrofit devices, inspections, and mass transit. On HOV lanes, see Gladwin Hill, "Bus Lane's Future Dimmer on Coast," *New York Times*, 15 Aug. 1976, 20. For an analysis of the mass transit defeats, see Peter Marcuse, "Mass Transit for the Few," *Society* 13:6 (1976): 43–50. Free rapid transit is from Chernow, "Implementing the Clean Air Act in Los Angeles," 568n, who cites the TRW study "Transportation Control Strategy Development for the Metropolitan Los Angeles Region," December 1972, 153–154, E-11.

28. For air quality, see "Ozone Trends Summary: South Coast Air Basin," California Air Resources Board website, www.arb.ca.gov. "Worst air" is from Bruce Newman, "From the Land of Private Freeways Comes Car Culture Shock," *New York Times*, 16 Oct. 1997, G-8.

29. For lung development, see Nicholas Bakalar, "Highway Exhaust Stunts Lung Growth, Study Finds," *New York Times*, 30 Jan. 2007, D6. For the study itself, see W. J. Gauderman et al., "Effect of Exposure to Traffic on Lung Development from Ten to Eighteen Years of Age: A Cohort Study," *Lancet*, 17 Feb. 2007, 571–577. Michelle L. Bell et al., "Ozone and Short-term Mortality in 95 US Urban Communities, 1987–2000," *Journal of the American Medical Association*, 17 Nov. 2004, 2372–2378.

30. John P. Holdren and Paul R. Ehrlich, "Human Population and the Global Environment," *American Scientist* 62:3 (1974): 282–292.

11. SMALL WAS BEAUTIFUL

1. Nixon explained his decision in his memoirs. See *RN: The Memoirs of Richard Nixon* (New York: Grosset and Dunlap, 1978), 920–922, 924, 926–927. For the oil implications, see Daniel Yergin, *The Prize: The Epic Quest for Oil, Money, and Power* (New York: Simon and Schuster, 1991), 588–612, which recounts in dramatic detail Nixon's decision to resupply the Israelis and the resulting Arab oil embargo. For an account of these events that focuses on the auto industry, see David Halberstam, *The Reckoning* (New York: William Morrow, 1986), 451–459. OPEC figures from William D. Smith, "Energy Crisis: Shortages amid Plenty," *New York Times*, 17 Apr. 1973, 26.

2. Nixon, *RN*, 927.

3. James E. Akins, "The Oil Crisis: This Time the Wolf Is Here," *Foreign Affairs*, April 1973, 462–490, identifies the 1920s, late 1930s, and post–World War II periods. For domestic production, Halberstam, *The Reckoning*, 453. Federal Highway Administration, U.S. Department of Transportation, *Highway Statistics Summary to 1995* (Washington, DC: GPO, 1997), Total Drivers in

Force in the United States, 1949–Present, Table DL-201, III-5-6, and Annual Vehicle Distance Traveled in Miles and Related Data, 1936–1995, Table VM-201A, V-17.

4. Stewart Udall, "The Last Traffic Jam," *Atlantic Monthly,* October 1972, 72.

5. On the environmental dimension, see John Noble Wilford, "The Long-Term Energy Crisis," *New York Times,* 19 Apr. 1973, 53. For "energy crisis," see Roger A. Guiles, "What Ya Gonna Do When the (Oil) Well Runs Dry?" *Iron Age,* 16 Nov. 1972, 23.

6. See Floyd G. Lawrence, "Is the Big Car Doomed?" *Industry Week,* 29 Apr. 1974, 30. Floyd quotes GM marketing vice president Mack W. Worden to the effect that Nixon added to the uncertainty with his message. Henry A. Kissinger, *Years of Upheaval* (Boston: Little, Brown, 1982), 885.

7. For figures, see "The Painful Change to Thinking Small," *Time,* 31 Dec. 1973, 18; and "More Gas—At Higher Prices," *Newsweek,* 1 Apr. 1974, 62. For the speed limit, see "Emergency Highway Energy Conservation Act," Public Law 93-239, *U.S. Statutes at Large,* vol. 87 (Washington, DC: GPO, 1974), 1046–1048; and "Nixon Approves Limit of 55 M.P.H.," *New York Times,* 3 Jan. 1974, 1.

8. Paul Lienert, "CB Fever Sweeping Auto Industry," *Automotive News,* 10 Nov. 1975, 1, 21.

9. "Consumer Price Index, All Urban Consumers–(CPI-U), U.S. City Average, All Items," www.bls.gov/cpi.

10. "Happy Days Here Again? Not Quite, Survey Finds," *Automotive News,* 14 June 1975, 19. For the Michigan time series from November 1952 through November 1977, see research.stlouisfed.org/fred2/data/UMCSENT1.txt. "The Squeeze on the Middle Class," *Business Week,* 10 Mar. 1975, 52.

11. For GM's problems, see "The Small Car Blues at General Motors," *Business Week,* 16 Mar. 1974, 76. The panic at GM was mentioned again by *Business Week* reporters Stephen Shepard and J. Patrick Wright in "The Auto Industry," *Atlantic Monthly,* December 1974, 18. Ending large cars is from "How GM Manages Its Billion-Dollar R&D Program," *Business Week,* 28 June 1976, 55. On the Ford-Iacocca power struggle, see Lee Iacocca, *Iacocca: An Autobiography* (New York: Bantam, 1984), 103–140; Halberstam, *The Reckoning,* 510–532; and Douglas Brinkley, *Wheels for the World: Henry Ford, His Company and a Century of Progress* (New York: Viking, 2003), 669–670.

12. The small car percentages are derived from data published in the annual market data issues of the *Automotive News* from 1979 and 1980. See "A Time to Think Small," *Consumer Reports,* April 1971, 197–198, which defined smaller cars—compacts and subcompacts—as those 196 inches in overall length or less. Adams is quoted in "Detroit Tries to Close Its Credibility Gap," *Business Week,* 2 Oct. 1971, 46.

13. On multiple car ownership, see "Detroit Must Retool Its Old Success Formula," *Business Week,* 8 Mar. 1976, 72. On changing tastes, see Dan Cordtz, "Autos: A Hazardous Stretch Ahead," *Fortune,* April 1971, 68–71+; "Americans Put the Car in Its Place," *Business Week,* 18 Sept. 1971, 66; "Autos 1970: A Good Year to Buy?" *Consumer Reports,* April 1970, 197; Maurine Christopher, "Volvo Continuous Car Buyer, Dodge Sheriff Tell All on TV," *Advertising Age,* 19 June 1972, 46; "Detroit Thinks Small," *Newsweek,* 1 Apr. 1974, 55; Robert M. Lienert, "Car of the Future to be Smaller and Safer," *Advertising Age,* 19 June 1972, 29; "The Little Ones Are Hotter Than Ever," *Business Week,* 11 Apr. 1970, 21; and E. B. Weiss, "Legislation and Inconvenience Will Limit Use of Private Car," *Advertising Age,* 12 Oct. 1970, 61.

14. On California, see "L.A. Shows Off Its Mini Mania," *Business Week,* 12 June 1971, 27; "Autos:

The Minicar Invasion," *Newsweek*, 23 Aug. 1971, 64; and Brinkley, *Wheels for the World*, 665. Robert M. Lienert, "Imports to Slip in '72, But Future Appears Strong," *Advertising Age*, 19 June 1972, 39.

15. "How the Japanese Blitzed the California Auto Market," *Forbes*, 15 Sept. 1971, 29. Frank is quoted in Robert M. Finlay, "What's Happening in Small Cars?" *Automotive News*, 3 Feb. 1975, 25.

16. For several perceptive contemporary analyses of the arrival of the Japanese in the American market, see Sherman M. Robards, "Here Come the Japanese," *Fortune*, 1 June 1967, 114–115; "How the Japanese Blitzed the California Auto Market," 28–29. For auto club quotation, see Henry R. Bernstein, "California Trend: Small Truck Sold as Second Car," *Advertising Age*, 19 June 1972, 28. One market at a time is from Finlay, "What's Happening in Small Cars?" 25.

17. Small car percentage is from "Subcompacts Set Records, Big Cars Dive," *Automotive News*, 16 July 1973, 1, 72. E. F. Schumacher, *Small Is Beautiful: Economics as If People Mattered* (New York: Harper and Row, 1973).

18. "The Converter Switch Stalemates Detroit," *Business Week*, 24 Mar. 1975, 37. On prices, see "Detroit Bucks a Buyer Rebellion," *Time*, 2 Dec. 1974, 36; "Price Changes, 1957–1975," *Automotive News*, 1975 Almanac Issue, 71; "Detroit's Dilemma on Prices," *Business Week*, 20 Jan. 1975, 82–83.

19. On the downturn as well as imports, see "Autos: Dragging Down the Recovery," *Business Week*, 28 July 1975, 47; "Detroit Bucks a Buyer Rebellion," 35; "A Slowing Market Scares Detroit," *Business Week*, 9 Jan. 1978, 38; "When GM Overlooked an Economy Model," *Business Week*, 7 July 1975, 21; Roger Rowand, "Economics Play Key Role as Imports Plan for '75," *Automotive News*, 7 Apr. 1975, 15; and "Autos: Invasion," *Newsweek*, 19 May 1975, 79.

20. William Tucker, "The Wreck of the Auto Industry," *Harper's*, November 1980, makes this point. See the nice graph of "Small Car Sales vs. Gasoline Prices" at 53. Significantly, the graph begins with 1973. Gertrude I. McWilliams, "Beleaguered Auto Industry: Victim of Simplistic Charges?" *Automotive News*, 3 Nov. 1975 43.

21. For mileage penalty, see John Quarles, *Cleaning Up America: An Insider's View of the Environmental Protection Agency* (Boston: Houghton Mifflin, 1976), 205. For the impact of options and weight, see "Options: The Gasoline Trade-Off," *Consumer Reports*, April 1974, 304, where the magazine reported that Consumers Union had found in its own tests that each five-hundred-pound increase in the weight of an automobile decreased fuel economy by half a mile per gallon. For average mileage, see Russell E. Train, as quoted in "Environment in an Energy Crisis," *Business Week*, 15 Dec. 1973, 54.

22. For the automakers' promise, see Helen Kahn, "MPG Standards Added to Auto Regulation List," *Automotive News*, 29 Dec. 1975, 1. For the number and type of bills as well as background on the legislation, see Richard R. John et al., "Mandated Fuel Economy Standards as a Strategy for Improving Motor Vehicle Fuel Economy," in Douglas H. Ginsburg and William J. Abernathy, eds., *Government, Technology, and the Future of the Automobile* (New York: McGraw-Hill, 1980), 120–121.

23. For EPCA, see Public Law 94-163, *United States Statutes at Large*, vol. 89 (Washington, DC: GPO, 1977), 871–969; and Philip Shabecoff, "Ford Signs Bill on Energy," *New York Times*, 23 Dec. 1975, 1. The *Times*'s quotation is from "The Energy Bill," *New York Times*, 15 Dec. 1975, 30. For background on the legislation, see Richard Corrigan, "Energy Report 'Compromise' Oil Bill Ends Up Pleasing Few," *National Journal*, 27 Dec. 1975, 1735; John et al., "Mandated Fuel Econ-

omy Standards," 120–127; and Stan Luger, *Corporate Power, American Democracy, and the Auto-mobile Industry* (New York: Cambridge, 2000), 94–95. For 1991 figure, see "Table Df383–412: Distance Traveled, Fuel Consumption, and Registered Vehicles, by Motor Type: 1936–1995," in Susan B. Carter et al., eds., *Historical Statistics of the United States*, vol. 4, pt. D, *Economic Sectors* (New York: Cambridge, 2006), 4-838.

24. *Consumer Reports* reported on the trade-off debate in "What's Happening to Auto Pollution Con-trols," *Consumer Reports*, April 1974, 348–349. On the UAW, see "The Environment: Jobs vs. Clean Air?" *Newsweek*, 3 Feb. 1975, 55, 57; "The Converter Switch Stalemates Detroit," *Business Week*, 24 Mar. 1975, 38; and Jack Doyle, *Taken for a Ride: Detroit's Big Three and the Politics of Pollution* (New York: Four Walls Eight Windows, 2000), 109–111. On Dingell, see ibid., 128–129.

25. For an inside perspective on EPA's woes, see Quarles, *Cleaning Up America*, esp. 196–215. For staffing, budget, and lawsuits, Gladwin Hill, "E.P.A. Turns from Cures to Prevention," *New York Times*, 2 Dec. 1975, 1. Russell E. Train, "Remarks . . . before the National Press Club, Washington, DC—September 18, 1973," in Russell E. Train Papers (hereinafter RET), Manuscripts Division, Library of Congress, box 48. Power is quoted by Robert M. Finlay, "Mileage Irritates Economy Buyers," *Automotive News*, 16 June 1975, 21.

26. Train recounts this effort in *Politics, Pollution, and Pandas: An Environmental Memoir* (Washing-ton, DC: Island, 2003), 171–175. On industry efforts, see Vogel, *Fluctuating Fortunes*, 130–132. On oil industry ads, see Walter Rugaber, "F.T.C. Urged to Ask Companies to Prove Image-Building Ads," *New York Times*, 10 Jan. 1974, 21; Les Brown, "Mobil Asks Limit on Backing Up Ads," *New York Times*, 5 Apr. 1974, 75; and Vogel, *Fluctuating Fortunes*, 217–220.

27. For a good example of public efforts, see "EPA's Position on the Energy Crisis," n.d. [but 1973], in RET, box 3, folder "Energy Policy—1973." "Remarks by the Honorable Russell E. Train . . . before Town Hall of California, December 18, 1973," in RET, box 48. On numbers, Russell E. Train, interview with the author. For examples of using economic data, see Alvin L. Alm, "Energy Impact of Environmental Controls," memorandum to the administrator, 11 Jan. 1974, in RET, box 15, folder "Press Conference—Nat. Press Club, January 16, 1974"; and "The Economic Impact of the Federal Environmental Program," EPA, November 1974, in RET, box 43, folder "Domestic Council."

28. For the proposals, see Russell E. Train, "Statement . . . with regard to Legislative Proposals to Amend the Clean Air Act of 1970, March 22, 1974," in RET, box 15. On economic impact of environmental regulation, see Hill, "E.P.A. Turns from Cures to Prevention," 1. Alvin L. Alm, interviews with Dennis Williams, 12 Apr., 23 June 1993, published by the EPA as *Alvin L. Alm, U.S. EPA Oral History Interview—3 January 1994*, EPA 202-K-94-005, 9. Train interview.

29. "Press Conference of Russell E. Train, EPA Administrator," 13 Sept. 1973, typewritten transcript, in RET, box 26, folder "EPA Press Conference." Eric Stork interview with the author. Public Law 94-163, 910–911.

30. Doyle, *Taken for a Ride*, 79. Doyle's case is made in relentless, dispiriting detail. It is borne out in the industry's trade newspaper *Automotive News*. Train interview. GM's Washington office is from Vogel, *Fluctuating Fortunes*, 197. Stan Luger, *Corporate Power, American Democracy, and the Automobile Industry* (New York: Cambridge University Press, 2000), 89, implied this need.

31. On the Clean Air Act fight of 1977, see J. Dicken Kirschten, "It's Washington Taking on Detroit in the Auto Pollution Games," *National Journal*, 1 Jan. 1977, 9–15; J. Dicken Kirschten, "Attempts to Curb Pollution Are Still Up in the Air," *National Journal*, 25 June 1977, 987–989; Dick Kirschten,

"The Clean Air Conference—Something for Everybody," *National Journal*, 13 Aug. 1977, 1261–1263; Vogel, *Fluctuating Fortunes*, 181–186; and Luger, *Corporate Power*, 91–93. Billings is quoted by Vogel, *Fluctuating Fortunes*, 184.

32. Joseph M. Colucci, interview with the author. Vogel, *Fluctuating Fortunes*, 148–192, concluded that big business had blunted efforts at further reform by 1978. Luger, *Corporate Power*, 76, came to the same conclusion, and the author concurs.

33. On loans, see Roger Rowand, "Over-36-Month Loans Climb toward 40 Pct.," *Automotive News*, 1 Sept. 1975, 8; and Howard Rudnitsky, "The Willy Loman Factor," *Forbes*, 1 Aug. 1976, 19. Public Law 94-163, 941-948. Tucker recounts this story in "The Wreck of the Auto Industry," 51–55.

34. This figure is from "A Look at the Cars of 1985," *Time*, 16 May 1977, 61. On rationing see Tucker, "The Wreck of the Auto Industry," 58; and Brinkley, *Wheels for the World*, 666.

35. On resistance, see "A Rougher Road," *Newsweek*, 22 Nov. 1976, 90; "Password for '78: 'Down-size,'" *Time*, 1 Aug. 1977, 32; and "Weaning the Buyer Away from Big Cars," *Business Week*, 24 Oct. 1977, 78.

36. On Ford's problems during this period see Halberstam, *The Reckoning*, 585–610. For market share by vehicle class, see "Market Class Comparisons," *Automotive News*, 1980 Market Data Book Issue, 16. This table broke out imports separately. Nearly all the imports could be classed as sub-compacts and compacts. For the ten most fuel-efficient makes, see "Big Surge in Smaller Cars," *Time*, 7 Jan. 1980, 88. On rationing, see Marshall Loeb, "How to Counter OPEC," *Time*, 9 July 1979, 23. Power is quoted in "Big Surge in Smaller Cars," *Time*, 7 Jan. 1980, 88. Government statistics are from "Selected Statewide Statistics," Table SSS-200, *Highway Statistics Summary to 1995* (Washington, DC: GPO, n.d.), Introduction-7.

37. "Consumer Price Index, All Urban Consumers-(CPI-U), U.S. City Average, All Items," www.bls. gov/cpi. On loan rates, see "U.S. Autos: Losing a Big Segment of the Market Forever?" *Business Week*, 24 Mar. 1980, 79; "Tight Credit Slams Car Dealers Two Ways," *Business Week*, 14 Apr. 1980, 29; and "Financing a Car at Sky High Rates," *Consumer Reports*, April 1981, 194. In the late 1970s American automakers typically earned seven hundred to eight hundred dollars on each small car and up to six times that amount on luxury cars (before fixed costs were apportioned). According to John Z. DeLorean, from 1950 to 1975 it rarely cost GM more than three hundred dollars more to build a Cadillac than it did a Chevrolet. For these figures, see "Luxury Car Sales Skid to New Lows," *Business Week*, 19 May 1980, 35; and Tucker, "The Wreck of the Auto Industry," 48. Quotation from "U.S. Autos: Losing a Big Segment of the Market Forever?" *Business Week*, 24 Mar. 1980, 79.

38. "Consumer Price Index, All Urban Consumers–(CPI-U), U.S. City Average, All Items," www.bls. gov/cpi. The developments in this paragraph were reported in "Detroit's Worsening Plight," *Time*, 26 May 1980, 42. On consumer sentiment, see "Autos: Thinking Smaller for Survival," *Business Week*, 14 Jan. 1980, 55; and "Table 1: The Index of Consumer Sentiment, 1960–2005," www. sca.isr.umich.edu/data-archive/gettable.php. Layoffs are from "Detroit Hits a Roadblock," *Time*, 2 June 1980, 56; and "Thinking the Unthinkable," *Forbes*, 1 Sept. 1980, 133. On Ford, see Brinkley, *Wheels for the World*, 659–678. Quotation from "Detroit's Uphill Battle," *Time*, 8 Sept. 1980, 48.

39. Japanese sales are from Charles K. Hyde, *Riding the Roller Coaster: A History of the Chrysler Corporation* (Detroit, MI: Wayne State University Press, 2003), 260. On the changed regulatory climate, see Luger, *Corporate Power*, 97–134.

40. For making wrong cars, see Tucker, "The Wreck of the Auto Industry," 46. *Time* quotation is from "Detroit Hits a Roadblock," 56. "U.S. Autos: Losing a Big Segment of the Market Forever?" *Business Week*, 24 Mar. 1980, 84. Ward's survey is from "Autos Hit Forty Miles of Bad Road," *Time*, 28 Apr. 1980, 41.

41. Henry Ford II, as quoted in Halberstam, *The Reckoning*, 462. Sperlich is paraphrased at 512.

42. For "no longer," see "Detroit Hits a Roadblock," 58. "Cheap energy" is from "U.S. Autos: Losing a Big Segment of the Market Forever?" 88. Lost tolerance is from Charles G. Burck, "A Comeback Decade for the American Car," *Fortune*, 2 June 1980, 52. Murphy is quoted in "Detroit's Uphill Battle," 50. Iacocca is quoted in "Autos Hit Forty Miles of Bad Road," 42. Allan Sloan, "Regulation That's Working," *Forbes*, 6 Aug. 1979, 28.

43. On V-8s in 1977, see "U.S. Autos: Losing a Big Segment of the Market Forever?" 81.

12. THE RIDDLE OF THE SPORT UTILITY VEHICLE

1. For efforts to encourage broader use, see Joseph C. Ingraham, "Sales of Jeep-Type Vehicles Are Forging Ahead in High Gear," *New York Times*, 29 Aug. 1965, F1; and Robert B. Konikow, "The Birth of a New Product," *Advertising and Sales Promotion*, October 1965, 13–16.

2. For suburban focus, see John Holusha, "For Drivers, the Fun Is Back," *New York Times*, 17 June 1986, D1.

3. "Business Bulletin," *Wall Street Journal*, 22 Mar. 1984, 1. For 1981–1987 growth, see John Holusha, "The Detroit Advantage in Trucks," *New York Times*, 28 Dec. 1987, D6. On personal use, see Joseph Bohn, "Light Trucks, Heavy Market," *Automotive News*, 22 Sept. 1986, E8. The sixfold jump is from Holusha, "For Drivers, the Fun Is Back," D13; and Stephen Buckley, "Sport-Utility Vehicles Tap Market Niches," *Wall Street Journal*, 2 Sept. 1987, 21. Charles E. Dole, "Full-Size Four-Wheel-Drive Workhorses Are Back in Brisk Demand," *Christian Science Monitor*, 5 Apr. 1984, 32.

4. On crossover buyers, see Paul A. Eisenstein, "Chrysler's AMC Buyout Fits Long History for Both Firms," *Christian Science Monitor*, 7 Aug. 1987, 11. The median age and Lutz quotation are from Holusha, "Detroit Advantage in Trucks," D1, D6. SUV personal use figure is from Joseph Bohn, "Compact Off-Roaders Get Big Marketing Push in '87," *Automotive News*, 22 Sept. 1986, E8. For masculinity, see Keith Bradsher, "Light Trucks Have Passed Cars on the Retail Sales Road," *New York Times*, 4 Dec. 1997, C-4.

5. "Table 1: The Index of Consumer Sentiment, 1960–2005," www.sca.isr.umich.edu/data-archive'gettable.php. "High Attitude," *Automotive News*, 13 Jan. 1986, 24; "Survey Puts Car-Buying Attitudes at Record High," *Automotive News*, 2 June 1986, 4.

6. The off-road figure is from John Holusha, "Car Makers Hop aboard the Non-Truck Truck," *New York Times*, 4 Jan. 1987, F8. On information from advertising, Keith Bradsher, *High and Mighty: SUVs—The World's Most Dangerous Vehicles and How They Got That Way* (New York: Public Affairs, 2002), 276–277. Ley is quoted in Matt DeLorenzo, "Emotions Rule Sales, Consultant Says in Book," *Automotive News*, 27 Oct. 1986, 36.

7. For a thoughtful piece on SUV advertising and the connection to nature, see William Rollins, "Reflections on a Spare Tire: SUVs and Postmodern Environmental Consciousness," *Environmental History* 11:4 (2006): 684–723. Timothy D. Wilson and Nancy Brekke, "Mental Contami-

nation and Mental Correction: Unwanted Influences on Judgments and Evaluations," *Psychological Bulletin* 116:1 (1994): 117–142. O'Grady is quoted in Holusha, "For Drivers, the Fun Is Back," D13.

8. On consumer boredom see John Skow, "High Ride and Handsome," *Time*, 5 Feb. 1996, 48. For "commodity product," see Jerry Flint, "Mine Is Bigger than Yours," *Forbes*, 26 Feb. 1996, 70.

9. Holusha, "For Drivers, the Fun Is Back," D1; Jack Kadden, "Samurai Success: Find Market and Conquer," *New York Times*, 19 Apr. 1987, AC WA4. Keller is quoted in Holusha, "Detroit Advantage in Trucks," D6. Ibid., D1.

10. Doron P. Levin, "Consumer Group Asks Recall of Suzuki Samurai as Unsafe," *New York Times*, 3 June 1988, A1; Philip Shabecoff, "Sharp Cut in Burning of Fossil Fuels Is Urged to Battle Shift in Climate," *New York Times*, 24 June 1988, A1. See Spencer R. Weart, *Discovery of Global Warming* (Cambridge, MA: Harvard University Press, 2003). Svante Arrhenius, "On the Influence of Carbonic Acid in the Air upon the Temperature of the Ground," *Philosophical Magazine* 41 (1896): 237–276. "How Can a Gallon of Gasoline Produce 20 Pounds of Carbon Dioxide?" www.fueleconomy.gov/feg/co2.shtml.

11. On use in the 1980s, see Levin, "Cars That Are Not Are Hot," *New York Times*, 24 Aug. 1993, D2. On the trend toward SUV family use, see "Sport-Utility Vehicles," *Consumer Reports*, September 1990, 594. On obscuring the family image, see Charles E. Ramirez, "Staying Power," *Automotive News*, 11 Oct. 1993, 1.

12. On the Explorer, see Paul C. Judge, "Trusty, Profitable Jeep Faces New Challenge," *New York Times*, 2 May 1990, D1; Adam Bryant, "A Bigger Jeep Wagon Lifts Hopes at Chrysler," *New York Times*, 26 Nov. 1991, D6; Keith Bradsher, "Light Trucks Increase Profits But Foul Air More Than Cars," *New York Times*, November 1997, 42; and Flint, "Mine Is Bigger," 70. For light truck share, see Doron P. Levin. "Helped by Friendly Trade Rules, Trucks Fly Off Lots," *New York Times*, 17 Jan. 1993, E6; and Matthew L. Wald, "Forget Sleek and Sexy: The Boxy and Bulky Look Is In," *New York Times*, 16 Jan. 1994, E5.

13. For women, see Bradsher, "Light Trucks Have Passed Cars," C-4; Jerry Edgerton, "I Want My SUV," *Money*, October 1999, 142; and John Greenwald, "Kings of the Road," *Time*, 6 June 1994, 57, which is also the source for the quotation. For a similar remark see Skow, "High Ride and Handsome," 49.

14. Trip Gabriel, "A Big Hulk Rolls in as the Latest Status Symbol," *New York Times*, 12 May 1996, 39. Jerry Flint, "All This and Luxury, Too," *Forbes*, 30 Aug. 1993, 76. On both Stewart and the CEOs, see Gabriel, "A Big Hulk Rolls in as the Latest Status Symbol," 40.

15. Figure for four-wheel-drive use is from William J. Cook, "Size Doesn't Matter," *U.S. News and World Report*, 19 Apr. 1999, 51. On towing, see Bradsher, "Light Trucks Increase Profits," 43. Skow, "High Ride and Handsome," 49, puts towing use at 15–18 percent. Flint, "All This," 77. Cedergren is quoted by Peter Passell in "How G.I. Joe's Little Jeep Grew into a Yuppy Hunk," *New York Times*, 16 Oct. 1997, G-7. "Fantasy machines" is from Skow, "High Ride and Handsome," 49.

16. Sue Zesiger, "Which Bigger Is Better?" *Fortune*, 25 Nov. 1996, 226. See also Laurie Freeman, "Trendy Sport-Utility Fords across Generation Gap," *Advertising Age*, 1 Apr. 1996, S24, who also argues for a connection between an uncertain economy, corporate downsizing, and the appeal of the SUV. Wald, "Forget Sleek and Sexy," E5.

17. Bradsher, *High and Mighty*, 170. See ibid., 149–165, for a thorough discussion of the SUV rollover problem. "U.S. to Investigate 21 Deaths Tied to Tires on Sport Utilities," *New York Times*, 3 Aug. 2000, C7.

18. For 5.5 times, see Bradsher, *High and Mighty*, xvii. The 1997 figures are from Bradsher, "Light Trucks Increase Profits," 42. Ozone reduction is from Marianne Lavelle, "New Rules for Those Fun Trucks," *U.S. News and World Report*, 15 Feb. 1999, 28.

19. Figures are from Bradsher, *High and Mighty*, 248.

20. For red-state Republicans, see Gina Chon, "Cars and Politics," *Wall Street Journal*, 17 Apr. 2006, R5.

21. Bill Maher, *When You Ride Alone You Ride with bin Laden: What the Government Should Be Telling Us to Help Fight the War on Terrorism* (Beverly Hills, CA: New Millennium, 2002), 6–8, 28–30. On the Huffington ads, see Katharine Q. Seelye, "TV Ads Say S.U.V. Owners Support Terrorists," *New York Times*, 8 Jan. 2003, A17; Arianna Huffington, letter to the editor, *New York Times*, 22 Feb. 2003, A16; and Dan DeLuca, "What Succeeds Like Excess?" *Philadelphia Inquirer*, 17 Jan. 2003, W3. For an excellent example of the literature deploring exaggeration, see Aaron Wildavsky, *But Is It True? A Citizen's Guide to Environmental Health and Safety Issues* (Cambridge, MA: Harvard University Press, 1995).

22. Figures and quotation from Patricia Leigh Brown, "Among California's S.U.V. Owners, Only a Bit of Guilt in a New 'Anti' Effort," *New York Times*, 8 Feb. 2003, A15.

23. On known suburban use, see Jack Doyle, *Taken for a Ride: Detroit's Big Three and the Politics of Pollution* (New York: Four Walls Eight Windows, 2000), 401.

24. See Bradsher, *High and Mighty*, 18–42, for a more extensive review of these decisions. For the light-truck CAFE standards, see "Agency Seeks to Require Higher Mileage in Vans," *New York Times*, 13 Dec. 1977, 16; Reginald Stuart, "Compromise Hinted on Proposed Truck Fuel Rules," *New York Times*, 18 Jan. 1978, A13; and Richard R. John et al., "Mandated Fuel Economy Standards as a Strategy for Improving Motor Vehicle Fuel Economy," in Douglas H. Ginsburg and William J. Abernathy, eds., *Government, Technology, and the Future of the Automobile* (New York: McGraw-Hill, 1980), 131–133.

25. The problems of the 1980s and 1990s are well covered by Luger, *Corporate Power*, 127–133, 154–171. On the 1985 and 1986 CAFE rollbacks, see Reginald Stuart, "U.S. Announces It Will Ease '86 Auto Fuel Economy Rule," *New York Times*, 19 July 1985, A12; and Reginald Stuart, "Fuel Economy Standard for Cars Said to be Eased for '87 and '88," *New York Times*, 1 Oct. 1986, B11.

26. For the Clean Air Act Amendments of 1990, see Public Law 101-549, in *United States Statutes at Large*, 1990, vol. 104, pt. 4 (Washington, DC: GPO, 1991). On ending the Partnership, see Neela Banerjee with Danny Hakim, "U.S. Ends Car Plan on Gas Efficiency; Looks to Fuel Cells," *New York Times*, 9 Jan. 2002, A1. For the California option, see Public Law 101-549, 2529.

27. The 1991 fuel economy figure is from "Table Df383–412: Distance Traveled, Fuel Consumption, and Registered Vehicles, by Motor Type: 1936–1995," in Susan B. Carter et al., eds., *Historical Statistics of the United States*, vol. 4, pt. D, Economic Sectors (New York: Cambridge University Press, 2006), 4-838.

28. As quoted in Bradsher, *High and Mighty*, 295.

29. Luger, *Corporate Power*, 195.

30. On motives and the Japanese, see Bradsher, *High and Mighty*, 101, 385.

31. Wilson paraphrases Daryl Bem in *Strangers to Ourselves: Discovering the Adaptive Unconscious* (Cambridge, MA: Harvard University Press, 2002), 205; see also 15–16, 97, 105, 165.

32. Bradsher, *High and Mighty*, 412. On the Mini Cooper, see Peter Bohr, "Is There Room in America for the Small Car?" *AAA World*, January–February 2003, 31–36; Micheline Maynard, "BMW Has Maxi Expectations for Its Next, Slightly Larger Mini Cooper," *New York Times*, 11 Oct. 2006, C1,

C4. On the overall trend, see Micheline Maynard, "Small-Car Nation," *New York Times*, 25 Oct. 2006, E1.

33. Most of these figures are from Jad Mouawad and Matthew L. Wald, "Prices Cause Concern, but Little Change in Behavior or Laws," *New York Times*, 12 July 2005, C1. On worries, see Kenneth S. Deffeyes, *Hubbert's Peak: The Impending World Oil Shortage* (Princeton, NJ: Princeton University Press, 2001); and Peter Maass, "The Breaking Point," *New York Times Magazine*, 21 Aug. 2005, 30–35, 50, 56–59.

34. Schlesinger is quoted by Mouawad and Wald, "Prices Cause Concern, but Little Change in Behavior or Laws," C1. On complacency, see Warren Brown, "Drivers Still Seek Horsepower, Not Fuel Efficiency," *Washington Post*, 25 Sept. 2005, G 02. Toyota dealer quoted in Sholnn Freeman, "Smaller Cars Enjoy New Chic; Dealers Notice SUV Demand Dropping Off," *Washington Post*, 28 Sept. 2005, D 01. Figures from Micheline Maynard, "A Comeback for the Car Species," *New York Times*, 5 Jan. 2006, C1. For Toyota's long-term focus especially, see Jon Gertner, "How Toyota Conquered the Car World," *New York Times Magazine*, 18 Feb. 2007, 34–41, 54, 58–60.

35. On electrics, see Lawrence M. Fisher, "G.M., in a First, Will Sell a Car Designed for Electric Power this Fall," *New York Times*, 5 Jan. 1996, A10; Micheline Maynard, "Ford Abandons Venture in Making Electric Cars," *New York Times*, 31 Aug. 2002, C1; Susan Ball, "For the Return of Electric Cars, The Reality Is Virtually Present," *New York Times*, 3 Mar. 1996, WC17; Matthew L. Wald, "Stumped, G.M. Recalls Half Its Electric Cars," *New York Times*, 4 Mar. 2000, A8; Danny Hakim, "Automakers Look Beyond Electric," *New York Times*, 22 Sept. 2002, L1; and Chris Dixon, "Carmakers Pull Plug on Electric Vehicles," *New York Times*, 28 Mar. 2004, 12:1. For the electricity-gasoline comparison, see Heather L. MacLean and Lester B. Lave, "Evaluating Automobile Fuel/Propulsion System Technologies," *Progress in Energy and Combustion Science* 29 (2003): 52, table 6.

36. Lindsay Brooke, "Challenging Toyota's Hybrid Hegemony," *New York Times*, 30 Apr. 2006, 12:1.

37. On flexible-fuel cars, see Micheline Maynard, "Ford Plans Shift in Focus Away from Hybrids," *New York Times*, 30 June 2006, C1; Alexei Barrionuevo, "A Range of Estimates on the Benefits," *New York Times*, 25 June 2006, 19; and Alexei Barrionuevo, "Fill Up on Corn If You Can," *New York Times*, 31 Aug. 2006, C1. For a skeptical view, see "The Ethanol Myth," *Consumer Reports*, October 2006, 15–19.

38. On hydrogen, see Danny Hakim, "Taking the Future for a Drive," *New York Times*, 2 Nov. 2005, C1; and James G. Cobb, "Hybrids, Hydrogen and Hype," *New York Times*, 1 Jan. 2006, 12:1.

39. Hosein Shapouri, James A. Duffield, and Michael Wang. *The Energy Balance of Corn Ethanol: An Update*. U.S. Department of Agriculture, Office of the Chief Economist, Office of Energy Policy and New Uses. Agricultural Economic Report No. 813 (July 2002).

40. For sophisticated life-cycle analyses of internal combustion automobiles and the alternatives, readers may consult the work of Mark Delucchi of the University of California at Davis and Michael Wang at the Argonne National Laboratory. For a recent review of such studies, see MacLean and Lave, "Evaluating Automobile Fuel/Propulsion System Technologies," 1–69.

INDEX

Page numbers in *italics* indicate illustrations